Lecture Notes in Physics

Volume 967

The Lecture Notes in Physics

The series Lecture Notes in Physics (LNP), founded in 1969, reports new developments in physics research and teaching-quickly and informally, but with a high quality and the explicit aim to summarize and communicate current knowledge in an accessible way. Books published in this series are conceived as bridging material between advanced graduate textbooks and the forefront of research and to serve three purposes:

- to be a compact and modern up-to-date source of reference on a well-defined topic
- to serve as an accessible introduction to the field to postgraduate students and nonspecialist researchers from related areas
- to be a source of advanced teaching material for specialized seminars, courses and schools

Both monographs and multi-author volumes will be considered for publication. Edited volumes should however consist of a very limited number of contributions only. Proceedings will not be considered for LNP.

Volumes published in LNP are disseminated both in print and in electronic formats, the electronic archive being available at springerlink.com. The series content is indexed, abstracted and referenced by many abstracting and information services, bibliographic networks, subscription agencies, library networks, and consortia.

Proposals should be sent to a member of the Editorial Board, or directly to the managing editor at Springer:

Dr Lisa Scalone
Springer Nature
Physics Editorial Department
Tiergartenstrasse 17
69121 Heidelberg, Germany
lisa.scalone@springernature.com

More information about this series at http://www.springer.com/series/5304

Yves Pomeau • Minh-Binh Tran

Statistical Physics of Non Equilibrium Quantum Phenomena

 Springer

Yves Pomeau
LadHyX
École Polytechnique
Palaiseau Cedex, France

Minh-Binh Tran
Department of Mathematics
Southern Methodist University
Dallas, TX, USA

ISSN 0075-8450   ISSN 1616-6361   (electronic)
Lecture Notes in Physics
ISBN 978-3-030-34393-4   ISBN 978-3-030-34394-1   (eBook)
https://doi.org/10.1007/978-3-030-34394-1

Mathematics Subject Classification (2010): 82C40, 82C10

This Springer imprint is published by the registered company Springer Nature Switzerland AG
The registered company address is: Gewerbestrasse 11, 6330 Cham, Switzerland

Foreword

There are a number of excellent books on the kinetic theory of classical and quantum gases, but this book is unique. The focus of the first part of the book is a kinetic theory that describes the interaction of single atoms with a radiation field. It deals with radiation from two- and three-level atoms and the statistical theory of *shelving*. Shelving refers to a situation where an excited atomic state, with a very long lifetime, can shut off the emission of light in a decay channel of another excited state.

The focus of the second part of the book is the kinetic theory of Bose–Einstein condensates (BEC) at ultra-low temperatures where the excitation spectrum is phonon-like, rather than particle-like. This brings a new type of contribution to collision processes in the BEC. The kinetic theory used to describe the hydrodynamic processes in the ultra-low temperature BEC uses an expansion of distribution functions in powers of the wave vector of hydrodynamic processes (an approach first suggested by Résibois). This approach to the kinetic theory of the BEC provides a very elegant way to compute the transport processes in the BEC.

The authors are well qualified to write this book. Yves Pomeau has a long history of exemplary (and award winning) work in kinetic theory, and Minh-Binh Tran is a mathematician whose work has focused on the derivation and mathematical properties of kinetic equations. The book provides a summary of the applications of kinetic theory to atomic systems interacting with radiation fields, and excitations interacting with very low temperature coherent phases of matter. The topics discussed in this book are not discussed in a similar manner elsewhere, and therefore the book provides an important addition to the literature on the kinetic theory of matter-radiation interaction and the kinetic theory of BECs.

Professor of Physics
Co-Director of the Center for Complex Quantum Systems
University of Texas at Austin, USA

Linda Reichl

Preface

The transition from Newtonian mechanics to quantum mechanics in the early years of the twentieth century was a major step in the progress of our understanding of the world. This transition involved more than just a change of equations – it led to a deep change in our understanding of the limits of human knowledge. It included from the very beginning a statistical interpretation of the theory.

Statistical concepts were originally introduced to provide a classical (i.e. not quantum-theoretic) description of complex systems with many degrees of freedom, such as a volume of fluid including a very large number of molecules. These large systems cannot be fully described and/or predicted since no human being has enough computational power to solve Newton's equations[1] where the initial data (position and velocity) involve too many particles.

In the limit of a dilute gas Boltzmann (cf. [41, 42]) found the right theory for describing the evolution of a large number of particles interacting by short range two-body forces. In this theory, kinetic equations are a primary tool. Kinetic equations modelize systems made up of a large number of particles (gases, plasma, etc.) by a distribution function in the phase space of particles, based on the modeling assumption that there are so many particles that the whole system can be treated as a continuum. The phase space under investigation includes not only microscopic variables, which describe the state of the particles, but also macroscopic variables, which depend on time and on position in physical space. As a consequence, kinetic equations accurately describe the dynamics of dilute gases and provide connections between macroscopic observable and microscopic properties. Kinetic equations take into account collisions and more generally interactions between particles, all phenomena which change their momentum and kinetic energy. After

---

[1] We reluctantly use the word "equations" here because Newton never wrote down ordinary differential equations of classical mechanics in the modern sense. As is well known, he solved dynamical problems by using elegant geometrical methods, replacing what we now call calculus. For instance, the solution of the two-body problem with a general spherical potential was obtained by replacing integrals by calculations of area.

a collision, on one hand, the number of particles with the previous momentum will decrease; on the other hand, the number of particles with the new momentum will increase. In order to modelize the effects of particle collisions on the probability distribution of momentum, one includes in the equations for this probability terms that are integrals over the distribution function, as was done by Boltzmann for dilute gases. These new terms are called collision integrals, which introduce conserved quantities that are used to construct hydrodynamic balance equations and parameterize the equilibrium distribution. Based on collision integrals, we can obtain the rates of dissipation and relaxation and expressions for the transport coefficients in the limit of a slow dependence of the hydrodynamic parameters following the general method invented by Enskog and Hilbert [85].

Quantum kinetic theory is developed in the second part of this book, which is devoted to the extension of Boltzmann's classical (non-quantum) kinetic theory of a dilute gas of quantum bosons. This is the source of many interesting fundamental questions, particularly because, if the temperature is low enough, such a gas is known to have at equilibrium a transition, the Bose–Einstein transition, where a finite portion of the particles stay in the quantum ground state. Chapter 9 is devoted to the Bose–Einstein transition at equilibrium in the limit of a small, but non-vanishing interaction between particles. This shows in particular that the transition is first order although it is second order without interaction. The existence of this transition obviously raises the question of the way this "condensate" grows dynamically from zero as the temperature is lowered and how it interacts with itself and with the rest of the particles. To be more precise, attempts to extend the Boltzmann equation, to study quantum gases, using arguments similar to those used to derive the classical Boltzmann equation, resulted in the Boltzmann–Nordheim kinetic equation. Nordheim derived [125] the kinetic equations for a dilute gas of bosons and of fermions by using methods similar to those of Boltzmann. In the literature the equation for bosons is sometimes called the Uehling–Uhlenbeck equation [159]. This equation is for the dynamics of the momentum distribution as the Wigner transform of the one-particle density matrix. Below the Bose–Einstein transition temperature T_{BE}, which depends on the density of the gas, this Wigner function becomes singular at zero momentum, which corresponds to a finite fraction of particles in the condensate. Therefore, one needs to modify the Boltzmann–Nordheim (B-N) equation to take into account this singular part in the Wigner transform. For a homogeneous system, deriving a coupled system of equations for the condensate fraction and the excited states is not too difficult. On the other hand, the non-homogeneous case is more subtle: at zero temperature the dynamics of a weakly interacting Bose gas is described by the Gross–Pitaevskii (G-P) equation when out of equilibrium (note that we use the word "temperature" a bit loosely since it does not strictly have a meaning in 'out of equilibrium' situations – it would be more accurate to replace it by an average energy per atom). Therefore, for a dilute Bose gas at finite temperature (a temperature between absolute zero 0K and the BE transition temperature), the system is described by a coupled system of B-N and G-P equations. In such a system of trapped Bose gases at finite temperature, the quantum Boltzmann equation corresponds to the evolution of the density distribution function of the thermal cloud and the G-P is the equation of the condensate.

However, this is not the only case where a kind of kinetic theory is necessary to describe the dynamics of a quantum system. In its early days quantum mechanics dealt almost exclusively with the interaction between atoms or molecules and the electromagnetic (EM) field. Even with a single isolated atom this introduces a system with infinitely many degrees of freedom, the modes of the radiated EM field if the atom is, as is often the case, in unbounded vacuum or at least in a big enough box. Following recent work by one of the authors (YP) with Martine Le Berre and Jean Ginibre, we introduce in the first part of this book a kinetic equation, of Kolmogorov type, which is needed to describe a rather standard physical situation, namely an isolated atom (actually, in experiments, an ion) under the effect of a classical pumping EM field which keeps it in its excited state(s) together with the random emission of fluorescence photons putting this atom back into its ground state. When reduced to a two-level system, this yields a rather simple Kolmogorov kinetic equation describing the fluctuations of the quantum state of the atom, whereas the case of a three-level system is far more complex in general and will be dealt with in a limit pertinent to some experiments.

This book provides an introduction to these systems for both mathematicians and theoretical physicists who are interested in the topic.

Yves Pomeau
Minh-Binh Tran

Acknowledgements

Y. Pomeau thanks Martine Le Berre and Jean Ginibre for many discussions and their invaluable help. M.-B. Tran would like to thank Ricardo Alonso, Alejandro Aceves, Stanislav Boldyrev, Gheorghe Craciun, J. Robert Dorfman, Miguel Escobedo, Pierre Germain, Alexandru D. Ionescu, Shi Jin, Sergey Nazarenko, Alan C. Newell, Toan T. Nguyen, Leslie M. Smith, Benno Rumpf, Avy Soffer, Herbert Spohn and Eitan Tadmor for their important help and fruitful discussions. He would especially like to express his gratitude to Linda E. Reichl and Enrique Zuazua for their illuminating advice and careful reading of the manuscript, and to Irene M. Gamba for her great support over the years. M.-B. Tran is supported by NSF Grants DMS-1814149, DMS-1854453 and a Sam Taylor Fellowship.

Contents

Part II Statistical Physics of Dilute Bose Gases   57

Part I
Statistical Physics of the Interaction of a Single Atom or Ion with Radiation

Chapter 1
Introduction

1.1 Overview

The use of statistical concepts to describe Nature was first introduced to describe classical (not quantum) systems with many degrees of freedom, such as a very large number of molecules or atoms in a macroscopic volume of fluid. However this is not the only situation where statistical concepts must be used to describe natural phenomena. We think of quantum physics. There the use of statistics met strong resistance, in part because it was unclear on which ensemble the statistics should be done. The starting point seems to be a deterministic set of equations – the Schrödinger equation – but after some discussion it loses its 'deterministic character' in connection with something called "reduction of the wave-packet". The meaning of this concept has been, and still is, a source of debate on the interpretation of quantum mechanics. The physical situation we shall discuss requires us to give a rather precise meaning to this interpretation of quantum mechanics. This is because the initial state of the atom we consider is not an eigenstate, but a linear superposition of an excited state and the ground state, a source of rather subtle differences with the standard situation considered in the emission-absorption process, as explained below.

The situation we shall deal with is the emission of photons by a single atom. Even though this is somewhat hidden, the randomness in the emission-absorption process is, as always in statistical physics, in the infinitely many degrees of freedom under consideration. It is due to the fact that the radiated field in an unbounded space has an infinite number of single photon states at a given frequency. This infinite number of photon states explains the randomness of the time of emission of a photon when the atom decays from the excited state. The link between these two aspects (randomness of the emission time and the infinite number of states of the photon field) is apparent in the calculation by Dirac [54] of the rate of emission of a photon by an atom in an excited state. Section 1.2 illustrates this remark.

In our non-trivial example, the fluorescence of a two-level atom, we explain how to relate explicitly statistical concepts to the quantum phenomena we

© Springer Nature Switzerland AG 2019
Y. Pomeau, M.-B. Tran, *Statistical Physics of Non Equilibrium Quantum Phenomena*,
Lecture Notes in Physics 967, https://doi.org/10.1007/978-3-030-34394-1_1

want to describe. In this example the Markovian dynamics of the atomic state is described by a Kolmogorov-like equation (2.2.19) which takes into account the quantum jumps associated to the emission of photons in a statistical way. One of the goals of our presentation is to draw attention to a theory of quantum phenomena using the full apparatus of statistical physics for the description of out-of-equilibrium phenomena, including in the seemingly simple situation of an atom emitting radiation when returning to its ground state.

We look at the fluorescence (namely the emission of photons) of a single two-level atom (or ion) illuminated by a monochromatic light beam at the frequency of resonance or quasi-resonance between two quantum levels (this beam is also called a pump). Later on we extend this to the three-level case. In the absence of spontaneous decay of the excited level(s) by the random emission of photons, the atomic system presents under the effect of a monochromatic pump regular oscillations, called optical Rabi oscillations, between the two atomic states. This name comes from the similarity of the relevant dynamical equations with those of a spin in a uniform magnetic field and a periodic electromagnetic field at Larmor frequency, which was imagined by Rabi [138]. The optical Rabi oscillations (just called Rabi oscillations later on) of a single atom get a statistical underpinning when one takes into account the possibility of spontaneous decay, namely the possibility of connecting the atom with the outside space where electromagnetic waves with all possible directions and locations can propagate. This spontaneous decay, or quantum jump, occurs typically with a very short duration but at random times.

The Kolmogorov equation displays this randomness together with the reversible dynamics of the Rabi oscillations. This equation is for the probability distribution of the state of the atom and it can be solved in various instances. The solutions give access to physical quantities like the spectrum of fluorescent light, and also to higher order time correlations allowing us to display the fundamentally irreversible character of fluorescence (see [134] on this matter).

The probability distribution p evolves as follows:

  i) In the coherent regime, namely between two successive photon emissions (or quantum jumps), the probability is transported by the Schrödinger flow describing the Rabi oscillations.
 ii) When a photon is emitted, the quantum state undergoes an instantaneous jump. The description of fluorescence by quantum jumps is standard in the subject. Our theory describes the quantum trajectories followed by the atom in the course of its history with the successive emission of photons.

In Section 2.1 we introduce the general framework of the Kolmogorov approach to the study of fluorescence. Chapter 2 is the main chapter of this part and is devoted to the two-level atom case.

We derive and solve the Kolmogorov equation for the exact resonant atom-laser interaction and also for a detuned laser (that is, not exactly resonant with the frequency of the atomic transition). Formally, the detuned case gives the same equation as the resonant case, but with coefficients depending on the detuning. In Chapter 3, we derive some statistical properties of the emitted light in the so-called shelving situation, namely in the three-level case when one of the excited level has a very long lifetime. The observed intermittent fluorescence is a quite spectacular manifestation of the intrinsic randomness of the quantum jumps, which is well described by the Kolmogorov equation in a convenient range of parameters.

Of course the theory of emission and absorption of photons by an atomic system (atom, ion or molecule) has been a topic of interest in atomic physics for a long time. The classical situation considered by Planck and Einstein in the early years of the twentieth century is different from the one we deal with. Naturally, all this concerns a weak interaction between the electromagnetic field and the atom (in practice, the electrons). However this does not mean that the changes made in the atomic system by radiation are small. Actually a transition from the excited state to the ground state is not a small perturbation of this atomic state, at least when seen from the point of view of the final state. What is small is the rate of transition compared to the frequencies of the atomic motion (in practice, the frequencies of the electrons in an atomic bound state). When considering the simple case of an atom in an excited state as the initial state, the rate of decay toward the ground state by emission of photons is small and constant. Therefore it is tempting to add to the "energy" of this excited state a small imaginary part tuned to represent a slow exponential decay of this excited state to the ground state. However this yields a representation of the decay which is not completely accurate, or more precisely that cannot be extended to more complex situations. The atom, after it has decayed to the ground state, is exactly in the ground state. If other things happen to the atom after the first decay, this atom will react completely differently if it is in the ground state (which has a certain probability) or if it has remained in the excited state. After the first emission a fundamental difference of behavior between the two atomic states (excited and ground state) shows up if the atom is pumped by a monochromatic beam, because pumping acts differently on the two states of the atom. Therefore the addition of a small imaginary part to the energy to represent the decay can describe the first decay, but not what will happen afterwards. Theories based on extending the idea of adding such a small imaginary part to the energies, or, in a more sophisticated way, non-Hermitian terms to the Hamiltonian operator of the atom, have been presented in the literature. They have difficulties with the conservation of the total probability and/or representing the fluctuations inherent to the emission process. We refer the interested reader to [130] and the references therein. The theory we present below is free of these difficulties and satisfies the basic constraints of probability theory, the total probability is conserved and it is positive. Moreover the dynamical equations are derived in a rational way from the fundamental physics of the atomic processes.

1.2 A Model Physical Problem: Emission of Photons by a Pumped Atom or Ion

Here we consider the emission of radiation by a single atom, a fairly standard topic in quantum physics. By conservation of energy an atom initially in an energy state above its ground state may return to this ground state by emitting a photon with an energy exactly equal to the energy difference between the excited and ground state. This is different from the classical (i.e. non-quantum) emission of radiation by a Hertz dipole in a number of ways: first a Hertz dipole is assumed to oscillate under the effect of an unspecified forcing, depending itself on an outside source of energy. Then the energy loss of the Hertz dipole is continuous in time, without any jump. The quantization of the radiation field yields discontinuities in the loss of energy of the emitting atom because the energy of each of the two systems (atom and electromagnetic radiation) can change by a discrete amount only, the quantum of energy of Planck's theory.

The spontaneous emission of a photon by an atom initially in an excited state was long ago considered by Einstein and by Dirac. This introduced a quantum process with a fundamental randomness showing up in the Markovian point process defined by the photo-emission times. In particular, the time intervals between successive emissions are independent and identical random variables. Thanks to the progress of experimental atomic physics the discreteness of the emission times has been observed. In the experiment proposed and realized by Dehmelt [52, 53] the fluorescence of a single three-level atom is observed to be intermittent because the atom is, so to speak, maintained in one of the two excited states for a long time. As in the two-level case, this atom is subject to two phenomena. First, under the effect of a pump field, it tends to make Rabi oscillations between the ground and the excited state. Once in this excited state it can also jump spontaneously to its ground state by emitting a photon. The process goes on forever provided the pump field remains on.

Chapter 2

The Kolmogorov Equation for a Two-Level System

2.1 The Kolmogorov Equation in General

The Kolmogorov equation is the dynamical equation for the probability distribution of a system (in our case a single atom or ion) under the effect of two processes: first a deterministic dynamics changes the parameters of the system in a deterministic way. Let $\Theta(t)$ be the set of time-dependent parameters of such a system. Under the deterministic part of the dynamics the time derivative $\frac{\partial}{\partial t}\Theta = v(\Theta)$ is a function of Θ, $\frac{\partial}{\partial t}$ being here and elsewhere the derivative with respect to time. For this deterministic dynamics the conservation of probability yields the equation of evolution of the probability distribution $p(\Theta, t)$,

$$\frac{\partial}{\partial t}p(\Theta, t) + \frac{\partial}{\partial \Theta}(v(\Theta)p(\Theta, t)) = 0, \qquad (2.1.1)$$

where $\frac{\partial}{\partial \Theta}$ is the derivative with respect to Θ. This is actually a gradient in general because Θ has more than one component, but this only complicates the writing in an unessential way. Moreover, the probability distribution is positive and normalized.

The Kolmogorov equation follows by adding to this equation a right-hand side representing instantaneous transitions occurring at random instants of time. This effect of the transitions (or jumps) is represented with the help of a positive-valued function $\Gamma(\Theta'|\Theta)$. During a small interval of time dt, if the system is in state Θ, it jumps with probability $\Gamma(\Theta'|\Theta)dt$ to state Θ'. The validity of the Kolmogorov equation for a given system implies that the duration of a jump from Θ to Θ' is much shorter than the time scale of the deterministic dynamics. The Kolmogorov equation describes the changes in $p(\Theta, t)$ under the joint effects of the deterministic dynamics and the random jumps. It reads [100]:

© Springer Nature Switzerland AG 2019

Y. Pomeau, M.-B. Tran, *Statistical Physics of Non Equilibrium Quantum Phenomena*,
Lecture Notes in Physics 967, https://doi.org/10.1007/978-3-030-34394-1_2

$$\frac{\partial}{\partial t}p(\Theta, t) + \frac{\partial}{\partial \Theta}(v(\Theta)p(\Theta, t)) = \tag{2.1.2}$$

$$\int \Gamma(\Theta|\Theta_1)p(\Theta_1, t)\mathrm{d}\Theta_1 - p(\Theta, t)\int \Gamma(\Theta'|\Theta)\mathrm{d}\Theta'.$$

In the right-hand side the first positive term describes the increase of probability of the Θ-state due to jumps from other states to Θ. The second term represents the loss of probability because of jumps from Θ to any other state Θ'. We assume that the jump probability $\Gamma(\Theta'|\Theta)$ is smooth enough near $\Theta = \Theta'$ that we do not have to care about jumps from Θ to almost identical states (as may happen, for instance, because of the grazing two-body collisions in Boltzmann's kinetic theory with a long range pair potential). Indeed, the very existence of the probability transition Γ implies that we are considering a Markov process where the transition rate depends only on the present state of the system.

By integration over Θ of equation (2.1.2) one finds that the L^1-norm

$$\int p(\Theta, t)\mathrm{d}\Theta$$

is constant (if it converges, as we assume). Moreover, because $\Gamma(\Theta'|\Theta)$ is positive, if positive initially, $p(\Theta, t)$ remains positive later. Below we shall derive an explicit Kolmogorov equation for a two-level atom subject both to an external electric field at quasi-resonance frequency between the two levels (the pump field giving the deterministic part of the dynamics) and to spontaneous decay by radiation (the process of fluorescence). We shall also consider the three-level case with two pump fields at the two frequencies defined by the resonances between the ground state and the two excited states. This makes a realistic limiting case corresponding to the phenomenon of shelving. Otherwise the general three-level case is fairly complex.

The validity of this approach requires the existence of two widely different time scales. The long time scale is associated to the deterministic dynamics, and so to the left-hand side of a Kolmogorov equation. From time to time there is a jump from one state to another state, a jump that is instantaneous at the time scale of the deterministic dynamics. This splitting into two time scales, and consequently the use of a Kolmogorov equation, is particularly well justified in the case of fluorescence of a single atom: the deterministic dynamics is due to the interaction with an external pump field, whereas the quick random jumps are due to the emission of a photon when the excited state decays to the ground state. This decay occurs rarely but with a time scale of the order of the period of the emitted photon, a time much shorter than the two other typical times of the problem, namely the average interval between two spontaneous jumps from the excited state and the period of the oscillations due to the interaction with a classical external pump field.

The physical results we shall derive from this Kolmogorov approach will rest on explicit calculations of time-dependent correlations of the fluctuations of the system around its steady state, given by a steady solution of a Kolmogorov equation. Let $F(\Theta, t)$ and $G(\Theta, t)$ be two functions of the state of the system at time t. We are interested in the computation of the time-dependent correlation $\langle F(\Theta, t)G(\Theta, t')\rangle$ where the average is taken on the steady state $P_{st}(\Theta)$, the time-independent solution of equation (2.1.2). When $t = t'$ the correlation $\langle F(\Theta, t)G(\Theta, t)\rangle$ is time-independent and is simply

$$\langle F(\Theta, t)G(\Theta, t)\rangle = \int P_{st}(\Theta)F(\Theta)G(\Theta)\mathrm{d}\Theta.$$

The time-dependent correlation is found as follows. Let $P(\Theta, t|\Theta_0)$ be the solution of the Kolmogorov equation (2.1.2) with variables (Θ, t) and initial condition

$$P(\Theta, t = 0) = \delta(\Theta_0 - \Theta),$$

$\delta(\cdot)$ being the Dirac delta function. The time-dependent correlation reads

$$\langle F(\Theta, 0)G(\Theta, t)\rangle = \iint P_{st}(\Theta_0)P(\Theta, t|\Theta_0)F(\Theta_0)G(\Theta)\mathrm{d}\Theta\mathrm{d}\Theta_0. \qquad (2.1.3)$$

2.2 The Kolmogorov Equation for a Two-Level System

Let us consider a two-level quantum system illuminated by a laser beam at resonance (in practice this system is an ion in a Paul's trap, but following the common use in this field we shall call this ion an "atom"). This system does two things: first, starting from the ground state, it will get to the excited state and back, making Rabi oscillations. Moreover, it will decay spontaneously from the excited to the ground state by random emission of photons, each one having a frequency very close to the difference between the atomic frequencies of the two atomic levels. This emission yields fluorescent light. This knowledge of the fluctuations of the emitted field, that is, what can be measured, requires us to introduce and solve a Kolmogorov equation for this two-level system, an equation for the evolution of the probability distribution of its quantum states. This probability distribution yields the statistical properties of the density matrix ρ_{at} reduced to the states of the atom.

The Kolmogorov equation is classical (i.e. non-quantum) in the sense that different histories of quantum jumps of the same atom yield quantum states of this atom which cannot interfere with one another because they bear a random phase. Therefore the contribution of each trajectory adds to that of the others in the sense of classical probability, without quantum interference between different histories. From the point of view of the interpretation of quantum mechanics, probabilities are added because the various trajectories belong to different Universes in the sense of Everett [67]. This is explained as follows. Let $(...t_j, t_{j+1}, ...)$ be the set of times of emission of a fluorescence photon with $t_j < t_{j+1}$. This defines what we shall call a trajectory of the atom. The quantity measuring the properties of the whole system, atom and radiated photons, is the full density matrix, which can be written symbolically

as $\rho(At, F; At', F')$, where At denotes the degrees of freedom of the atom (in
the present case this will be a finite discrete set representing the indices of
the quantum levels involved) and where F denotes the degrees of freedom of
the field of the emitted photons. The primes indicate another choice of the
variables, either for the degrees of freedom of the atom or for the field. The
rules of quantum mechanics are such that the state of the atom is given by
the reduced density matrix derived from the overall density matrix by tracing
it over the degrees of freedom of the field,

$$\rho_{at}(At; At') = \Sigma_F \, \rho(At, F; At', F). \tag{2.2.1}$$

Consider the contribution to the reduced density matrix arising from different
histories of quantum jumps. Those contributions correspond to a different set
of field coordinates F, because there is no overlap of the fields emitted by
the atom at different times: the emitted field propagates at the speed of light
out of the atom and has no overlap if the emissions are at different times.
Therefore the contributions of different histories to this density matrix add
to each other in the ordinary classical sense. In particular, no contribution
to the off-diagonal part of the density matrix may arise from the addition
of contributions of different trajectories which have no correlation in the
quantum sense. This explains why one can add contributions to the reduced
density matrix with a probability distribution in the ordinary sense for each
contribution. For this problem this average, with a probability distribution
in the classical sense, is the translation in this specific case of the summation
over the degrees of freedom of the field, as done in the general formula defining
a reduced density matrix.

2.2.1 Deterministic Regime: Rabi Oscillations

We consider here the deterministic regime which begins at the instant when a
photon is emitted by the atom, and goes on until another photon is emitted.
Let us consider the two coupled equations for the quantum amplitudes of the
ground state $|0\rangle$ and the excited state $|1\rangle$ (in Dirac's notation) of a two-level
atom, of energy E_0 and E_1, respectively ($E_0 < E_1$). In the absence of an
external field, the wave function reduces to

$$|\Psi\rangle = a_0(t)e^{-iE_0t/\hbar}|0\rangle + a_1(t)e^{-iE_1t/\hbar}|1\rangle, \tag{2.2.2}$$

with complex coefficients $a_{0,1}$ which do not depend on time if there is no
pumping field. The quantity $\hbar$ is Planck's constant divided by 2π. With such
a quasi-resonant pumping (electrical) field $\mathcal{E}\cos(\omega_L t)$, of frequency ω_L close
to the difference of Bohr frequencies of the atom, $\omega_0 = (E_1 - E_0)/\hbar$, and
assuming that all other atomic levels play no role, the decomposition (2.2.2)
of the wave function is still valid but with coefficients depending on time.
The interaction representation gets rid of the free evolution by introducing

coefficients c_0 and c_1 such that $a_0 = c_0$ and $a_1 = c_1 e^{i\omega_L t}$. Those coefficients are associated with the slow evolution of the state vector. In the absence of any relaxation by fluorescence, the dynamics is purely deterministic and the two complex amplitudes obey the two coupled differential equations

$$\frac{\partial}{\partial t} a_0 = -i\frac{\Omega}{2} e^{i(t\delta - \xi)} a_1 \qquad (2.2.3)$$

and

$$\frac{\partial}{\partial t} a_1 = -i\frac{\Omega}{2} e^{-i(t\delta - \xi)} a_0, \qquad (2.2.4)$$

where $\delta = \omega_L - \omega_0$ is the detuning between the laser and the frequency difference of the two atomic levels, ξ is the phase of the matrix element of the electric dipole moment $d = e\langle 0|x|1\rangle$ of the atom and Ω is the Rabi frequency of the atom illuminated by the laser, proportional to the amplitude $\mathcal{E}$ of the electric field in the incoming pump field, assumed to be classical (without quantum fluctuations, that is, with many photons in the same quantum state), and to the modulus of d,

$$\Omega = -\frac{|d|\mathcal{E}}{\hbar}. \qquad (2.2.5)$$

The dynamics described by equations (2.2.3) and (2.2.4) is unitary because the sum $(a_1 a_1^* + a_0 a_0^*)$ is constant in time, a^* being the complex conjugate of a. We consider the case of a real dipolar moment, or $\xi = 0$, for simplicity. A complex dipolar moment changes nothing essential.

2.2.1.1 Laser at Exact Resonance

In addition to the real dipole condition, we consider first the simplest case of a laser at exact resonance with the atomic transition, which gives $\delta = 0$. Therefore we have:

$$\delta = \xi = 0. \qquad (2.2.6)$$

Let us introduce the scaled time $\tilde{t} = \Omega t/2$, the spinor A and the Pauli matrix σ_1, defined as

$$A = \begin{pmatrix} a_0 \\ a_1 \end{pmatrix} \qquad (2.2.7)$$

and

$$\sigma_1 = \begin{pmatrix} 0 & 1 \\ 1 & 0 \end{pmatrix}. \qquad (2.2.8)$$

The system (2.2.3)–(2.2.4) can be written as

$$\frac{\partial}{\partial \tilde{t}} A = -i\sigma_1 A.$$

The solution

$$A(t) = \exp(-i\tilde{t}\sigma_1)A(0)$$

can be written as

$$A(\tilde{t}) = (\cos\tilde{t} - i\sigma_1\sin\tilde{t})A(0). \tag{2.2.9}$$

Because the deterministic regime begins just after the emission of a photon by transition from state $|1\rangle$ to $|0\rangle$, the amplitude of state $|1\rangle$ vanishes exactly after this emission. Therefore the post-decay initial conditions are $|a_0(0)| = 1$ and $|a_1(0)| = 0$, and equation (2.2.9) becomes

$$A(\tilde{t}) = e^{i\phi}\begin{pmatrix} \cos\tilde{t} \\ i\sin\tilde{t} \end{pmatrix}. \tag{2.2.10}$$

Equation (2.2.10) shows that the two amplitudes evolve in quadrature between two jumps. We have introduced the constant phase ϕ, however in this section it plays no role in the derivation of the Kolmogorov equation and can be ignored for the moment. However, this parameter will play an important role when deriving the statistical properties of the light emitted by the atom because this phase is associated with the random times of emission, therefore ϕ changes after each quantum jump. Returning to the original variable and introducing the angle variable

$$\theta(t) = \frac{\Omega}{2}t, \tag{2.2.11}$$

we obtain

$$a_0(t) = e^{i\phi}\cos\theta(t); \quad a_1(t) = ie^{i\phi}\sin\theta(t), \tag{2.2.12}$$

which describes the Rabi nutation of the atom between the fundamental state and the excited state.

Equation (2.2.11) shows that the amplitudes a_0 and a_1 remain in quadrature during the deterministic motion, an important result which characterizes the resonant case only. This explains why we should be able to derive a Kolmogorov equation in the resonant case from the dynamics of the complex amplitudes. Therefore in the deterministic regime without fluorescence the dynamical equation for the probability $p(\theta, t)$ is

$$\frac{\partial}{\partial t}p + \frac{\Omega}{2}\frac{\partial}{\partial\theta}p = 0. \tag{2.2.13}$$

Equation (2.2.13) conserves the norm defined below in equation (2.2.17). We point out the peculiar status of the resonant situation, for which the quantum flow can be studied directly by using the dynamics of the amplitudes $a_{0,1}$, as done here.

Let us observe that other variables may be used to describe the trajectory in another phase space. For example, taking the variables (r, θ) used in the next subsection and defined in Chapter 9, which are quadratic combinations of the amplitudes a_0 and a_1, it can be shown that the quantum flow rotates in the plane $(r\cos 2\theta, r\sin 2\theta)$ at constant angular velocity $\Omega/2$ along a circle of radius unity. With these variables the probability density p is such that

$$p(r, \theta + \Omega t, t) = \delta_D(r - 1)p(\theta, 0),$$

where δ_D is the Dirac distribution (not to be confused with the detuning δ). Moreover, we show below that the probability is still carried by equation (2.2.13), where the variable θ does not have the same meaning (it differs by a factor of 2 because the function r is quadratic with respect to the amplitudes).

2.2.1.2 Laser Off-Resonance

When the laser is not at exact resonance (this is what we also call a detuned laser) with the atomic transition, the solution of equations (2.2.3)–(2.2.4) can be expressed in terms of two time-dependent phases, $\theta(t) = \Omega t/2$ and $\Psi(t) = t\delta/2$, see the Appendix, equations (4.2.2)–(4.2.3). But it appears that when one describes the motion in terms of the amplitudes $a_{0,1}$, the trajectory is complicated because their relative phase evolves with time, in contrast to the resonant case where the amplitudes are in quadrature all along the deterministic trajectory.

We propose in the Appendix to use a pertinent set of variables (r, θ) such that the situation becomes very much like that of the resonant case. These variables are the modulus and the phase of a complex function $u + iv$, where u and v are quadratic functions of new amplitudes $c_{0,1}$ defined in terms of $a_{0,1}$ by equation (4.2.14). In the phase space ($u = r \cos 2\theta, v = r \sin 2\theta$) we show that the quantum flow displays a rotation at constant angular velocity $\omega = \Omega_\delta/2$, where

$$\Omega_\delta = \sqrt{\Omega^2 + \delta^2} \qquad (2.2.14)$$

is the effective Rabi frequency in the presence of detuning. We show that the rotation takes place along a circle of radius

$$r_\delta = \frac{\Omega}{\Omega_\delta}, \qquad (2.2.15)$$

smaller than unity. At exact resonance the trajectory is a circle of radius unity in this phase space, as written above. After each emission of one photon, the trajectory starts again on the circle from the same point $\theta = 0$. Therefore one may forget the inside and outside of the circle and restrict attention to the circle of radius r_δ given by equation (2.2.15). On this trajectory one may put a probability density $p(r, \theta, t)$ which obeys a partial differential equation having the same form as (2.2.13),

$$\frac{\partial}{\partial t}p + \frac{\Omega_\delta}{2}\frac{\partial}{\partial \theta}p = 0. \qquad (2.2.16)$$

2.2.2 Rabi Oscillations and Emission of Photons

The dynamics of the system results from two physical phenomena, first the Rabi oscillation described by the simple equation (2.2.11), which is an oscillation between the quantum amplitudes of the two states, and secondly, spontaneous decay (not described by the equations above) of the state $|1\rangle$ toward the ground state $|0\rangle$. The rate of transition is proportional to γ, a quantity with the physical dimension of an inverse time which is proportional to the square of d, the atomic dipolar moment, and to ω_L^3. Any statistical theory is based on a probability distribution. In the present case, we could consider the probability distribution of the values of $\cos(\theta)$ or $\sin(\theta)$, both in the interval $[-1, 1]$, or more simply of the variable θ, all measuring the probability of the atom to be in the excited or the ground state. Let $p(\theta, t)$ be the probability distribution describing this system, a wrapped probability distribution because it is a periodic function of θ, of period π instead of 2π because a change of sign in front of both a_0 and a_1 does not change the state of the system. The probability $p(\theta, t)$ is normalized by the constraint

$$\int_{-\pi/2}^{\pi/2} p(\theta, t)\mathrm{d}\theta = 1. \qquad (2.2.17)$$

As explained earlier (at the beginning of the present section) we consider probabilities with their classical (non-quantum) meaning. This relies on the assumption that the various quantum trajectories (or histories) are independent, that is, for different values of the angle θ at a *given time* no quantum correlations exist. Such correlations would forbid us to consider trajectories with different θ at a given time as statistically independent. This statistical independence in the quantum sense is there because what we call different histories are different sets of times of emission of fluorescence photons.

Another issue must also be discussed: why is the photon emission treated as a quantum jump? It implies that this process is much faster than any other physical process in the problem, so that its time duration is so short that it can be seen as instantaneous, making the Kolmogorov equation valid. The typical time scale for the duration of a "quantum jump" is on the time scale of atomic processes. Roughly speaking, it is about of the order of magnitude it takes for the emitted photon to leave the close neighborhood of an atom. This time is almost independent of the weak interaction with the laser field, for instance. Therefore the only relevant time scale for such an atomic process is the period of the electromagnetic wave emitted by the transition from the excited to the ground state. In the situation of the kind of experiment we are thinking of [52, 53], this period of the electromagnetic waves is much shorter than the interval between two photon emissions (or, equivalently than the lifetime of the excited state) and much shorter too than the period of Rabi oscillations. Therefore it is legitimate to consider this "duration of the quantum jump" as negligibly small compared to any other relevant time scale.

Let us derive the equation of motion of the probability $p(\theta, t)$, taking into account the joint effect of Rabi oscillations and the spontaneous decay toward the ground state. The probability distribution is expected to be a periodic function of the variable θ, of period π. Let us consider the domain $-\pi/2 < \theta < \pi/2$.

2.2.2.1 Laser at Resonance

The probability of the event "the atom decays at time t from the excited state toward the ground state" is proportional to the square modulus of the amplitude of the excited state, namely to $\sin^2(\theta)$ times the decay rate γ, a data of the problem with the physical dimension of an inverse time. Let us now explicitly write the right-hand side of equation (2.1.2). The probability $\Gamma(\theta|\theta')$ for the atom to make a quantum jump from the state θ towards the state θ' is proportional to $\delta_D(\theta')$ because any jump lands on $\theta' = 0$ in the interval $[-\pi/2, \pi/2]$ and this probability is proportional to $\gamma \sin^2 \theta$ because it comes from the state a_1 with the squared amplitude $\sin^2 \theta$. Therefore

$$\Gamma(\theta|\theta') = \gamma \, \sin^2 \theta \, \delta_D(\theta'). \qquad (2.2.18)$$

This introduces a parameter γ, which is an inverse time or the rate of emission of photons by the atom when in the excited state. This rate is small compared with the atomic frequencies, which justifies that it can be derived by explicit calculations from the quantum theory of the interaction of an atom and the electromagnetic field. This calculation was done by Dirac in 1925 at the age of 24 in a masterpiece of ingenuity. Although this does not make any difference in the interval $[-\pi/2, \pi/2]$, we shall sometimes replace the Dirac function $\delta_D(\theta)$ by $\delta_D(\sin(\theta))$ to make more obvious the periodicity with respect to θ of a given expression. Using the above expression for the jump probability one obtains the following Kolmogorov equation for the dynamics of the π-periodic function $p(\theta, t)$,

$$\frac{\partial}{\partial t} p + \frac{\Omega}{2} \frac{\partial}{\partial \theta} p = \gamma \left(\delta_D(\sin(\theta)) \left(\int_{-\pi/2}^{\pi/2} p(\theta', t) \sin^2(\theta') d\theta' \right) - p(\theta, t) \sin^2(\theta) \right).$$

$$(2.2.19)$$

2.2.2.2 Laser Off-Resonance

In the presence of detuning, we have shown in the previous subsection that the trajectory in the phase space (r, θ) is confined to the circle of radius r_δ given by (2.2.15). After each emission of a photon, in the plane $u = r \cos 2\theta$, $v = r \sin 2\theta$ schematized in Figure 4.2-(b), the quantum flow starts from the same point $u = r$, $v = 0$, and rotates with angular velocity $\Omega_\delta/2$ until the next emission. To derive the right-hand side of the Kolmogorov equation, we must express the transition function $\Gamma(\theta|\theta')$ in this space. We start naturally from the expression

$$\Gamma(\theta|\theta') = \gamma |a_1(\theta)|^2 \, \delta_D(\theta'). \tag{2.2.20}$$

The amplitude $a_1(\theta)$ is obtained via the amplitudes $c_{0,1}$ expressed in terms of the functions (u, v), as explained in the Appendix. We get an expression $|a_1|^2$ which is surprisingly proportional to $\sin^2(\theta)$. This gives a transfer function (2.2.20) which becomes exactly (2.2.18) for zero detuning, as expected. The probability $p(\theta, t)$ obeys the following equation

$$\frac{\partial}{\partial t} p + \frac{\Omega_\delta}{2} \frac{\partial}{\partial \theta} p = \gamma \left(\delta_D(\theta) \left(\int_{-\pi/2}^{\pi/2} p(\theta', t) f(\theta') \mathrm{d}\theta' \right) - p(\theta, t) f(\theta) \right),$$
$$\tag{2.2.21}$$

where

$$f(\theta) = \left(\frac{\Omega}{\Omega_\delta} \right)^2 \sin^2 \theta. \tag{2.2.22}$$

After a simple change of γ, this equation is the same as the one in the resonant case:

$$\frac{\partial}{\partial t} p + \frac{\Omega}{2} \frac{\partial}{\partial \theta} p = \gamma \left(\delta_D(\sin \theta) \left(\int_{-\pi/2}^{\pi/2} p(\theta', t) \sin^2 \theta' \mathrm{d}\theta' \right) - p(\theta, t) \sin^2 \theta \right).$$
$$\tag{2.2.23}$$

Summarizing, the Kolmogorov equation for the quasi-resonant case and for the exactly resonant case are formally identical. They only differ by two coefficients, one concerns the Rabi frequency in the left-hand side, the other is the factor $\gamma(\frac{\Omega}{\Omega_\delta})^2$ which is in front of $\sin^2 \theta$ in the right-hand side. When rescaling the time, as done below, these coefficients disappear formally. In the following we shall deal with equation (2.2.19) only.

2.2.3 The Physical Meaning of the Kolmogorov Equation

The Kolmogorov equation describes the wave function of the two-level atom with wave function:

$$\Psi_{at}(t) = \left(\cos(\theta(t)) |g\rangle + i e^{i\omega t} \sin(\theta(t)) |e\rangle \right) e^{i\varphi}. \tag{2.2.24}$$

This equation is for the probability distribution $p(\theta, t)$ of the atomic states indexed by a single variable θ, and has a built-in conservation law of the total probability at any time. Without emission of photons the dynamics reduces itself to $\dot{\theta} = \Omega/2$, where Ω is the Rabi frequency. Because it is linear and autonomous, the Kolmogorov equation (2.2.19) for a given initial condition can be solved using the Laplace transform, as shown in the next section. We shall discuss below some physical properties derived from it without using this formal solution.

The probabilities for the atom to be in the excited or in the ground state are respectively

$$\rho_1(t) = \int_{-\pi/2}^{\pi/2} p(\theta', t) \sin^2 \theta' d\theta' \tag{2.2.25}$$

and

$$\rho_0(t) = \int_{-\pi/2}^{\pi/2} p(\theta', t) \cos^2 \theta' d\theta'. \tag{2.2.26}$$

Their sum is one, as it should be, if $p(\theta, t)$ is normalized to one. From equation (2.2.19) one can derive an equation for the time dependence of $\rho_1(t)$ and $\rho_0(t)$ by multiplying (2.2.19) by $\sin^2 \theta$ and by $\cos^2 \theta$ and integrating the result over θ. This gives

$$\dot{\rho}_1 = -\frac{\Omega}{2} \int_{-\pi/2}^{\pi/2} \sin^2 \theta' \frac{\partial p}{\partial \theta} d\theta' - \gamma \left(\int_{-\pi/2}^{\pi/2} p(\theta', t) \sin^4 \theta' d\theta' \right) \tag{2.2.27}$$

and

$$\dot{\rho}_0 = -\frac{\Omega}{2} \int_{-\pi/2}^{\pi/2} \cos^2 \theta' \frac{\partial p}{\partial \theta} d\theta' + \gamma \left(\int_{-\pi/2}^{\pi/2} p(\theta', t) \sin^4 \theta' d\theta' \right). \tag{2.2.28}$$

In both equations the dots indicate time derivatives.

The two equations (2.2.27)–(2.2.28) yield the rate of evolution of the quantum probabilities of the two levels. The first term, proportional to the Rabi frequency Ω, describes the effect of the Rabi oscillations, whereas the second term, proportional to γ, displays the effect of the quantum jumps responsible for the photo-emission. After integration by parts it becomes

$$\dot{\rho}_1(t) = \int_{-\pi/2}^{\pi/2} p(\theta, t) \left(\frac{\Omega}{2} \sin 2\theta - \gamma \sin^4 \theta \right) d\theta. \tag{2.2.29}$$

This set of equations is not closed, that is, it *cannot* be mapped into equations for $\rho_1(t)$ and $\rho_0(t)$ only: their right-hand sides depend on higher momenta of the probability distribution $p(\theta, t)$, momenta that cannot be derived from the knowledge of $\rho_1(t)$ and $\rho_0(t)$. This is a rather common situation. To name a few cases, the BBGKY hierarchy of non-equilibrium statistical physics makes [144] an infinite set of coupled equations for the distribution functions of systems of interacting (classical) particles: the evolution of the one-body distribution depends explicitly on the two-body distribution, which itself depends on the three-body distribution, etc. This can be reduced to a closed equation for the one-body distribution function (the Boltzmann equation) for dilute gases only. In the theory of fully developed turbulence the average value of the velocity depends on the average value of the two-point correlation of the velocity fluctuations, depending itself on the three-points correlations, etc.

Fortunately, one can solve the Kolmogorov equation (2.2.19) via an implicit integral equation [134] and there is generally no need to manipulate an infinite hierarchy of equations as in those examples. In the present case one can say, following Everett, that the probability distribution $p(\theta, t)$ allows us to make averages over the states of the atom in different Universes, each being labeled by a value of θ at a given time t. As mentioned above, physical phenomena, like the observation of a quantum state decay measured by emission of a photon, are relative to the measurement apparatus which takes place in the Universe associated to the observer. At every emission of a photon a new history begins. The creation of new Universes at each step defines a Markov process, which is well described by the Kolmogorov statistical picture, and could not be considered as a deterministic process depending on averaged quantities such as those on the instantaneous values of the quantum probabilities of each state. In particular, $\dot{\rho}_1(t)$, the time derivative of the population $\rho_1(t)$, cannot be expressed in terms of $\rho_1(t)$, which is an averaged value over all the Universes.

To illustrate how one can use the Kolmogorov equation, we derive the time-dependent probability of photo-emission by a single atom, first without any pump field, then in the presence of a laser. We consider first an isolated atom initially in the pure state $\Psi_{at}(0)$ given by (2.2.24) with $\theta(0) = \theta_0$. We search for the evolution of the probability

$$\rho_1(t) = \int_{-\pi/2}^{\pi/2} p(\theta, t) \sin^2 \theta d\theta$$

that the atom is in an excited state at time t. From (2.2.19) with $\Omega = 0$ (no pump), the emission of a photon occurs randomly in time with rate

$$\dot{\rho}_1 = \gamma \sin^2 \theta(t) \rho_1(t). \tag{2.2.30}$$

Once the atom jumps to its ground state it cannot emit another photon, then the emission of a photon, if recorded, is a way to measure the state of the atom. The solution of (2.2.30) leads to the probability of the excited state

$$\rho_1(t) = \sin^2 \theta_0 e^{-\gamma \sin^2 \theta_0 \, t}$$

when taking into account the initial condition, and the photo-emission rate is

$$\dot{\rho}_1(t) = -\gamma \sin^4 \theta_0 e^{-\gamma \sin^2 \theta_0 \, t}. \tag{2.2.31}$$

The probability of one photo-emission in the time interval $(0, \infty)$ is the integral of $\dot{\rho}_1$

$$\int_0^\infty \gamma \sin^4 \theta_0 \, e^{-\gamma \sin^2 \theta_0 \, t} dt = \sin^2 \theta_0, \tag{2.2.32}$$

which implies that the final state of the coupled system "atom+emitted photon field" is

$$\Psi(\infty) = \sin\theta_0|g,1\rangle + e^{i\phi'}\cos\theta_0|g,0\rangle, \qquad (2.2.33)$$

where the numbers (1,0) in the ket vectors correspond to the one and zero photon state, respectively.

The relation (2.2.32) implies that if we consider N atoms initially prepared in a given pure state with $\theta(0) = \theta_0$, namely with total energy $N\sin^2\theta_0\hbar\omega$, we get at infinite time, when all the atoms in an excited state have decayed, N atoms in the ground state and $N\sin^2\theta_0$ emitted photons, each one of energy $\hbar\omega$. In the final state only a fraction of them, $N\sin^2\theta_0$, jumped from the excited state to the ground state, with the emission of a photon. The other part, $N\cos^2\theta_0$, did not emit any photons because those atoms are initially in the ground state $|g\rangle$.

A stimulating discussion between Yves Pomeau, Martine Le Berre and Jean Ginibre with C. Cohen-Tannoudji, J. Dalibard and S. Reynaud motivated the above derivation.

In the case of an atom illuminated by a resonant pump field, this atom, when in the excited state, will emit photons at random times t_i. The set of those times makes what is called a point process in statistical theory. In the present case this process is Markovian because the probability of emission only depends on the present state of the atom. This remark is used in Section 2.2.7 to find the probability distribution of time intervals between two emitted photons.

2.2.4 Explicit Solutions of the Kolmogorov Equation

Solutions of equation (2.2.19) keep constant the L^1-norm

$$\int_{0_-}^{\pi} p(\theta,t)\mathrm{d}\theta$$

for any π-periodic distribution function of the variable θ.

Below we look at various questions related to this Kolmogorov equation. We first give its formal explicit solution and derive some interesting properties, then its steady solution reached from arbitrary initial conditions. Afterwards we look at various limiting cases, namely those of large and of small damping. Lastly, we explain how to obtain various physical quantities from solutions of this Kolmogorov equation, in particular for the spectral properties of the emitted light.

2.2.4.1 Derivation of the Probability Density and Average Value of $\sin^2(\theta)$

The calculation given below is due to Jean Ginibre. Let us introduce the auxiliary function

$$b(t) = \int_{-\pi/2}^{\pi/2} p(\theta', t) \sin^2(\theta') d\theta'. \tag{2.2.34}$$

The equation we are trying to solve becomes

$$\frac{\partial p}{\partial t} + \frac{\Omega}{2} \frac{\partial p}{\partial \theta} + \gamma \sin^2(\theta) p = \gamma \delta_D(\sin(\theta)) b(t). \tag{2.2.35}$$

Let us take $\frac{2}{\Omega}$ as the unit of time and introduce the dimensionless parameter $\gamma' = \frac{2\gamma}{\Omega}$. Equation (2.2.35) becomes

$$\frac{\partial p}{\partial t} + \frac{\partial p}{\partial \theta} = g(\theta, t) - f(\theta) p, \tag{2.2.36}$$

with

$$g(\theta, t) = \gamma' \delta_D(\sin(\theta)) b(t), \tag{2.2.37}$$

where $b(t)$ is still given by equation (2.2.34) and

$$f(\theta) = \gamma' \sin^2(\theta). \tag{2.2.38}$$

From equation (2.2.36), the function

$$h(\theta, t) = p(\theta + t, t)$$

satisfies the following differential equation:

$$\frac{\partial}{\partial t} h(\theta, t) = (g(\theta + t, t) - f(\theta + t)) h(\theta, t). \tag{2.2.39}$$

This can be solved as an initial value problem as follows. Take

$$s(\theta, t) = h(\theta, t) e^{\int_0^t f(\theta + t') dt'}.$$

Therefore

$$s(\theta, t = 0) = h(\theta, t = 0).$$

The auxiliary function $s(\theta, t)$ is a solution of

$$\frac{\partial}{\partial t} s = g(\theta + t, t) e^{\int_0^t f(\theta + t') dt'}.$$

This has the solution

$$s(\theta, t) = s(\theta, t = 0) + \int_0^t g(\theta + t', t') e^{\int_0^{t'} f(\theta+t'')dt''} dt'.$$

The equivalent result for the function $h(\theta, t)$ reads

$$h(\theta, t) = h(\theta, 0) e^{-\int_0^t f(\theta+t')dt'}$$

$$+ \int_0^t g(\theta + t', t') \times \exp\left(\int_0^{t'} f(\theta + t'')dt'' - \int_0^t f(\theta + t'')dt'' \right) dt'.$$

Tracing back the path from this explicit solution to the original equation, one finds the general solution of equation (2.2.36):

$$p(\theta, t) = p(\theta - t, 0)\alpha(\theta, t) + \int_0^t \alpha(\theta, t')g(\theta - t', t - t')dt', \qquad (2.2.40)$$

where

$$\alpha(\theta, t) = e^{\left(-\int_0^t f(\theta-t')dt'\right)}, \qquad (2.2.41)$$

and $g(\cdot)$ is given by equation (2.2.37). Multiplying both sides of this equation by

$$f(\theta) = \gamma' \sin^2(\theta)$$

and integrating the result over one period for θ, one finds the following Fredholm integral equation for $b(t)$,

$$b(t) = m(t) + \int_0^t b(t')l(t - t')dt'. \qquad (2.2.42)$$

In (2.2.42) we have

$$m(t) = \int_T p(\theta, 0)f(\theta + t)\alpha(\theta + t, t)d\theta, \qquad (2.2.43)$$

where T stands for the period of the function f (here $T = (0, \pi)$), and

$$l(t) = f(t)\alpha(t), \qquad (2.2.44)$$

where $\alpha(t)$ is the reduction to $\theta = t$ of the function of two variables $\alpha(\theta, t)$ (hopefully no confusion will arise from the use of the same notation, α for $\alpha(\theta, t)$ and $\alpha(t, t) = \alpha(t)$):

$$\alpha(t) = \alpha(t, t) = e^{-\int_0^t f(t')dt'}. \qquad (2.2.45)$$

Note the relation

$$l(t) = -\frac{\partial}{\partial t}\alpha(t). \qquad (2.2.46)$$

For $m(t)$ given, $b(t)$ can be derived from equation (2.2.42) either by iterations or by Laplace transforming both sides. In the numerics we use the iteration method, which gives the result drawn in Figure 2.1, where the initial condition for the probability is chosen as

$$p(\theta, t = 0) = \delta(\theta - \theta_0). \tag{2.2.47}$$

This choice is made in view of the derivation of correlation functions. It follows that any mean value calculated by using the probability (2.2.40) with initial condition (2.2.47) is actually a conditional average. It should depend on the parameter θ_0 and should be labeled $b(t, \theta_0)$ as done below. Figure 2.1 shows this function $b(t)$ for two values of θ_0, both converging at long times towards the stationary value $\langle b_{st} \rangle$ (see below and the Appendix for a sketch of a proof of this property). With initial condition (2.2.47) the probability $p(\theta, t)$ in (2.2.40) becomes the conditional probability,

$$p(\theta, t | \theta_0, 0) = \delta(\theta - t - \theta_0)\alpha(\theta_0, t) + \int_0^t \alpha(\theta, t')g(\theta - t', t - t')dt', \tag{2.2.48}$$

where $g(\theta, t)$ is given by equation (2.2.37).

Formally the Laplace transform method is simple: Let $\Phi_L(z)$ be the Laplace transform of a function $\Phi(t)$ of time, defined as

$$\Phi_L(z) = \int_0^\infty \Phi(t)e^{-zt}dt. \tag{2.2.49}$$

To ensure the convergence of the integral in this definition of the Laplace transform, the real part of z must be big enough, depending on how $\Phi(t)$ behaves at infinity (assuming it is a smooth function otherwise). The Fredholm equation (2.2.42) has a simple solution using the Laplace transform

$$b_L(z) = m_L(z)/(1 - l_L(z)),$$

or using equation (2.2.46)

$$b_L(z) = -\frac{m_L(z)}{z\alpha_L(z)}, \tag{2.2.50}$$

but this result is not very useful because inverting this Laplace transform requires us to know the singularities of the right-hand side.

2.2.4.2 Properties of the Kolmogorov Solution

Several properties of the solution (2.2.40)–(2.2.42) of the Kolmogorov equation deserve to be mentioned. They are valid for any periodic function $f(\cdot)$ which is positive and bounded. An important result is

$$M(t) = B(t), \tag{2.2.51}$$

where $M(t)$ is the integral of $m(t)$,

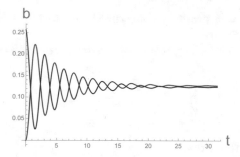

Fig. 2.1 Solution of equation (2.2.42) for small damping $\gamma = 1/7.8$. The initial condition is given by equation (2.2.47) with $\theta_0 = 0$ (red curve) and $\theta_0 = \pi/2$ (blue curve).

$$M(t) = \int_0^t m(t')dt', \qquad (2.2.52)$$

and

$$B(t) = \int_0^t b(t')\alpha(t - t')dt'. \qquad (2.2.53)$$

The proof goes as follows. Using equation (2.2.46), equation (2.2.34) becomes

$$m(t) = b(t) + \int_0^t b(t')\frac{\partial}{\partial t}\alpha(t - t')dt' = \frac{\partial}{\partial t}B(t). \qquad (2.2.54)$$

Integration of (2.2.54) with respect to time gives (2.2.51).

Another striking result concerns the evolution of several functions as time increases. Let us consider for example the function $\alpha(t)$ for any positive π-periodic function $f(\cdot)$, and define its particular value $\alpha(\pi)$,

$$\bar{\alpha} = e^{-\int_0^\pi f(t)dt}. \qquad (2.2.55)$$

From (2.2.45) we have, by definition:

$$\alpha(t + \pi) = \bar{\alpha}\alpha(t). \qquad (2.2.56)$$

Moreover, equation (2.2.43) can be written as

$$m(t) = \int_T p(\theta, 0)\left(-\frac{d}{dt}(e^{-\int_0^t f(\theta+t')dt'})\right)d\theta, \qquad (2.2.57)$$

which gives by integration with respect to time

$$M(t) = 1 - \int_T p(\theta, 0)e^{-\int_0^t f(\theta+t')dt'}d\theta = 1 - k(t). \qquad (2.2.58)$$

The periodicity of f leads to the relation between $k(t+\pi)$ and $k(t)$ defined in (2.2.58),

$$k(t + \pi) = \bar{\alpha}\, k(t). \qquad (2.2.59)$$

Finally, an interesting property concerns the evolution of the function $B(t)$. Defining the difference

$$\beta(t) = B(t + \pi) - \bar{\alpha}\, B(t) = \int_t^{t\,|\,\pi} b(t')\alpha(t + \pi - t')\mathrm{d}t', \qquad (2.2.60)$$

we can show that

$$\beta(t) = 1 - \bar{\alpha}, \qquad (2.2.61)$$

for $t \geqslant 0$. Here is the proof:

$$\beta(t) = M(t + \pi) - \bar{\alpha}M(t) = 1 - k(t + \pi) - \bar{\alpha}(1 - k(t)) = 1 - \bar{\alpha}. \quad (2.2.62)$$

The relations established above are useful when checking the numerical results. In particular, we have verified the property (2.2.61). Finally, we make the conjecture that, as t tends to $+\infty$, $b(t)$ tends to its average value, $\langle b_{st} \rangle$ of $\gamma' \sin^2(\theta)$ calculated with the stationary probability given in equation (2.2.72):

$$\lim_{t \to +\infty} b(t) = \langle b_{st} \rangle. \qquad (2.2.63)$$

This statement is expected intuitively and agrees with the numerics. We haven't established it rigorously, but made a step towards the proof by considering the Fourier transform of $H(t)b(t)$, H being the Heaviside function, equal to unity for $t \geq 0$ and zero for $t < 0$. Defining the Laplace transform of Hb by

$$\widehat{Hb}(z) = \int_0^\infty b(t)e^{-izt}\mathrm{d}t$$

we show in the Appendix that

$$\widehat{Hb} = \frac{\langle b_{st} \rangle}{iz} + \frac{\tilde{b}(z)}{\widehat{H_\pi \alpha}(z)}, \qquad (2.2.64)$$

where H_π is the characteristic function of $[0, \pi]$, and $\tilde{b}(z)$ is an analytical function for any bounded value of z. The full proof of convergence toward the steady state at infinite positive times requires some knowledge of the location of the zeros of $\widehat{H_\pi a}(z)$ in the complex plane.

2.2.5 Steady Solutions of the Kolmogorov Equation

The steady distribution $p_{st}(\theta)$, if it exists, must satisfy the integro-differential equation

$$\frac{\partial}{\partial \theta}p_{st} = \gamma'\left(\delta_D(\theta)\left(\int_{-\pi/2}^{\pi/2} p_{st}(\theta)\sin^2(\theta)\mathrm{d}\theta\right) - p_{st}((\theta))\sin^2(\theta)\right), \quad (2.2.65)$$

which shows that it depends on the single dimensionless parameter $\gamma' = \frac{2\gamma}{\Omega}$. It can be checked that the solution of equation (2.2.65) is a periodic function of θ of period π. By integrating the equation from $-\pi/2$ to $+\pi/2$ one obtains zero on the right-hand side, whereas the left-hand side is proportional to the difference $p_{st}(\pi/2) - p_{st}(-\pi/2)$, which is zero as well.

For θ different from zero, the integro-differential equation (2.2.65) reduces to

$$\frac{\partial}{\partial \theta}\hat{p}_{st}(\theta) = -\gamma' \sin^2(\theta)\,\hat{p}_{st}(\theta),\qquad(2.2.66)$$

which can be formally integrated as

$$\hat{p}_{st}(\theta) = \hat{p}_{st}(0_+)\alpha(\theta),\qquad(2.2.67)$$

where, as before, $\alpha(\cdot)$, with only one argument, is the restriction to the line $\theta = t$ of the function $\alpha(\theta, t)$ defined in equation (2.2.41). It gives

$$\alpha(t) = e^{\left(-\int_0^t\, f(t-t')\mathrm{d}t'\right)},$$

or

$$\alpha(\theta) = e^{-\frac{\gamma'}{4}(2\theta - \sin(2\theta))},\qquad(2.2.68)$$

for the particular case of f given in (2.2.38). This solution is formally not convenient a priori because the exponent of α is not periodic with respect to θ . However the periodicity is restored by noticing that the solution has a jump at $\theta = 0$. This jump is such that the value of $p_{st}(\theta)$ for $\theta = 0_-$ is just equal to $\hat{p}_{st}(\theta)$ for $\theta = \pi_-$, as given by the solution of equation (2.2.67). It remains to find the constant of integration $\hat{p}_{st}(0_+)$, which is derived from the norm constraint (2.2.17). Defining

$$\mathcal{I}_x = \int_0^x \alpha(\theta)\mathrm{d}\theta,$$

the norm condition gives

$$\hat{p}_{st}(0_+) = (\mathcal{I}_\pi)^{-1},\qquad(2.2.69)$$

which depends on the dimensionless parameter γ', as illustrated in Figure 2.2-(a). The stationary probability distribution, see Figure 2.2-(b), is the wrapped periodic function built by translating the solution $\hat{p}_{st}(\theta)$ just found, given by equations (2.2.67)–(2.2.69) and defined on the interval $[0_+, \pi_-]$. This can be written formally as

$$p_{st}(\theta) = \sum_{k=-\infty}^{\infty} \hat{p}_{st}(\theta - k\pi),\qquad(2.2.70)$$

where k is an integer. This function displays discontinuities for any value $\theta = k\pi$. In the interval $[0, \pi]$ the derivative $p_{st,\theta}$ calculated with the meaning of a distribution is equal to

$$\frac{\partial p}{\partial \theta} + \mu \delta_D(\theta),$$

where μ is the size of the jump of p_{st} at $\theta = 0$, $\mu = \hat{p}(0_+) - \hat{p}(\pi_-)$. This derivative is identical to the right-hand side of (2.2.65) and we have the relation

$$\langle b_{st} \rangle = \gamma' \langle \sin^2 \theta \rangle = \hat{p}_{at}(0_+) - \hat{p}_{st}(\pi_-) = \hat{p}_{st}(0_+)(1 - \alpha(\pi)).$$

From equations (2.2.67)–(2.2.69) we derive the two relations

$$\gamma' \int_0^\pi d\theta = 1 - \alpha(\pi) \qquad (2.2.71)$$

and

$$\langle b_{st} \rangle = \frac{1 - \alpha(\pi)}{\mathcal{I}_\pi}, \qquad (2.2.72)$$

where α is given by equation (2.2.68).

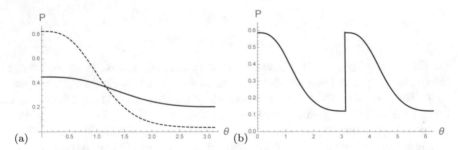

Fig. 2.2 The stationary probability distribution $p_{st}(\theta)$, a solution of equation (2.2.65). (a) In the interval $[0, \pi]$, solid line for $\gamma' = 0.5$, dashed line for $\gamma' = 2$. The periodicity of the wrapped distribution is illustrated in (b) for $\gamma' = 1$ on the interval $[0, 2\pi]$.

2.2.6 The Kolmogorov Equation in Two Limits

Below we look at what kind of approximation can be made in the opposite limits of a large or a small dimensionless ratio $\frac{\gamma}{\Omega}$.

2.2.6.1 Large Damping: $\frac{\gamma}{\Omega} \gg 1$

When this ratio is large it means that the damping due to the random emission of photons is large compared to the driving by the external pump field. Therefore the atom will be mostly in its ground state. This means that the angle θ remains close to zero. Therefore one can replace in this limit $\sin(\theta)$ by

θ. The Kolmogorov equation (without introducing any new symbol) becomes:

$$\frac{\partial p}{\partial t} + \frac{\Omega}{2}\frac{\partial p}{\partial \theta} = \gamma\left(\delta_D(\theta)\left(\int_{0_-}^{\pi} p(\theta',t)(\theta')^2 d\theta'\right) - p(\theta,t)\theta^2\right). \qquad (2.2.73)$$

This can be transformed into a parameterless equation by taking $\left(\frac{\Omega}{\gamma}\right)^{1/3}$ as the unit for θ and $1/2\left(\Omega^2\gamma\right)^{-1/3}$ as the unit of time. Because this is a small number in the limit of large γ we can ignore the condition of periodicity with respect to θ and take the new transformed scaled variable replacing θ as going from minus to plus infinity. With this variable the Kolmogorov equation becomes:

$$\frac{\partial p}{\partial t} + \frac{\partial p}{\partial \theta} = \left(\delta_D(\theta)\left(\int_{-\infty}^{+\infty} p(\theta',t)(\theta')^2 d\theta'\right) - p(\theta,t)\theta^2\right). \qquad (2.2.74)$$

The normalization condition becomes:

$$\int_{-\infty}^{+\infty} p(\theta',t)d\theta' = 1.$$

This yields a universal parameterless problem where all the scaling laws are exhibited as a consequence of the transformations made to obtain this parameterless equation.

2.2.6.2 Small Damping: $\frac{\gamma}{\Omega} \ll 1$

This limit is interesting because it gives an idea of the general solution, which is more involved. In this limit it is easy to check that, if one neglects the right-hand side of equation (2.2.65), the solution for the steady distribution $p_{st}(\theta)$ is the constant $1/\pi$.

The case of a small γ can be dealt with as follows. The Rabi oscillations make a fast motion. Therefore, during those oscillations, the function $\sin^2(\theta)$ as it appears in equation (2.2.19) can be replaced by its average $1/2$. Therefore, in this limit the Kolmogorov equation becomes:

$$\frac{\partial p}{\partial t} + \frac{\partial p}{\partial \theta} = \gamma'\left(\delta_D(\sin\theta) - p(\theta,t)\right). \qquad (2.2.75)$$

In the equation above, $2/\Omega$ is taken as the time unit and $\gamma' = \frac{\gamma}{\Omega}$ is a small parameter. Let us introduce the function

$$\hat{p} = e^{-\gamma't}p(\theta,t).$$

It satisfies the equation:

$$\frac{\partial \hat{p}}{\partial t} + \frac{\partial \hat{p}}{\partial \theta} = g(\theta,t), \qquad (2.2.76)$$

where

$$g(\theta, t) = \gamma' e^{-\gamma' t} \delta_D(\sin\theta) = \gamma' e^{-\gamma' t} \sum_k \delta_D(\theta - k\pi),$$

where the index k is any integer, positive, zero or negative. The solution of equation (2.2.76) with initial condition $\hat{p}(\theta, t = 0) = p_0(\theta)$ reads:

$$\hat{p}(\theta, t) = p_0(\theta - t) + \int_0^t g(\theta - t', t - t')dt'. \tag{2.2.77}$$

The integral over t' can be carried out and yields

$$\int_0^t g(\theta - t', t - t')dt' = \sum_k e^{-\gamma'(\theta - k\pi)} H(\theta - k\pi)H(t - \theta + k\pi),$$

where $H(\cdot)$ is the Heaviside function. This yields at once the final result for the solution of the initial value problem in the limit of very small γ':

$$p(\theta, t) = e^{-\gamma' t} p(\theta - t, t = 0) + \gamma' \sum_k e^{-\gamma'(\theta - k\pi)} H(\theta - k\pi)H(t - \theta + k\pi).$$
$$\tag{2.2.78}$$

Because of the Heaviside function it is not completely obvious how to find the limit of this expression of $p(\theta, t)$ for t tending to plus infinity. The sum over k is limited by the condition expressed by the product of Heaviside functions. If one takes $\theta = 0$ to simplify the algebra, k should be such that $0 > k > -t/\pi$. In the limit as t tends to plus infinity the lower bound on k' can be taken as minus infinity because the sum over k' converges in this limit of large negative values. The result is:

$$\gamma' \sum_k e^{-\gamma'(-k\pi)} H(-k\pi)H(t + k\pi) \to \gamma' \sum_{k<0} e^{\gamma' k\pi} = \gamma' \frac{\gamma'}{1 - e^{-\gamma'\pi}}.$$

One can check that, in the limit of small γ', this is also the same limit for any value of θ in the interval $(0, \pi)$ and that this limit is equal to the constant $1/\pi$, the same as the value of $p_{st}(\theta)$ derived before.

2.2.7 Between Two Jumps: Statistics of the Time Intervals

Here we study the statistics of the point process formed by the quantum jumps. It has the following properties:

(i) The number of jumps in non-overlapping intervals are independent variables.

(ii) In a small interval $[t, t + dt]$ there is at most one quantum jump.

These two properties define a Poisson process. We show below that this Poisson process is non-stationary, in the sense that its density (or intensity) is not constant versus time.

Let us look at a simple problem, namely the calculation of the probability distribution of the time intervals between two photons emitted consecutively by an atom (or an ion) resonantly pumped. This question has an interesting answer and it also displays in a non-trivial way the statistical principles of quantum mechanics. The random process in the present case is the emission of a photon by an atom in an excited state. Let a_0 and a_1 be the amplitude of the wave functions of the ground and excited state. Their coupled dynamical equations read, with the pumping field on:

$$i\dot{a}_0 = \Omega \cos(\omega t) a_1$$

and

$$i\dot{a}_1 = -\omega a_1 + \Omega \cos(\omega t) a_0,$$

where $\hbar\omega$ is the energy difference between the two eigenstates of the atom and where $\omega/(2\pi)$ is the frequency of the resonant pump. The energy of the ground state is set to zero to simplify the writing. The pump field is represented in the equations by the time-dependent terms with coefficients $\Omega \cos(\omega t)$. We assume the (realistic) limit $\Omega \ll \omega$, namely a pump field far weaker than the electric field binding the electrons to the atom. In this limit one can solve the equations by perturbation with the result that a_0 and a_1 display optical Rabi oscillations, such that $a_0(t) = A_0(t)$ and $a_1(t) = A_1(t)e^{i\omega t}$, where A_0 and A_1 depend slowly on time.

$$A_0(t) = \cos\theta(t) \tag{2.2.79}$$

and

$$A_1(t) = i\sin\theta(t), \tag{2.2.80}$$

with $\dot{\theta} = \Omega/2$. Therefore:

$$A_0 = \cos(\Omega t) \tag{2.2.81}$$

and

$$A_1 = i\sin(\Omega t). \tag{2.2.82}$$

Note that $|A_0|^2 + |A_1|^2 = 1$, a normalization condition easy to impose in the two-level case, but more problematic in the three-level case. This picture of the pumped two-state atom is not complete because it does not include the emission of photons when the atom is in the excited state. When the atom is exactly in the excited state, it may decay toward the ground state by emitting a photon at a rate γ per unit time. The emission of a photon is a very quick "quantum jump" with a duration of order $1/\omega$, which is small compared to any other time scale in the process. When the state of the atom is a linear superposition of the excited and ground state, the rate of emission of a photon is less than γ because it has to tend to zero as the angle θ tends to zero. As was shown by Dirac [54], the rate γ is proportional to a matrix element of V, the interaction operator between the electromagnetic field and the atom:

$$\Gamma = k \, |\langle i|V|f\rangle|^2,$$

k being a coefficient independent of the quantum states. The state $\langle i|$ is the initial quantum state of the *full* system (atom + electromagnetic (EM) field) before the transition and $|f\rangle$ is the final state after the transition. Because Γ is a rate of transition, it is found as the coefficient of a secular term in the expansion of the solution of the quantum equations of motion for small interaction between the EM field (namely the photons) and the electrons in the atom, a secular term due to the fact that the energies of the final and initial state are the same.

The state $\langle i|$ is a linear superposition of the ground and excited state of the atom whereas the field is in a photon-less state. This is written symbolically as:

$$\langle i| = (\sin(\theta)\langle e| + \cos(\theta)\langle g|) \otimes \langle 0_{ph}|,$$

where $\langle e|$ is the excited state and $\langle g|$ the ground state, whereas $\langle 0_{ph}|$ and (later) $\langle 1_{ph}|$ are the zero-photon (above the zero point energy) and one-photon state. In this notation $\langle i|$ is obviously not an eigenstate of the energy operator, because $\langle g|$ and $\langle e|$ have different energies. This does not matter because the state $\langle g|$ will not contribute to $\langle i|V|f\rangle$. Our writing of $\langle i|$ is to emphasize that the initial state is normalized to one, with the sum of two contributions which are not eigenstates with the same energies. This initial state comes from the Rabi oscillations which mix the ground and excited state. The quick quantum jump brings back the atom to the ground state and the balance of energy is satisfied by the emission of the photon.

The final state, with the same energy as the initial state, reads:

$$|f\rangle = |g\rangle \otimes (\sin(\theta)|1_{ph}\rangle + \cos(\theta)|0_{ph}\rangle).$$

The contributions of $|f\rangle$ and $\langle i|$ to Γ that are proportional to $\cos(\theta)$ do not matter because the transition does not change them. Moreover, the coefficients $\sin(\theta)$ and $\cos(\theta)$ ensure that the energy of the final and initial states are the same. The coefficient $\sin(\theta)$ of the contribution $|g\rangle \otimes |1_{ph}\rangle$ to the final state is the same as the coefficient of $\langle e| \otimes \langle 0_{ph}|$ in the initial state, because this part of the initial state is the only one changed by the interaction, which has to conserve the norm, proportional to $\sin^2(\theta)$. Once the expressions of the initial and final state are inserted into the above expression for the rate of transition Γ one obtains two contributions to the matrix element $\langle i|V|f\rangle$. Because V is time-independent, the only contributions to this matrix element that survive (that is, which do give at the end secular contributions to the amplitude of the ground state and whence contribute to Γ) are those with the same energies of the initial and final state, namely:

$$\langle i|V|f\rangle = \sin^2(\theta)(\langle e| \otimes \langle 0_{ph}|V|g\rangle \otimes |1_{ph}\rangle) + \cos^2(\theta)\langle g| \otimes \langle 0_{ph}|V|g\rangle \otimes |0_{ph}\rangle.$$
$$(2.2.83)$$

Only the first term on the right-hand side is relevant. The second one includes a matrix element of V proportional to the dipole operator and diagonal with respect to the ground state. Such a matrix element is zero by symmetry because the wave-function of the ground state is isotropic, a so-called s-state. The part of $\langle i|V|f\rangle$ proportional to $\sin^2(\theta)$ is the one derived by considering the decay of pure excited states only (namely for values of the angle θ such that $\sin^2(\theta) = 1$). Putting this into Γ one finds that for any value of θ, $\Gamma = \gamma \sin^4(\theta)$, where γ is the rate computed for eigenstates of the Hamiltonian without mixing between excited and ground states, namely with $\sin^2(\theta) = 1$.

This allows us to find the probability distribution of the time intervals between two photons in the case of a two-level atom pumped resonantly. Let us first compute the probability distribution of the time t of emission of a photon, assuming the atom is in the excited state without pumping at $t = 0$. Let $K(t)$ be the probability that the atom is still in the excited state at time t. Between t and $t + dt$, dt small, the atom will jump back to the ground state by emitting a photon with probability $K(t)$ times the probability of emitting a photon, namely γdt. The factor $K(t)$ is there to represent the constraint that the atom is still in the excited state at time t. To obtain $K(t + dt)$, one must subtract $K(t)\gamma dt$ from the probability $K(t)$. Therefore

$$K(t + dt) = K(t) - K(t)\gamma dt.$$

This becomes the differential equation for a Poisson process:

$$\dot{K}(t) = -\gamma K(t),$$

with the solution

$$K(t) = e^{-\gamma t}.$$

This solution satisfies the constraint that $K(t = 0) = 1$, which expresses that immediately after the emission of a photon, the atom is in the ground state. Let us apply the same reasoning to the case of a two-level atom resonantly pumped. This atom will emit photons. Let $L(t)$ be the probability distribution for the atom to have emitted no photon in the interval $(0, t)$, $t = 0$ being again the instant of emission of a photon. A first-order differential equation for $L(t)$ is derived in the same way as that for $K(t)$. We have the same kind of relation between $L(t + dt)$ and $L(t)$:

$$L(t + dt) = L(t) - \gamma r(t)L(t)dt.$$

Compared to the equation for $K(t)$ there is a new term in this equation, the factor $r(t)$ on the right-hand side. It is there to take into account that γ is changed into the θ-dependent Γ. In the case of the decay of an atom initially in the excited state and without pump field, the atom remains in the excited state all the time before the quantum jump and $r = 1$. In the present case this is not so because between time 0 and t the atom makes Rabi oscillations. Therefore $\gamma r(t)$ is the probability that the atom, in an excited state with a photon-less field, will jump to the ground state and a one-photon state of the EM field. We just showed that this probability of jump per unit time is $\Gamma = \gamma \sin^4(\theta)$, where θ is the angle permitting us to compute the balance of probability between the excited and ground state, as it appears in equations (2.2.79) and (2.2.80). In the present case this angle θ drifts at constant speed under the effect of the Rabi oscillations. The angle $\theta(t)$ starts at $t = 0$ with the value zero because just after the quantum jump the atom is in the ground state, which corresponds to $\theta = 0$. Therefore at any positive time, $\theta = \Omega t$, after emission of a photon. By simple algebra one has

$$r(t) = \sin^4(\theta) = \frac{1}{4}\left(\frac{3}{2} + \frac{1}{2}\cos(4\Omega t) - 2\cos(2\Omega t)\right).$$

By integration:
$$L(t) = e^{-\gamma q(t)},$$

where

$$q(t) = \int_0^t r(t')dt' = \frac{1}{4}\left(\frac{3t}{2} + \frac{1}{8\Omega}\sin(4\Omega t) - \frac{1}{\Omega}\sin(2\Omega t)\right).$$

Let us turn now to the physically measurable quantity, namely the statistics of the time intervals between two successive emission of photons. Let $M(t)$ be this probability, which must be normalized to one, because, unlike $K(t)$ and $L(t)$, it is a probability in the usual sense and not a conditional probability. The probability of no emission between time 0 and time t is the integral from time t to infinity of $M(t)$, namely of the statistical weight of all intervals longer than t:

$$L(t) = \int_t^\infty M(t')dt'.$$

Therefore by simple algebra $\dot{M}(t) = -L(t)$. In the present case this yields:

$$M(t) = \gamma r(t)e^{-\gamma \int_0^t r(t')dt'}. \tag{2.2.84}$$

As is easily checked, this is a probability distribution normalized to 1.

Note that the probability distribution of time intervals $M(t)$ shows "anti-bunching" of the jumps at different times: the function $M(t)$ decreases periodically to zero at times $\Omega t = k\pi$, k positive integers, when the Rabi oscillations bring the atom back to the ground state. This anti-bunching is explained by the property that, starting from the ground state, it takes some time to build up the amplitude of the excited state by interaction with the pump field. Later on there is also no emission of a photon when the atom is back in its ground state because of the Rabi oscillations.

A limit of interest is when the rate of spontaneous decay is far bigger than the period of the Rabi oscillations, namely the limit $\Omega << \gamma$. In this limit one must consider Ωt as small in the exponential of equation (2.2.84). This yields

$$r(t) \approx \Omega^4 t^4$$

so that

$$q(t) \approx \frac{\Omega^4}{5} t^5.$$

Therefore

$$\frac{\gamma}{2}(t - \frac{\sin \Omega t}{\Omega}) \approx \frac{\gamma \Omega^2}{12} t^3,$$

and

$$L(t) \approx \gamma \Omega^4 t^4 e^{-\frac{\gamma \Omega^4}{5} t^5}. \tag{2.2.85}$$

In the limit $\Omega << \gamma$ this makes a fair approximation for the bulk of the probability distribution of the intervals between two emitted photons. It shows that this mean time interval is of order

$$\tau_L = (\Omega^4 \gamma)^{-1/5}. \tag{2.2.86}$$

This "typical" time for the interval between two emissions is between the long time $1/\Omega$ (period of Rabi oscillations) and the short time $1/\gamma$ (lifetime of the excited state). It shows in particular that between two jumps, on average, a complete Rabi oscillation cannot take place because it takes too long. This time is also longer than $1/\gamma$. This illustrates the anti-bunching effect which prevents the occurrence of very close quantum jumps, because it takes some time after a jump for the amplitude of the excited state to grow enough until an emitted photon brings the atom back to its ground state.

Chapter 3
The Statistical Theory of Shelving

The idea of shelving was presented by Dehmelt [52, 53]. It is a way to provide evidence for fundamental principles of quantum theory and also to measure frequencies of photons with an unprecedented accuracy. This uses the property of what is called Paul's trap, a device that keeps almost motionless a single charged particle like an ion. This trap was invented for accurate mass spectroscopy, but it can also be used in studies of the interaction of light with a single ion. For that purpose one illuminates a trapped ion and observes its fluorescence. The shelving process requires an ion with three (electronic) states, called 0 (the ground state), 1 (the "regular" excited state) and 2 (the long lived excited state). State 2 may even take minutes to decay spontaneously by emission of photons. The idea of shelving is to put the ion in the long lived state. When this occurs, and if the ion is pumped resonantly at the transition frequency between level 0 and 1, the fluorescence light between these two levels is interrupted by a dark period because the ion is "on the shelf", that is, in state number 2. The challenge is to find the statistical properties of the average duration of emission of fluorescence light between shelving and the average duration of dark periods when the ion is on the shelf.

The shelving process is concerned with the limit where all time scales associated to level 2 (except its Bohr frequency) are much longer than those associated to the other two "regular" levels, 0 and 1. Because of this, one can consider instead of the full dynamics of the 0 and 1 levels the "average" oscillations between them, so that everything implying level 2 will involve only such averaging. This is a case of adiabatic perturbation. Things work rather differently depending on the case of shelving one considers. The case of the so-called Λ shelving is rather straightforward as it implies only simple averages of the fast oscillations between level 0 and 1. The V shelving is more complex because it introduces the time dependence of the fast fluctuations of the amplitude of level 0, in a similar way to how the motion of a heavy Brownian particle depends on the fast fluctuations of the Langevin force. In the coming two subsections we deal with the two cases one after the other.

© Springer Nature Switzerland AG 2019

Y. Pomeau, M.-B. Tran, *Statistical Physics of Non Equilibrium Quantum Phenomena*,
Lecture Notes in Physics 967, https://doi.org/10.1007/978-3-030-34394-1_3

3.1 The Statistical Theory of Shelving: The Λ Case

In this so-called Λ case, state 1 is at the highest energy level. It is excited by pumping from the ground state to level 1. Rarely it jumps from state 1 to state 2, which has an energy between the ground state and the energy of level 1. When this happens the fluorescence of the jumps between level 1 and 0 stops, which can be observed because state 2 is long-lived. This dark period stops when state 2 decays to 0 by emitting a photon.

We split the probability distribution of the amplitude of the three levels into two pieces, one is the probability for the atom to be either in state 0 or 1. Let P_1 be this probability. The equation of motion of P_1 is nothing but the Kolmogorov equation for a two-level system with an optical Rabi circular frequency Ω_1 and a rate of emission of photons γ_1. The equation for the probability distribution P_1 is equation (2.2.19), the dynamical equation for a probability distribution depending on an angle θ such that the amplitude of state 0 is proportional to $\cos\theta$ whereas that of state 1 is proportional to $\sin\theta$. Compared to the two-level case of equation (2.2.19) we shall put an index 1 to mean that only the pair of states 0 and 1 is involved.

Like the general Kolmogorov equation, the equation for $P_1(\theta_1, t)$ includes on its left-hand side a streaming term representing the Rabi oscillation, whereas the right-hand side includes two terms, both representing the effect of random emission of photons leading to a return to the ground state. This right-hand side has the familiar structure of the Kolmogorov equations for Markov processes with a gain and a loss term. The final result is,

$$\frac{\partial}{\partial t} P_1 + \frac{\Omega_1}{2} \frac{\partial}{\partial \theta_1} P_1 = \tag{3.1.1}$$

$$\gamma_1 \left(\delta_D(\sin\theta_1) \left(\int_{-\pi/2}^{\pi/2} P_1(\theta_1', t) \sin^2\theta_1' d\theta_1' \right) - P(\theta_1, t) \sin^2\theta_1 \right).$$

This equation obviously neglects the small terms (small with respect to the terms kept explicitly) representing the interaction with the shelf level 2, namely the filling of this level by the unlikely transition from 1 to 2 and the return to the ground state 0 by emission of a photon. Because this concerns the Λ case, there is a priori no pumping by a resonant EM field at the transition frequency between 2 and 0. Therefore one does not have to introduce any dependence of the probability distribution linked to state 2 with respect to an angle θ_2. It is sufficient to consider a simpler quantity, namely the norm of state 2, denoted Q_2, a number between 0 and 1. The total probability is now the sum

$$Q_2(t) + \int_{-\pi/2}^{\pi/2} P_1(\theta_1, t) d\theta_1.$$

To take into account the quantum jumps from 1 to 2 and from 2 to 0 one has to add new terms to equation (3.1.1) and to write another equation for $Q_2(t)$. Because the transition rates involving the shelf state 2 are much smaller than both Ω_1 and γ_1 one can assume that P_1 remains close to the steady solution of equation (3.1.1), up to a slowly varying norm, denoted Q_1 and defined formally as

$$Q_1(t) = \int_{-\pi/2}^{\pi/2} P_1(\theta_1, t) \mathrm{d}\theta_1.$$

In the adiabatic limit, $P_1(\theta_1, t)$ is very close to the function $Q_1(t)P_{st,1}(\theta_1)$, where $P_{st,1}$ is the steady solution of equation (3.1.1) normalized to 1 and derived in Section 2.2.5. The slow evolution of $Q_1(t)$ and $Q_2(t)$ is due to the rare transitions from state 1 to 2 with rate Γ_2 and from state 2 to the ground state with rate γ_3, the adiabatic approximation being valid if $\gamma_i \ll \gamma_1$ with $i = 2$ and 3. The equations for Q_1 and Q_2 read:

$$\frac{\partial Q_1}{\partial t} = \gamma_3 Q_2 - \Gamma_2 Q_1 \tag{3.1.2}$$

and

$$\frac{\partial Q_2}{\partial t} = \Gamma_2 Q_1 - \gamma_3 Q_2. \tag{3.1.3}$$

The rate Γ_2 was computed before, it is the average rate of transition from 1 to 2 during the fast oscillations between the ground state and state 1. It is

$$\Gamma_2 = \gamma_2 \langle \sin^4(\theta_1) \rangle,$$

where γ_2 is the small rate of decay of the state 1 to state 2 and where the average is made over the steady state distribution of θ_1 given by the steady solution of the Kolmogorov equation (3.1.1) normalized to one.

Those are the familiar equations for a two-state Poisson process. Their solution directly gives all the statistical properties of the time intervals of darkness (when the atom is on the shelf) and of emission of light (when the transitions 1 to 0 and back are operating).

3.2 The Statistical Theory of Shelving: The V Case

This case is more complex than the previous one because the shelf level is filled by either absorption of random photons near the 0 to 2 transition or by (slow) Rabi oscillations between levels 0 and 2. The first case can be dealt with almost in the same way as the Λ case, even though it obviously corresponds to a different physics. The other one, namely the filling of level 2 by Rabi oscillations between level 0 and 2, is different. The origin of this difference lies in the fact that, a priori, the oscillations from 0 to 2 are very slow. Therefore during these oscillations the state 0 has time to change, in particular to reach a different phase because of the random emission of photons from state 1 to state 0. Because of this randomness of the phase the oscillations are not coherent and the changes of the amplitude of 2 are just the random addition or subtraction of small contributions to this amplitude.

To see this last point, let us consider the equation for the amplitude of the 2 state in the presence of weak pumping resonant with the frequency difference of level 0 and 2. Following the same line of thinking as when deriving equation (2.2.80), we obtain the equation for the amplitude $A_2(t)$:

$$\frac{\partial A_2}{\partial t} = i\Omega_2 A_0(t). \tag{3.2.1}$$

In this equation, because of the smallness of the coefficients associated to the shelf mode, the amplitude A_0 can be seen as a given fluctuating quantity depending on the transitions from level 0 to 1, either by Rabi oscillations or by random emission of photons, as described by equation (3.1.1). The solution of equation (3.2.1) is:

$$A_2(t) - A_2(0) = i\Omega_2 \int_0^t A_0(t')\mathrm{d}t'. \tag{3.2.2}$$

We are looking for the value of

$$|A_2(t)|^2 = |A_2(t)A_2^*(t)|,$$

where A^* is the complex conjugate of A. It is given by:

$$|A_2(t)|^2 - |A_2(0)|^2 = \Omega_2^2 \int_0^t \int_0^t A_0(t')A_0^*(t'')\mathrm{d}t''\mathrm{d}t'. \tag{3.2.3}$$

Because Ω_2 is small, $|A_2(t)A_2^*(t)|$ changes much more slowly than $A_0(t)$. Therefore one can find the leading order value of the right-hand side of equation (3.2.3) by substituting for the product $A_0(t')A_0^*(t'')$ its averaged value resulting over the fast dynamics of the levels 0 and 1. This yields

$$|A_2(t)|^2 - |A_2(0)|^2 \approx 2t\Omega_2^2 \int_0^\infty \langle A_0(0)A_0^*(t')\rangle \mathrm{d}t'. \tag{3.2.4}$$

This shows that the effect of the fluctuations of A_0 is to induce a Brownian-like dynamics of A_2. This is equivalent to substituting for the streaming term on the left-hand side of equation (3.1.1) a diffusion term when dealing with the probability distribution $P_2(\theta_2, t)$, which is the equivalent of $P_1(\theta_1, t)$ but for the pair of levels 0 and 2. The result is the equation:

$$\frac{\partial P_2}{\partial t} = \frac{\partial}{\partial \theta_2}\left(D\cos^2(\theta_2)\frac{\partial P_2(\theta_2, t)}{\partial \theta_2}\right) \tag{3.2.5}$$

$$+ \gamma_2\left(\delta_D(\sin\theta_2)\left(\int_{-\pi/2}^{\pi/2} P_2(\theta_2', t)\sin^2\theta_2'\mathrm{d}\theta_2'\right) - P_2(\theta_2, t)\sin^2\theta_2\right),$$

where

$$D = 2\Omega_2^2 \int_0^\infty \langle A_0(0)A_0^*(t')\rangle \mathrm{d}t'.$$

In this expression of D the average is taken over the fluctuations in the steady state of the two-level system, without any third partner (no level 2), so that the full probability is one. To map it into an equation for P_2 which involves a third level, a level of index 2, one has to multiply the product $\langle A_0(0)A_0^*(t')\rangle$ by $\cos^2(\theta_2)$.

Chapter 4

Summary, Conclusion and Appendix of Part 1

4.1 Summary and Conclusion

We intended to show how helpful the concepts of non-equilibrium statistical mechanics are in the understanding of the phenomenon of fluorescence of an atom or ion submitted to an electromagnetic wave at the frequency of resonance between two or three quantum levels. The coherent part of the dynamics of this system is well understood and is standard quantum physics. The spontaneous decay brings randomness into this system, a phenomenon requiring statistical methods for its understanding. This leads quite naturally to a Kolmogorov equation where the randomness inherent to the decay process and the determinism linked to the interaction with the light beam are put together in a coherent picture. Thanks to this, one recovers the physically reasonable result that, if the natural lifetime of the excited state decreases, the time-lag between two emissions of photons by the atom (or ion) also decreases. This shows how statistical theory can be successfully used to understand complex quantum phenomena of interaction of single atoms (or ions) with radiation.

4.2 Appendix of Part 1

This Appendix explains how to extend the theory of fluorescence from the two-level case to the three-level case in general. We recall first the case of a detuned pump and then use part of the results for the three-level case.

The matter of this Appendix is mostly due to Jean Ginibre, who is thanked for this masterpiece of science.

© Springer Nature Switzerland AG 2019
Y. Pomeau, M.-B. Tran, *Statistical Physics of Non Equilibrium Quantum Phenomena*,
Lecture Notes in Physics 967, https://doi.org/10.1007/978-3-030-34394-1_4

4.2.1 The Kolmogorov Equation for a Detuned Atom-Laser Transition: The 2-Level Case

For simplicity of presentation let us set $\xi = 0$ (see below the solution for $\xi \neq 0$). With $A = (a_0, a_1)^\dagger$, equations (2.2.3)–(2.2.4) can be written as

$$2i\frac{\partial}{\partial t}A = \Omega \begin{pmatrix} 0 & e^{i(t\delta-\xi)} \\ e^{-i(t\delta-\xi)} & 0 \end{pmatrix} A \tag{4.2.1}$$

in the general case of non-zero detuning δ. The solution corresponding to the initial condition $a_0(0) = 1$ and $a_1(0) = 0$ is given by the expressions

$$a_0(t) = e^{i\frac{t\delta}{2}} \left(\cos\omega t - i\frac{\delta}{\Omega_\delta} \sin\omega t \right) \tag{4.2.2}$$

and

$$a_1(t) = -ie^{-i(\frac{t\delta}{2}-\xi)} \frac{\Omega}{\Omega_\delta} \sin\omega t, \tag{4.2.3}$$

where $\omega = \Omega_\delta/2$, with Ω_δ the effective Rabi frequency in the presence of detuning,

$$\Omega_\delta = \sqrt{\Omega^2 + \delta^2}. \tag{4.2.4}$$

Using the variables $a_{0,1}$ the evolution equation for the probability (or any other function of the amplitudes $a_{0,1}$) should have partial derivatives not only with respect to t and $\theta = \omega t$ but also with respect to $\psi = \delta t$ and ξ, which leads to awful expressions for the Kolmogorov equation. We are going to show that a fair change of variables allows us to obtain an unexpected result, namely a Kolmogorov equation of the same form as the one for the resonant case. The choice of appropriate phase space comes from the one used for the three-level case treated in Section 4.2.3. In the next subsection we assume that the atomic dipole moment is real, or $\xi = 0$, in order to simplify the presentation. The solution for $\xi \neq 0$ is given at the end of this appendix.

4.2.1.1 New Amplitudes

Let us define a parameter η which obeys the relations

$$\sin 2\eta = \frac{\delta}{\Omega_\delta} \qquad \cos 2\eta = \frac{\Omega}{\Omega_\delta}. \tag{4.2.5}$$

To get rid of the phase factors, let us define the spinor B (which has nothing to do with the function $B(t)$ defined in equation (2.2.53)),

$$B = \begin{pmatrix} b_0 \\ b_1 \end{pmatrix} = \begin{pmatrix} e^{-it\delta/2}a_0 \\ e^{it\delta/2}a_1 \end{pmatrix}. \tag{4.2.6}$$

The system (4.2.1) becomes

$$i\frac{\partial}{\partial t}B = \frac{\Omega_\delta}{2}\begin{pmatrix} \sin 2\eta & \cos 2\eta \\ \cos 2\eta & -\sin 2\eta \end{pmatrix}B, \qquad (4.2.7)$$

with the initial condition $B(0) = A(0)$. Introducing the two Pauli matrices

$$\sigma_1 = \begin{pmatrix} 0 & 1 \\ 1 & 0 \end{pmatrix} \qquad \sigma_3 = \begin{pmatrix} 1 & 0 \\ 0 & -1 \end{pmatrix}, \qquad (4.2.8)$$

the system (4.2.7) becomes

$$i\frac{\partial}{\partial t}B = \frac{\Omega_\delta}{2}\left(\sigma_3 \sin 2\eta + \sigma_1 \cos 2\eta\right)B. \qquad (4.2.9)$$

Using the relation

$$\left(\sigma_3 \sin 2\eta + \sigma_1 \cos 2\eta\right)^2 = 1, \qquad (4.2.10)$$

the general solution of (4.2.9) is

$$B(t) = \left(\cos\frac{\Omega_\delta}{2}t - \left(\sigma_3 \sin 2\eta + \sigma_1 \cos 2\eta\right)\sin\frac{\Omega_\delta}{2}t\right)B(0). \qquad (4.2.11)$$

Using (4.2.6) we get the general solution for A, which gives back (4.2.2)–(4.2.3) for $a_1(0) = 0$.

Let us introduce the rotation matrix,

$$R(\eta) = \begin{pmatrix} \cos\eta & -\sin\eta \\ \sin\eta & \cos\eta \end{pmatrix}. \qquad (4.2.12)$$

The right-hand side of equation (4.2.7) can be written as a matrix product and equation (4.2.9) becomes

$$i\frac{\partial}{\partial t}B(t) = \frac{\Omega_\delta}{2}R^{-1}\sigma_1 RB. \qquad (4.2.13)$$

Using the spinor $C = (c_0, c_1)^\dagger$ defined by the relation

$$C(t) = R(\eta)B, \qquad (4.2.14)$$

we get the system

$$i\frac{\partial}{\partial t}C(t) = \frac{\Omega_\delta}{2}\sigma_1 C. \qquad (4.2.15)$$

Lastly the equations for the amplitudes (c_0, c_1) are identical to those of the amplitudes (a_0, a_1) in the resonant case. They are

$$i\frac{\partial}{\partial t}c_0 = \frac{\Omega_\delta}{2}c_1$$

and

$$i\frac{\partial}{\partial t}c_1 = \frac{\Omega_\delta}{2}c_0.$$

4.2.1.2 Change of Variables

To find a pertinent phase space we adapt the procedure detailed in Section 4.2.3 to the three-level atom. We define the new variables

$$u = 2|c_0|^2 - 1 = 1 - 2|c_1|^2 \qquad (4.2.16)$$

and

$$v = 2I(c_0 c_1^*), \qquad (4.2.17)$$

where $I(f)$ denotes the imaginary part of f. They obey the same differential equations as (4.2.61) with $2\omega = \Omega_\delta$. Using the polar coordinates (r, θ) associated to the variables (u, v) by the relation

$$u + iv = r \exp(2i\theta),$$

we are able to describe the motion in a 2D phase space (in the three-level case the phase space is $3D$, with variables (r, s, θ)). In this phase space the dynamics is governed by the equations

$$\frac{\partial}{\partial t}r = 0 \qquad (4.2.18)$$

and

$$\frac{\partial}{\partial t}\theta = \frac{\Omega_\delta}{2}. \qquad (4.2.19)$$

It follows that the deterministic motion is a rotation around a circle of constant radius $r = \sqrt{u^2 + v^2}$, with angular velocity $\Omega_\delta/2$. Note that the angular variable θ does not have the same meaning as the one used to describe the motion in the phase space of the amplitudes $a_{0,1}$. Here θ is twice that of the resonant case, because we consider in this subsection the motion of the functions $(u(t), v(t))$ which are quadratic with respect to the amplitudes (a_0, a_1).

4.2.1.3 Motion After an Emission

Just after the emission of a photon (at time indexed by a +), the initial conditions for the deterministic motion are $b_{1+} = a_{1+} = 0$ and $|b_{0+1}| = |a_{0+}| = 1$. In terms of the amplitudes $c_{0,1}$ this gives

$$c_{0+} = b_{0+} \cos\eta = a_{0+} \cos\eta$$

and

$$c_{1+} = b_{0+} \sin\eta = a_{0+} \sin\eta.$$

The two amplitudes $c_{0,1}$ are then in phase. It follows that $v_+ = \theta_+ = 0$, $R(c_0 c_1^*)_+ = \sin \eta \cos \eta$ and $r_+ = 2 \cos^2 \eta - 1$. In summary, the motion starts from the same initial point

$$\theta = 0$$

on the trajectory defined by the circle of radius

$$r_\delta = \cos 2\eta = \Omega/(2\omega). \tag{4.2.20}$$

4.2.1.4 The Kolmogorov Equation

We have proved that the phase space reduces to a single trajectory. On this trajectory one may put a probability $p(\theta, t) d\theta$. The left-hand side of the kinetic equation is that of equation (2.2.16). The right-hand side is built as in the resonant case, thanks to the transition function

$$\Gamma(\theta|\theta') = \gamma \delta_D(\theta') |a_1(\theta)|^2, \tag{4.2.21}$$

where δ_D stands for the wrapped Dirac distribution of period π. To calculate the amplitude $a_1(\theta)$ we use equations (4.2.16)–(4.2.17) together with the relation

$$R(c_0 c_1^*) = k, \tag{4.2.22}$$

where k is a constant. This constant can be deduced from the initial conditions $a_1 = b_1 = 0$ and the expressions

$$c_0 = b_0 \cos \eta - b_1 \sin \eta, \quad c_1 = b_0 \sin \eta - b_1 \cos \eta$$

taken from (4.2.14). This gives

$$k = |b_0|^2 = \sin \eta \cos \eta.$$

Using (4.2.6), we have $|a_1|^2 = |b_1|^2$. In terms of $c_{0,1}$, we have

$$|a_1|^2 = |-c_0 \sin \eta + c_1 \cos \eta|^2.$$

Introducing equation (4.2.16) and the relation $u = r \cos 2\theta$ with $r = r_\delta$ given in (4.2.20), we obtain:

$$|a_1|^2 = \frac{1}{2}(1 + u) \sin^2 \eta + \frac{1}{2}(1 - u) \cos^2 \eta - 2 \sin^2 \eta \cos^2 \eta = \cos^2(2\eta) \sin^2 \theta. \tag{4.2.23}$$

Finally, the transition function becomes

$$\Gamma(\theta|\theta') = \gamma (\frac{\Omega}{\Omega_\delta})^2 \delta_D(\theta') \sin^2 \theta. \tag{4.2.24}$$

The transport equation, or Kolmogorov equation, is given in (2.2.21). It takes the same form as in the resonant case when using the same variables, but with different coefficients. In the left-hand side the Rabi frequency is given in (4.2.4) and in the right-hand side the function $f(\theta)$ is given in (2.2.22).

4.2.2 Asymptotic Value of $b(t)$

We try to solve equation (2.2.34) by Fourier transform. Using relations (2.2.51) and (2.2.46) we get

$$Hb \star H\alpha = HM, \tag{4.2.25}$$

where $H(x)$ is the Heaviside function defined in Section 2.2.4.2, and $\star$ denotes convolution. The function $b(t)$ is positive and bounded, $0 \leq b \leq \|f\|_\infty$, and $0 \leq H\alpha \leq 1$, which gives the solution in the Fourier space

$$\widehat{Hb} = \frac{\widehat{HM}}{\widehat{H\alpha}}. \tag{4.2.26}$$

By definition, the Fourier transform of $H\alpha$ as a function of $z = x + iy$ is given by the integral

$$\widehat{H\alpha}(z) = \int_0^\infty \alpha(t)e^{-itz}\mathrm{d}t, \tag{4.2.27}$$

which can be written as

$$\widehat{H\alpha}(z) = \sum_{n \geq 0} \int_{n\pi}^{(n+1)\pi} \alpha(t)e^{-itz}\mathrm{d}t. \tag{4.2.28}$$

Putting relation (2.2.55) into (4.2.28) we get

$$\widehat{H\alpha}(z) = \sum_{n \geq 0} \int_0^\pi \alpha(t)e^{-itz}e^{-(\bar{f}+iz)n\pi}\mathrm{d}t, \tag{4.2.29}$$

where

$$\bar{f} = (1/\pi) \int_0^\pi f(t)\mathrm{d}t$$

is equal to $\gamma'/2$ in the particular case treated here $(f(x) = \sin(x))$, but we introduce the parameter $\bar{f}$ to be more general. For $y < \bar{f}$, equation (4.2.29) becomes

$$\widehat{H\alpha}(z) = \frac{1}{1 - e^{-(\bar{f}+iz)\pi}} \int_0^\pi \alpha(t)e^{-itz}\mathrm{d}t, \tag{4.2.30}$$

where the first r.h.s. factor is a meromorphic function of z having simple poles at $z = 2k + i\bar{f}$, k integer. The latter term is holomorphic for $y < \bar{f}$ and never vanishes. The second factor in equation (4.2.30) is an entire function of z. Its imaginary part,

$$I \int_0^\pi \alpha(t) e^{-itz} \mathrm{d}t = - \int_0^\pi \alpha(t) e^{ty} \sin tx \mathrm{d}t, \qquad (4.2.31)$$

is negative and has no zero for $y < 0$ because in this domain $\alpha(t)e^{yt}$ is decreasing. The question is: does this factor in (4.2.31) vanish in the domain $0 < y < \bar{f}$? Let us return to the original problem, namely the asymptotic solution of $b(t)$ from equation (4.2.25). The function M defined by

$$M(t) = 1 - \int_0^\pi p(\theta, 0) e^{- \int_0^t f(\theta + t') \mathrm{d}t'} \mathrm{d}\theta = 1 - k(t) \qquad (4.2.32)$$

has a Fourier transform given by

$$\widehat{HM}(z) = \widehat{H} - \widehat{Hk}. \qquad (4.2.33)$$

Treating the term $\widehat{Hk}$ in the same way as we did for $\widehat{H\alpha}$, we get

$$\widehat{Hk}(z) = \int_0^\infty k(t) e^{-itz} \mathrm{d}t = \frac{1}{1 - e^{-(\bar{f}+iz)\pi}} \int_0^\pi k(t) e^{-itz} \mathrm{d}t, \qquad (4.2.34)$$

where the second factor

$$\int_0^\pi k(t) e^{-itz} \mathrm{d}t = \int_0^\pi p(\theta, 0) \int_0^\pi e^{(- \int_0^t f(\theta + t') \mathrm{d}t' - itz)} \mathrm{d}\theta \qquad (4.2.35)$$

is still an entire function of z. Defining $H_\pi(t)$ as the characteristic function on the domain $0 \le t \le \pi$ (equal to unity in this domain and null outside), we finally get

$$\widehat{Hb}(z) = \left(\frac{1 - e^{-(\bar{f}+iz)\pi}}{iz} - \widehat{H_\pi k}(z) \right) \frac{1}{\widehat{H_\pi a}(z)}. \qquad (4.2.36)$$

Finally, let us write $\langle b_{st} \rangle$ in terms of the above functions. The stationary distribution is given by $p_{st}(\theta) = p_{st}(0)\alpha(\theta)$ in the domain $0 < \theta < \pi$. The norm condition

$$p_{st}(0) \int_0^\pi \alpha(\theta) \mathrm{d}\theta = 1 \qquad (4.2.37)$$

gives

$$p_{st}(\theta) = \frac{\alpha(\theta)}{\widehat{H_\pi \alpha}(0)}, \qquad (4.2.38)$$

with

$$\widehat{H_\pi \alpha}(0) = \int_0^\pi \alpha(t) \mathrm{d}t.$$

The average stationary value of b is defined by

$$\langle b_{st} \rangle = \int_0^\pi f(\theta) p_{st}(\theta) \mathrm{d}\theta = - \frac{1}{\widehat{H_\pi \alpha}(0)} \int_0^\pi \alpha_\theta(\theta) \mathrm{d}\theta, \qquad (4.2.39)$$

which gives the relation

$$\langle b_{st} \rangle = \frac{1 - \bar{\alpha}}{\widehat{H_\pi \alpha}(0)}. \tag{4.2.40}$$

Returning to equation (4.2.36) we get

$$\widehat{Hb}(z) = \frac{\langle b_{st} \rangle}{iz} + \frac{\tilde{b}(z)}{\widehat{H_\pi a}(z)}, \tag{4.2.41}$$

where $\tilde{b}(z)$ is an entire function of z, which allows us to make the following conjecture

$$Hb(t) = H\langle b_{st} \rangle + F(t), \tag{4.2.42}$$

where $F(t)$ tends to zero as time tends to infinity. To conclude with more arguments we have to control the zeros of $\widehat{H_\pi a}(z)$, although we have only stated that $\widehat{H_\pi a}(z)$ doesn't vanish for $y \leq 0$.

4.2.3 The Kolmogorov Equation for a Three-Level Atom

Let us consider a three-level system illuminated by two laser beams, each one at the frequency of transition between the ground state and one of the two excited states, the configuration imagined by Dehmelt [52, 53], named the V-configuration. In the absence of spontaneous decay of the excited states, the equation of motion connects three complex amplitudes, $a_0(t)$ for the ground and $a_1(t), a_2(t)$ for the two excited states, see Figure 4.1. These equations are discussed in detail in [134].

$$\frac{\partial}{\partial t} a_0 = -i\Omega_1 a_1 - i\Omega_2 a_2, \tag{4.2.43}$$

$$\frac{\partial}{\partial t} a_1 = -i\Omega_1 a_0 \tag{4.2.44}$$

and

$$\frac{\partial}{\partial t} a_2 = -i\Omega_2 a_0. \tag{4.2.45}$$

In these equations, $2\Omega_1$ and $2\Omega_2$ are Rabi frequencies associated to the transitions from level 1 to zero (index 1) and from level 2 to zero (index 2), each one being proportional to the amplitude of an electromagnetic wave at the frequency of the atomic transition between level 1 or 2 and the ground state. In the following, we shall assume, as in Dehmelt's proposal, that the transition $|0\rangle \rightarrow |1\rangle$ is saturated, contrary to the transition $|0\rangle \rightarrow |2\rangle$, and that level 2 has a very long lifetime (such that spontaneous emission from level 1 is frequent while it is very rare from level 2, and such that stimulated emission from 2 is still more rare), namely we shall consider the situation where the parameters fulfill the conditions

$$\Omega_1 \gg \gamma_1 \gg \gamma_2 \gg \Omega_2. \tag{4.2.46}$$

Within this framework we shall describe the intermittent fluorescence observed in the experiments [120], where clear periods of darkness were appearing in the fluorescent signal.

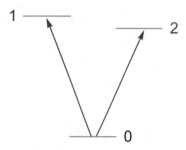

Fig. 4.1 Scheme of the three-level atom in the V-configuration.

4.2.4 Solution in the Deterministic Regime

Setting

$$A(t) = \begin{pmatrix} a_0 \\ a_1 \\ a_2 \end{pmatrix}, \tag{4.2.47}$$

equations (4.2.43)–(4.2.45) can be written as

$$i\frac{\partial}{\partial t} A(t) = \begin{pmatrix} 0 & \Omega_1 & \Omega_2 \\ \Omega_1 & 0 & 0 \\ \Omega_1 & 0 & 0 \end{pmatrix} A(t). \tag{4.2.48}$$

The eigenvalues of the matrix in (4.2.48) are $(0, \pm\Omega)$ with

$$\Omega^2 = \Omega_1^2 + \Omega_1^2.$$

We introduce the parameter ϵ defined by the relations $\cos\epsilon = \Omega_1/\Omega$ and $\sin\epsilon = \Omega_2/\Omega$ (ϵ will be small in the situation considered below). It is interesting to introduce two new amplitudes which are linear combinations of a_1 and a_2,

$$\begin{aligned} a_3 &= \cos\epsilon\, a_1 + \sin\epsilon\, a_2, \\ a_4 &= \sin\epsilon\, a_1 - \cos\epsilon\, a_2, \end{aligned} \tag{4.2.49}$$

and fulfill the condition

$$|a_1^2| + |a_2^2| = |a_3^2| + |a_4^2|.$$

The system (4.2.48) becomes

$$i\frac{\partial}{\partial t}\begin{pmatrix} a_0 \\ a_3 \\ a_4 \end{pmatrix} = \Omega \begin{pmatrix} a_3 \\ a_0 \\ 0 \end{pmatrix}. \tag{4.2.50}$$

In this system the amplitude a_4 remains constant, and the first and second equation in (4.2.50) give

$$\left(\frac{\partial^2}{\partial t^2} + \Omega^2\right) a_0 = 0$$

and

$$\left(\frac{\partial^2}{\partial t^2} + \Omega^2\right) a_3 = 0.$$

Therefore the dynamics for the original three amplitudes (4.2.47) reduces itself to the evolution of only two. We have

$$a_0(t) = \cos \Omega t\, a_0(0) - i \sin \Omega t\, a_3(0),$$
$$a_3(t) = \cos \Omega t\, a_3(0) - i \sin \Omega t\, a_0(0), \tag{4.2.51}$$
$$a_4(t) = a_4(0).$$

Note that the amplitude a_4 enters in the norm condition but plays no role in the dynamics.

Let us return to the original amplitudes and look at their evolution, assuming that at time $t = 0$ the atom emits a photon by the transition $|1\rangle \to |0\rangle$. Because this transition does not concern the state $|2\rangle$, the amplitude a_2 should be the same just before and just after this quantum jump. In contrast, the amplitudes a_0 and a_1 should change at $t = 0$, in particular we have to set

$$a_1(0_+) = 0, \tag{4.2.52}$$

because the emitted photon comes from the state $|1\rangle$ which becomes empty at $t = 0+$. Using equations (4.2.49) and (4.2.51) and setting $a_{i\pm} = a_i(0\pm)$, the solution in terms of $a_{2+} = a_{2-} = a_{20}$ and a_{0+} is given by the equations

$$a_0(t) = \cos \Omega t\, a_{0+} - i \sin \epsilon \sin \Omega t\, a_{20}, \tag{4.2.53}$$

$$a_1(t) = \sin \epsilon \cos \epsilon (\cos \Omega t - 1)\, a_{20} - i \cos \epsilon \sin \Omega t\, a_{0+} \tag{4.2.54}$$

and

$$a_2(t) = (\sin^2 \epsilon \cos \Omega t + \cos^2 \epsilon)\, a_{20} - i \sin \epsilon \sin \Omega t\, a_{0+}. \tag{4.2.55}$$

The relation (4.2.52) is satisfied, as well as the norm constraint

$$|a_0^2| + |a_1^2| + |a_2^2| = 1.$$

At time 0+ the two complex amplitudes a_{0+} and a_{20} are linked by a single relation

$$|a_{0+}|^2 + |a_{20}|^2 = 1,$$

which allows us to write

$$\begin{pmatrix} a_{0+} \\ a_{20} \end{pmatrix} = e^{i\phi} \begin{pmatrix} \cos\varphi \\ \sin\varphi e^{i\xi} \end{pmatrix}. \tag{4.2.56}$$

Finally, we have three unknown quantities, the angular variable φ (which describes the amplitudes of levels $|2\rangle$ and $|0\rangle$ after the jump), plus the two phases ϕ (in the factor) and ξ, the phase difference between the amplitudes a_{20} and a_{0+}. Note that only ξ is useful if we want to derive the dynamics of squared amplitudes. In contrast to the two-level atom, where we pointed out that the two amplitudes evolve in quadrature, here we have no reason to assume this and must consider the relative phase ξ as a random variable (with uniform distribution, for instance). Moreover, the phase ϕ which appears in the factor in the right-hand side of (4.2.56) enters into play for the calculation of the correlation functions of the complex amplitudes, as for the two-level atom case.

4.2.5 Change of Coordinates for the Schrödinger Flow

Let us consider the solution (4.2.51) of the pseudo-two-level atom and try to find the pertinent phase space to describe this deterministic dynamics. We first introduce the variable

$$s = |a_4|^2, \tag{4.2.57}$$

which fulfills the relation

$$|a_0|^2 + |a_3|^2 = 1 - s.$$

This relation differs from the true two-level case where one should have $s = 0$. We look for a description of the dynamics which takes the form of a rotation on a circle with angular velocity proportional to Ω. Let us consider the relation

$$i\frac{\partial}{\partial t}(a_0^* a_3) = \Omega(2|a_0|^2 - 1 + s) \tag{4.2.58}$$

and define the functions

$$u = 2|a_0|^2 - 1 + s \tag{4.2.59}$$

and

$$v = 2I(a_0 a_3^*), \tag{4.2.60}$$

where I denotes the imaginary part.

The amplitudes (u, v) obey the system

$$\frac{\partial}{\partial t} u = -2\Omega\, v,$$
$$\frac{\partial}{\partial t} v = 2\Omega\, u. \tag{4.2.61}$$

The time-dependent variables $(u(t), v(t))$ may be associated to the variables $(r, \theta(t))$ defined by

$$u = r\,\cos\theta,$$
$$v = r\,\sin\theta. \tag{4.2.62}$$

The radius

$$r = \sqrt{|u|^2 + |v|^2}$$

is constant during the deterministic motion, and the angular variable $\theta(t)$ obeys the equation

$$\frac{\partial}{\partial t}\theta = 2\Omega. \tag{4.2.63}$$

Recall that (u, v) are defined in terms of the amplitudes (a_0, a_3) by equations (4.2.59)–(4.2.60). In this picture the variable r is the radius of a circle along which the coordinates (u, v) or (r, θ) evolve with the angular velocity 2Ω, see Figure 4.2-(b).

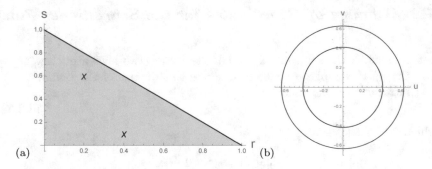

Fig. 4.2 (a) Domain of the phase space r, s bounded by the relation (4.2.64). (b) Trajectory of the coupled u, v or r, θ functions for the two different initial conditions corresponding to the red and blue points in (a).

The phase space (s, r) is bounded by the frontier

$$r + s \leq 1, \tag{4.2.64}$$

because we have

$$v^2 = 4|I(a_0 a_3^*)|^2 \leq 4|a_0|^2 |a_3|^2,$$

which gives

$$u^2 + v^2 \leq (1 - s)^2.$$

At time zero after a given jump, the trajectory starts from one particular point r, s of the 2D-phase space shown inside the filled region of Figure 4.2-(a). The motion is known as soon as the initial value of (r, s) is given, this pair of variables changing after each quantum jump.

In summary, one may consider a probability distribution $p(r, s, \theta; t)$ taken inside the domain

$$(0 \leq r \leq r + s \leq 1, 0 \leq \theta \leq 2\pi).$$

Doing this one has the transport equation

$$\frac{\partial}{\partial t} p + 2\Omega \frac{\partial}{\partial \theta} p = 0, \tag{4.2.65}$$

because the two variables (r, s) are constant during the deterministic dynamics, which depends on the initial conditions. The initial conditions in the phase space (r, s) can be expressed in terms of the initial values (4.2.56). Using (4.2.51) with $a_{1+} = 0$, and the definitions of (u, v) with (4.2.59)–(4.2.60), we get $a_{3+} = a_{20} \sin \epsilon$ and $a_{4+} = -a_{20} \cos \epsilon$, which gives

$$s_+ = |a_{4+}|^2 = |a_{20}|^2 \cos^2 \epsilon. \tag{4.2.66}$$

Using the relations (4.2.56) deduced from the norm constraint, we get

$$u_+ = 1 - |a_{20}|^2(1 + \sin^2 \epsilon) = \sin^2 \varphi(1 + \sin^2 \epsilon) \tag{4.2.67}$$

and

$$v_+ = 2 \sin \epsilon \, I(a_{0+} a_{20}^*) = -\cos \varphi \sin \varphi \sin \xi. \tag{4.2.68}$$

To figure out how the shelving process occurs, we imagine an experiment with a single laser, as done in [120]. An intense laser of frequency Ω_1 induces the transitions between the states $|0\rangle$ and $|1\rangle$. At this stage the atom makes Rabi nutations between these two states and emits photons "1" by stimulated and spontaneous emission. At a given time, taken as the origin and coinciding with a quantum jump from $|1\rangle$ and $|0\rangle$, the second laser is switched on. At this instant we have the initial conditions

$$a_{0+} = 1; \qquad a_{1+} = a_{2+} = 0, \tag{4.2.69}$$

or $r_+ = 1$ and $s_+ = 0$, which corresponds to a rotation on the circle of unit radius in the phase plane (u, v), the situation of the true two-level atom. For positive time, the trajectory in the plane (u, v) follows the circle of radius unity until the next quantum jump. At this time the initial conditions change from the coordinates $(r_0 = 1, s_0 = 0)$ in the plane (r, s) towards new values (r_1, s_1). We show below that the distance between $(1, 0)$ and the new initial conditions is of order ϵ^2 in each direction. Therefore the new trajectory is a circle of radius slightly smaller than unity.

To be more precise, let us consider the change of the amplitudes a_i. With initial conditions (4.2.69), the solution (4.2.53)–(4.2.55) shows that the amplitude a_2 evolves (before the next emission of a photon) as

$$a_2(t) = -i \sin \epsilon \sin \Omega t$$

(the two other amplitudes as $a_0 = \cos \Omega t$ and $a_1 = -i \sin \epsilon \sin \Omega t$). Recall that we have assumed that ϵ is much smaller than unity, because $\Omega_1 \ll \Omega_2$. Let $t_1 > 0$ be the emission time of the next photon "1". We have

$$a_2(t_1) = -i \sin \epsilon \sin \Omega t_1,$$

which gives the initial condition for the next deterministic stage,

$$a_{0+} = \cos \varphi^{(1)} e^{i\phi}; \qquad a_{1+} = 0; \qquad a_{2+} = -i \sin \varphi^{(1)}, \qquad (4.2.70)$$

where

$$\sin \varphi^{(1)} = \sin \epsilon \sin \Omega t_1$$

is of order ϵ. By comparing the initial conditions (4.2.69) and (4.2.70) we see that the population of level 2 increases during the first deterministic stage.

Let us describe now the shift of the initial coordinates in the phase space (r, s). Using expressions (4.2.70) and relations (4.2.67)–(4.2.68), the initial conditions after time t_1 are

$$u_+^{(1)} = 1 - \sin^2 \varphi^{(1)}(1 + \sin^2 \epsilon)$$

and

$$v{+}^{(1)} = \sin \epsilon \cos \phi \sin(2\varphi^{(1)}).$$

At first order with respect to the parameter ϵ, this gives

$$u_+^{(1)} = 1 - (\phi^{(1)})^2$$

and

$$v_+^{(1)} = \sin \epsilon \cos \xi_0 \varphi^{(1)}.$$

The radius of the next deterministic stage becomes equal to

$$1 - (\varphi^{(1)})^2.$$

Moreover, from (4.2.66) we have

$$s_+ = \cos^{2\epsilon}(\sin \varphi^{(1)})^2,$$

which gives

$$r^{(1)} \sim 1 - (\varphi^{(1)})^2; \qquad s^{(1)} \sim (\varphi^{(1)})^2, \qquad (4.2.71)$$

at first order with respect to $\phi^{(1)} \sim \epsilon \sin \Omega t_1$. This relation describes a small shift of the initial conditions along the frontier in Figure 4.2-(a), of order ϵ. The deterministic motion between t_1 and t_2 (emission time of the next photon from level $|1\rangle$), leads to more complicated expressions, that we shall not write. It appears that setting $\sin \Omega t_1 / \sin \Omega t_2 = 1$ and $\cos \phi = 0$, the modulus of a_2 gets a new increment equal to $\varphi^{(1)}$ at first order. This gives a new shift of the initial conditions r, s along the frontier. We infer that step by step, the modulus of the amplitude a_2 will grow with time sufficient to allow the transition of the atom from state $|0\rangle$ to $|2\rangle$ after a given number of successive deterministic stages interrupted randomly by emissions of photons 1.

Part II
Statistical Physics of Dilute Bose Gases

Chapter 5
Introduction

The most important equation in the kinetic theory of classical gases is the
Boltzmann equation [42], describing the time evolution of distribution functions of gas particles. To study quantum gases, attempts to extend the Boltzmann equation resulted in the Boltzmann–Nordheim (Uehling–Uhlenbeck)
equation for bosons and fermions (cf. [125, 159]). Let us explain in detail the
formulation of the equation.

Bosons of mass m at temperature T can be considered as quantum-
mechanical wavepackets with a spatial extent of the order of a thermal de
Broglie wavelength $\lambda_{dB} = \left(\frac{2\pi\hbar^2}{mk_BT}\right)^{\frac{1}{2}}$, in which k_B is the Boltzmann constant,
$\hbar$ is Planck's constant divided by 2π and m is the mass of the particle. The
de Broglie wavelength λ_{dB} shows the position uncertainty associated with
the mean thermal momentum. If the temperature of the gas is high, the de
Broglie wavelength λ_{dB} is very small; and thus, the weakly interacting gas
can be treated as a system of "billiard balls" (cf. [96]). The dynamics of the
gas is described by the Boltzmann–Nordheim equation

$$\frac{\partial}{\partial t} f(t,r,p) \ + \ r \cdot \nabla_r f(t,r,p) \ = \ Q_{BN}[f](t,r,p), \qquad (5.0.1)$$

whose operator reads

$$Q_{BN}[f](t,r,p_1)$$
$$= \frac{g^2}{m\hbar^3} \iiint_{\mathbb{R}^3 \times \mathbb{R}^3 \times \mathbb{R}^3} \delta(p_1 + p_2 - p_3 - p_4)\delta(\omega_{p_1} + \omega_{p_2} - \omega_{p_3} - \omega_{p_4})$$
$$\times \ [(1+f_1)(1+f_2)f_3f_4 - f_1f_2(1+f_3)(1+f_4)]\mathrm{d}p_2\mathrm{d}p_3\mathrm{d}p_4,$$

where $f(t,r,p)$ is the number of particles in a cell of volume $\hbar^3$ of the space
$dpdr$, at time t, position r and momentum p, $(t,r,p) \in \mathbb{R}_+ \times \mathbb{R}^3 \times \mathbb{R}^3$. In the
classical limit most cells are empty, so that $f \ll 1$, which gives back the
classical Boltzmann equation where Q is quadratic with respect to f. In this
limit the dimensionless quantity f, the number of particles in a cell of phase
space, yields the classical density in phase space (r,p) defined as $\varphi = f/\hbar^3$.

© Springer Nature Switzerland AG 2019
Y. Pomeau, M.-B. Tran, *Statistical Physics of Non Equilibrium Quantum Phenomena*,
Lecture Notes in Physics 967, https://doi.org/10.1007/978-3-030-34394-1_5

This gets rid of Planck's constant, as it should, in the classical Boltzmann equation. As the temperature T, at constant number density, becomes lower and lower, λ_{dB} becomes larger and larger. As the temperature of the gas is lowered, there is a critical temperature T_{BE} at which λ_{dB} becomes comparable to the distance between atoms. As a result, the atomic wavepackets "overlap" and the indistinguishability of atoms becomes important. A phase transition then occurs (Bose–Einstein condensation) and bosons begin to condense into the ground state (lowest energy state) of the system, which is the so-called Bose–Einstein condensate (BEC) (cf. [96]). This occurs because at equilibrium the decrease of entropy as temperature is lessened can only done by piling a finite proportion of particles in the ground state, which makes the condensate. However there are non-equilibrium energy states, even very close to absolute zero, where no condensate exists. Those states are precisely the ones we choose as initial conditions for the kinetic equation to study the onset of build-up of condensate and its growth.

After the first experiments of BEC in trapped atomic vapors of ^{87}Rb, ^{7}Li, and ^{23}Na [8, 10], a new period of intense experimental and theoretical research began. Despite the fact that the equilibrium properties of those systems was quite well understood, several open questions remained concerning their non-equilibrium behavior. One of the most important open problems concerns the behavior of the system after cooling a non-degenerate trapped Bose gas below the critical temperature. Up to now, while the experimental research has focused mainly on the initial formation of BECs, the theoretical behavior at finite temperatures remains a frontier of many-body physics. Such a theory has to take into account the coupled non-equilibrium dynamics of both the condensed and non-condensed components of the gas under investigation.

In order to understand this coupling dynamics, let us consider the number of particles in the condensate n_c as a function of the temperature T. Following the classical example considered in [51, 127, 128, 136], we suppose that the ideal gas under consideration has number density N and is confined in a harmonic potential well of the form

$$V_{ext}(x, y, z) = m \left(W_x^2 x^2 + W_y^2 y^2 + W_z^2 z^2 \right).$$

Then

$$\frac{n_c}{N} = 1 - \frac{T^3}{k_B^3},$$

where $k_B \approx 0.94 \hbar W_{h_0} N^{\frac{1}{3}}$ with (cf. [51, 127, 128, 136])

$$W_{h_0} = \left(W_x W_y W_z \right)^{\frac{1}{3}}.$$

Despite being very simple, this example is qualitatively useful. We observe that at $T = T_{BE}$, $n_c = 0$, but as T decreases, the condensate fraction $\frac{n_c}{N}$ increases, until at $T = 0K$, $N = n_c$ all the atoms are in the condensate. Therefore, at finite temperatures $0 < T < T_{BE}$, a BEC co-exists with non-condensed particles (normal fluid or thermal cloud). At these temperatures, the coupling of the condensate and non-condensate degrees of freedom leads to a two-component condensate-thermal cloud system. In a finite-temperature system, the low-density cloud of thermally excited atoms is spread over a larger spatial region whilst the high-density condensate is localized at the center of the trapping potential. In such systems, both components can exhibit coherent collective behavior, leading to many new phenomena.

Moreover, it is important to note that experiments can only be conducted at finite temperatures. In such situations, a thermal cloud/normal fluid is always present, and as the temperature increases towards T_{BEC} the influence of the thermal cloud/normal fluid becomes more and more significant. In the problem of condensate growth, or the heating of the gas under strong external perturbations, the role of the thermal cloud is the most important. An important implication of the coupling system is that under certain conditions, a trapped Bose gas at finite temperatures can be regarded as a two-fluid system, precisely analogous to the well-known macroscopic quantum properties of the superfluid ^{4}He.

The properties of a dilute Bose gas with repulsive interactions and possibly a condensate have been pioneered by Bogoliubov [28] and by Huang, Lee, Luttinger and Yang [86, 87, 104]. The latter authors used a regular perturbation method, considering the interaction as small and computing the first-order corrections, the unperturbed state being given by the usual Bose–Einstein theory, for a perfect gas. Another approach, used later by Lee and Yang [104], amounts to diagonalizing the Hamiltonian, which, thanks to a Bogoliubov transformation [1], includes the interaction between particles at finite momentum and the condensate. Lee and Yang [104] studied the model with a Bogoliubov spectrum as well, but the results differ from ours for the reason given below. The Bogoliubov renormalization makes the transition to condensation first order, which can be understood as follows. If the density of the condensate is finite, the excitation spectrum becomes a Bogoliubov spectrum, with a linear dependence of energy on momentum at low energies, although energy is a quadratic function of this momentum for free particles. Therefore, for a given momentum, the particle energy is larger than the free-particle energy and it is more difficult at a given temperature to thermally excite particles than free particles with the same momentum. This depletes the density of thermal particles to the benefit of the condensate, and yields a feedback making the transition first order. The thermodynamics of a Bose gas interacting weakly is explained in Chapter 9.

Many years after the work of Nordheim there was a renewal in the quantum kinetic theory of bosons, as shown by the work of Kirkpatrick and Dorfman [97, 98, 99]. Their work showed how the use of perturbation theory facilitates a generalization beyond the static thermal cloud approximation, after imposing the mean-field approximations. This approach was then employed and extended in the work of Zaremba, Nikuni and Griffin [165], where they introduced the full coupling system of a quantum Boltzmann equation for the density function of the normal fluid/thermal cloud and a Gross–Pitaevskii equation for the wavefunction of the condensate.

The theory is the consistent time-dependent extension of the Hartree–Fock–
Bogoliubov–Popov theory which additionally includes collisions within the
thermal cloud and particle-exchange collisions between condensate and thermal atoms. Independently, the same model was also derived by Pomeau,
Brachet, Métens and Rica in [118], using a completely different technique:
the quantum BBGKY hierarchy argument. The terminology "Quantum Kinetic Theory" was later introduced in a series of papers by Gardiner, Zoller
and collaborators [72, 73, 74, 91]. Indeed, at the limits, the Gardiner–Zoller
model becomes the model of Zaremba et al. and Pomeau et al. Moreover, the
two theories were tested against each other using the MIT experiment and
good agreements were found [119, 136].

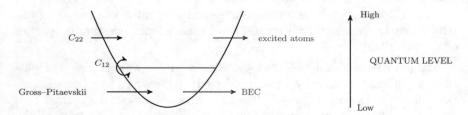

Fig. 5.1 The Bose–Einstein Condensate (BEC) and the excited atoms.

Let us rewrite below the model derived by Pomeau, Brachet, Métens and
Rica (cf. [118]) and by Zaremba, Nikuni and Griffin (cf. [165]):

The wavefunction $\Phi(t,r)$ of the condensate is a complex-valued function of time t and position r. The evolution of this function follows the
nonlinear Schrödinger (or Gross–Pitaevskii) equation coupled to the kinetic
Boltzmann–Nordheim equation:

$$i\hbar\frac{\partial}{\partial t}\Phi(t,r) = \left(-\frac{\hbar^2\Delta_r}{2m} + g[n_c(t,r) + 2n_n(t,r)] \right.$$

$$\left. - i\Lambda_{12}[f](t,r) + V(r) \right)\Phi(t,r), \quad (t,r) \in \mathbb{R}_+ \times \mathbb{R}^3,$$

$$\Lambda_{12}[f](t,r) = \frac{\hbar}{2n_c}\Gamma_{12}[f](t,r),$$

$$\Gamma_{12}[f](t,r) = \int_{\mathbb{R}^3} C_{12}[f](t,r,p)\frac{\mathrm{d}p}{(2\pi\hbar)^3},$$

$$n_n(t,r) = \int_{\mathbb{R}^3} f(t,r,p)\mathrm{d}p,$$

$$\Phi(0,r) = \Phi_0(r), \tag{5.0.2}$$

where $n_c(t,r) = |\Phi|^2(t,r)$ is the condensate density, g is the interaction cou-
pling constant proportional to the s-wave scattering length and $V(r)$ is the
trapping potential.

The atoms not belonging to the condensate are described by a probability distribution $f(t, r, p)$ that depends on time t, position r and momentum p. This function follows the quantum Boltzmann equation:

$$\frac{\partial}{\partial t} f(t, r, p) + \frac{p}{m} \cdot \nabla_r f(t, r, p) - \nabla_r U(t, r) \cdot \nabla_p f(t, r, p) \qquad (5.0.3)$$

$$= Q[f](t, r, p) := C_{12}[f](t, r, p) + C_{22}[f](t, r, p), \quad (t, r, p) \in \mathbb{R}_+ \times \mathbb{R}^3 \times \mathbb{R}^3,$$

$$C_{12}[f](t, r, p_1) :=$$

$$\lambda_1 n_c(t, r) \iint_{\mathbb{R}^3 \times \mathbb{R}^3} \delta(m v_c + p_1 - p_2 - p_3) \delta(\omega_c + \omega_{p_1} - \omega_{p_2} - \omega_{p_3})$$

$$\times [(1 + f_1) f_2 f_3 - f_1(1 + f_2)(1 + f_3)] \mathrm{d}p_2 \mathrm{d}p_3 \qquad (5.0.4)$$

$$- 2\lambda_1 n_c(t, r) \iint_{\mathbb{R}^3 \times \mathbb{R}^3} \delta(m v_c + p_2 - p_1 - p_3) \delta(\omega_c + \omega_{p_2} - \omega_{p_1} - \omega_{p_3})$$

$$\times [(1 + f_2) f_1 f_3 - f_2(1 + f_1)(1 + f_3)] \mathrm{d}p_2 \mathrm{d}p_3,$$

$$C_{22}[f](t, r, p_1) := \lambda_2 \iiint_{\mathbb{R}^3 \times \mathbb{R}^3 \times \mathbb{R}^3} \delta(p_1 + p_2 - p_3 - p_4) \qquad (5.0.5)$$

$$\times \delta(\omega_{p_1} + \omega_{p_2} - \omega_{p_3} - \omega_{p_4})$$

$$\times [(1 + f_1)(1 + f_2) f_3 f_4 - f_1 f_2 (1 + f_3)(1 + f_4)] \mathrm{d}p_2 \mathrm{d}p_3 \mathrm{d}p_4,$$

$$f(0, r, p) = f_0(r, p), (r, p) \in \mathbb{R}^3 \times \mathbb{R}^3,$$

where for the sake of simplicity, we impose the abbreviations $f_1 = f(t, p_1)$, $f_2 = f(t, p_2)$, $f_3 = f(t, p_3)$, $f_4 = f(t, p_4)$ and $\omega_1 = \omega(p_1)$, $\omega_2 = \omega(p_2)$, $\omega_3 = \omega(p_3)$, $\omega_4 = \omega(p_4)$. Moreover, $\lambda_1 = \frac{2g^2}{(2\pi)^2 \hbar^4}$, $\lambda_2 = \frac{2g^2}{(2\pi)^5 \hbar^7}$, m is the mass of the particles, and ω_p is the Hartree–Fock energy (cf. [26])

$$\omega_p = \omega(p) = \frac{|p|^2}{2m} + U(t, r). \qquad (5.0.6)$$

Writing

$$\Phi = |\Phi(t, r)| e^{i\phi(t, r)}, \qquad (5.0.7)$$

the condensate velocity can be defined as

$$v_c(t, r) = \frac{\hbar}{m} \nabla \phi(t, r), \qquad (5.0.8)$$

and the condensate chemical potential is then

$$\mu_c = \frac{1}{\sqrt{n_c}} \left(-\frac{\hbar^2 \Delta_r}{2m} + V + g[2n_n + n_c] \right) \sqrt{n_c}. \qquad (5.0.9)$$

We write

$$U(t, r) = V(r) + 2g[n_c(t, r) + n_n(t, r)] \qquad (5.0.10)$$

and

$$\omega_c(t, r) = \mu_c(t, r) + \frac{m v_c^2(t, r)}{2}. \qquad (5.0.11)$$

For the sake of simplicity, we suppose that $V \equiv 0$.

In the above system, the collision operator (5.0.4) describes collisions between the condensate and the non-condensate atoms (condensate growth term) and the collision operator (5.0.5) describes collisions between non-condensate atoms.

The model (5.0.2)–(5.0.3) is based on certain approximations of the excitations, for instance, they are single-particle-like (Hartree–Fock). The Bogoliubov dispersion law (cf. [55, 97, 99, 136])

$$\mathcal{E}_p = \mathcal{E}(p) = \sqrt{\kappa_1 |p|^2 + \kappa_2 |p|^4}, \qquad \kappa_1 = \frac{gn_c}{m} > 0, \quad \kappa_2 = \frac{1}{4m^2} > 0,$$
$$(5.0.12)$$

could be used instead of the Hartree–Fock energy, when the two terms $\kappa_1 p^2$ and $\kappa_2 p^4$ become of the same order for a thermal particle, that is for $p^2 \approx mk_B T'$, where k_B is the Boltzmann constant and T' is the temperature of the system. This gives $T' \approx gn_c/k_B$, a temperature well below T_{BE} in most physical situations.

As mentioned earlier, the study of quantum phase transitions has been a central research field of physics. In the context of ultracold Bose gases, the most important issue is the understanding of the formation of condensates and the associated evolution of coherence, leading to the establishment of off-diagonal long-range order. The pioneering MIT experiment [119] introduced the technique of "shock cooling". In the experiment, a trapped thermal cloud is suddenly quenched below the transition point, and then one studies the evolution of the resulting supersaturated cloud into a condensate. It was observed that the condensate grew slowly initially, before the bosonic enhancement set in, and then had an exponential growth. Let us quote [13]: *"The growth of a condensate is an interesting dynamical process – atoms must find the lowest energy state of the system, and long-range coherence has to be established. Experimentally, this process is observed after fast evaporative cooling, which cools the gas below the transition temperature for Bose–Einstein condensation, but is faster than the growth of the condensate to its equilibrium size. A full theoretical description must include the condensate and its elementary excitations, and the interactions with the cloud of thermal atoms (those not part of the condensate)."* The first numerical studies of the dynamics of the condensate growth were done by Gardiner, Zoller, Ballagh and Davis [74] and they are in qualitative agreement with existing experiments [119]. Parallel work was done by Bijlsma, Zaremba and Stoof using the model (5.0.2)–(5.0.3); these authors arrived at similar conclusions [26] to those of Davis, Gardiner and Ballagh, with the same initial parameters. The two theories were compared against each other using a particular set of experimental data from the MIT group. These two numerical implementations were found to be in very good agreement with each other [119, 136].

The above models are based on a picture of excitations with a particle-like spectrum, yielding a precise description for high temperature ranges $T_{BE} > T \geq 0.5T_{BE}$ [90]. However, to describe collective phonon-like excitations – the so-called bogolons – which become important at ultra low temperatures

[3], the models need to be modified. An original idea has been introduced in the PhD Thesis of Gust under the guidance of Reichl. In [77, 78, 79, 80, 142, 143], a modified kinetic equation which takes into account the non-conservation of the bogolon number during collisions and the phonon-like spectrum of bogolons at very low temperature has been introduced. In the model, a contribution to the collision operator $\mathbf{C}_{31}$ appears that takes into account $1 \leftrightarrow 3$ type collisions between the excitations, in addition to the $1 \leftrightarrow 2$ and $2 \leftrightarrow 2$ type collisions already known to occur. This work has its origins in a paper by Peletminskii and Yatsenko [126], in which they derived a more traditional kinetic equation that could incorporate a mean field description of relaxation processes in superfluids. In the model considered in these works, the collision operator $C_{12} + C_{22}$ is replaced by

$$\mathbf{G} \ = \ \mathbf{C}_{12} + \mathbf{C}_{22} + \mathbf{C}_{31}, \qquad (5.0.13)$$

where

$$
\begin{aligned}
\mathbf{C}_{12}[f_1] =& \lambda_1 n_c \iint_{\mathbb{R}^3 \times \mathbb{R}^3} (W^{12}_{1,2,3})^2 \delta(p_1 - p_2 - p_3)\delta(E_{p_1} - E_{p_2} - E_{p_3}) \\
& \times [(1 + f_1)f_2 f_3 - f_1(1 + f_2)(1 + f_3)]\mathrm{d}p_2 \mathrm{d}p_3 \\
& - 2\lambda_1 n_c \iint_{\mathbb{R}^3 \times \mathbb{R}^3} (W^{12}_{1,2,3})^2 \delta(p_2 - p_1 - p_3)\delta(E_{p_2} - E_{p_1} - E_{p_3}) \\
& \times [(1 + f_2)f_1 f_3 - f_2(1 + f_1)(1 + f_3)]\mathrm{d}p_2 \mathrm{d}p_3,
\end{aligned}
$$
$$(5.0.14)$$

$$
\begin{aligned}
\mathbf{C}_{22}[f_1] =& \lambda_2 \iiint_{\mathbb{R}^3 \times \mathbb{R}^3 \times \mathbb{R}^3} \delta(p_1 + p_2 - p_3 - p_4)(W^{22}_{1,2,3,4})^2 \\
& \times \delta(E_1 + E_2 - E_3 - E_4) \\
& \times [(1 + f_1)(1 + f_2)f_3 f_4 - f_1 f_2(1 + f_3)(1 + f_4)]\mathrm{d}p_2 \mathrm{d}p_3 \mathrm{d}p_4
\end{aligned}
$$
$$(5.0.15)$$

and

$$
\begin{aligned}
\mathbf{C}_{31}[f_1] =& \\
& \frac{2g^2 n_c}{3(2\pi)^2 \hbar^4} \iiint_{\mathbb{R}^3 \times \mathbb{R}^3 \times \mathbb{R}^3} \delta(p_1 - p_2 - p_3 - p_4)\delta(E_1 - E_2 - E_3 - E_4) \\
& \times (W^{31}_{1,2,3,4})^2 [f_1(f_2 + 1)(f_3 + 1)(f_4 + 1) - (f_1 + 1)f_2 f_3 f_4]\mathrm{d}p_2 \mathrm{d}p_3 \mathrm{d}p_4 \\
& - \frac{2g^2 n_c}{(2\pi)^2 \hbar^4} \iiint_{\mathbb{R}^3 \times \mathbb{R}^3 \times \mathbb{R}^3} \delta(p_2 - p_1 - p_3 - p_4)\delta(E_2 - E_1 - E_3 - E_4) \\
& \times (W^{31}_{1,2,3,4})^2 [f_2(f_1 + 1)(f_3 + 1)(f_4 + 1) - (f_2 + 1)f_1 f_3 f_4]\mathrm{d}p_2 \mathrm{d}p_3 \mathrm{d}p_4.
\end{aligned}
$$
$$(5.0.16)$$

The momenta considered in (5.0.14)–(5.0.16) are quasiparticle momenta in the "local" rest frame (i.e. the frame in which the superfluid is not moving). These momenta are connected to the momenta used in (5.0.3) by the formulation

$$p_i' = p_i + mv_c.$$

In the presentations of $\mathbf{C}_{12}, \mathbf{C}_{22}, \mathbf{C}_{31}$, the quasiparticle momenta in the "local" rest frame p_i' are still denoted by p_i, for the sake of simplicity. In the original works by Kirkpatrick and Dorfman [97, 98, 99], the quasiparticle momenta in the "local" rest frame are also used.

If we define

$$\epsilon(p) = \frac{\hbar^2 |p|^2}{2m},$$

and let Λ_0 be the equilibrium condensate order parameter, then

$$E_p = E(p) = \sqrt{(\epsilon(p) + \Lambda_0)^2 - \Lambda_0{}^2}$$

is the bogolon (excitations at ultra low temperature) energy. Moreover, E_1, E_2, E_3, E_4 stand for $E(p_1)$, $E(p_2)$, $E(p_3)$, $E(p_4)$.

The collision kernels are defined as

$$W_{1,2,3}^{12} = u_1 u_2 u_3 - u_1 v_2 u_3 - v_1 u_2 u_3 + u_1 v_2 v_3 + v_1 u_2 v_3 - v_1 v_2 v_3, \quad (5.0.17)$$

$$W_{1,2,3,4}^{22} = u_1 u_2 u_3 u_4 + u_1 v_2 u_3 v_4 + u_1 v_2 u_3 v_4 + v_1 u_2 u_3 v_4 + v_1 u_2 v_3 v_4, \quad (5.0.18)$$

and

$$W_{1,2,3,4}^{31} = u_1 u_2 u_3 v_4 + u_1 v_2 v_3 u_4 + u_1 v_2 u_3 v_4 + v_1 v_2 v_3 u_4 + v_1 u_2 v_3 v_4, \quad (5.0.19)$$

in which

$$u_i^2 = \frac{\epsilon(p) + gn_c + E_i}{2E_i}, \quad u_i^2 - v_i^2 = 1.$$

The new model has been tested by several experiments. For example, the speed and lifetime of the first and second sound modes in mono-atomic BECs computed by a simplified version of the model are in agreement with many experimental results. In the following, we will present this simplified version. The computation of the hydrodynamics modes and the experimental comparisons will be discussed in detail in Chapter 8. This model is indeed an equation for the spatial Fourier transform of the distribution of the excitations, normally called bogolons at this temperature regime. We denote this Fourier transform by

$$f(t,q,p) = \int_{\mathbb{R}^3} e^{iq \cdot r} f(t,r,p) \mathrm{d}r, \quad \forall q \in \mathbb{R}^3.$$

Notice that the notation of the Fourier transform usually comes with the hat symbol $\hat{f}$. However, to avoid unnecessary complexities in doing the computations, we remove this symbol, as an abuse of notation. At ultra-low temperatures, the Fourier transforms of the distribution of the bogolons and the microscopic phase of the condensate satisfy the system of equations:

$$\frac{\partial f(t,q,p)}{\partial t} = i\frac{\hbar}{m}p\cdot q\frac{\epsilon(p) + \Lambda_0}{E_p}f(t,q,p) + iq\cdot v_c(t,q)\mathcal{N}_p^{eq} - \mathbf{G}[f](t,q,p),$$

$$\text{on } (q,p,t) \in \mathbb{R}^3 \times \mathbb{R}^3 \times \mathbb{R}_+, \tag{5.0.20}$$

$$f(0,q,p) = f_0(q,p), \quad (q,p) \in \mathbb{R}^3 \times \mathbb{R}^3, \tag{5.0.21}$$

$$\frac{\partial^2 \phi(t,q)}{\partial t^2} = -i\frac{g}{m}\frac{1}{(2\pi)^3}\int_{\mathbb{R}^3} q\cdot p f(t,q,p)\mathrm{d}p - i\frac{g}{\hbar}q\cdot v_c(t,q)n^{eq},$$

$$\text{on } (t,q) \in \mathbb{R}^3 \times \mathbb{R}_+, \tag{5.0.22}$$

$$\phi(0,q) = \phi_0(q), \quad q \in \mathbb{R}^3. \tag{5.0.23}$$

Notice again that $\phi(t,q)$ and $v_c(t,q)$ are the Fourier transforms of $\phi(t,r)$ and $v_c(t,r)$ in (5.0.8). The hat symbols are omitted to simplify the mathematical expressions. For the same reason, $\mathbf{G}[f](t,q,p)$ is the Fourier transform in r of $\mathbf{G}[f](t,r,p)$, described in (5.0.13). The Gross–Pitaevskii equation (5.0.2) of $\Phi(t,r)$ is now replaced by the Hugenholtz–Pines equation (5.0.22) of $\phi(t,q)$. Moreover, q is the wave vector for spatial variations of the bogolon density, and

$$\mathcal{N}_p^{eq} = (e^{E(p)/k_B T} - 1)^{-1} \tag{5.0.24}$$

is the equilibrium Bose–Einstein distribution for bogolons at temperature T where k_B is the Boltzmann constant, $g = 4\pi\hbar^2 a/m$ is the coupling constant, a is the s-wave scattering length of the atoms in the gas, and n^{eq} is the total particle number density. The distribution $\mathcal{N}^{eq}(p)$ is a stationary state of Equation (5.0.20). The derivation of this simplified model will be given in Section 6.3.

The equilibrium particle density n^{eq} that appears in Equation (5.0.22) can be written

$$n^{eq} \approx N_c + \frac{1}{(2\pi)^3}\int_{\mathbb{R}^3} \frac{\epsilon(p) + \Lambda_0}{E_{\mathbf{k}_1}}\mathcal{N}_p^{eq}\mathrm{d}p, \tag{5.0.25}$$

where N_c is the density of bosons that have condensed into the ground state $p = 0$. This form of the equilibrium particle density is called the "Popov approximation" [89].

The superfluid velocity is determined by the spatial variation of the macroscopic phase $\phi(t,r)$ and is given by $v_c(t,r) = \frac{\hbar}{m}\nabla_r\phi(t,r)$. Therefore,

$$v_c(t,q) = -i\frac{\hbar}{m}q\,\phi(t,q). \tag{5.0.26}$$

These works have shown that, at low temperatures, not only are the $2 \leftrightarrow 2$ collisions as strong as the $1 \leftrightarrow 2$ collisions, but also the $1 \leftrightarrow 3$ collisions come into play. Recent theoretical and experimental findings support the Reichl et al. model to be a possible complement of the Zaremba et al./Pomeau et al. model at ultra-low temperatures ($0 < T < 0.3T_{BE}$).

We can see that n_c appears in the expressions of both $\mathbf{C}_{12}$ and $\mathbf{C}_{31}$. When $n_c = 0$, the two operators vanish and the system is reduced to the Boltzmann–Nordheim equation. However, we remark that when the system is closed to equilibrium, treating n_c as a dynamical variable has little effect on the evolution of the kinetic equation. This observation will be discussed later in Chapter 6.

Chapter 6
Quantum Boltzmann Equations

6.1 The Boltzmann–Nordheim Equation

This section is devoted to the derivation of the Boltzmann–Nordheim kinetic equation. It follows the work of Balescu [18] and is split into several steps. A different derivation could be found in the book of Akhiezer and Peletminskii [1].

6.1.1 Step 1: The Wigner Functions

Let us define an operator which destroys a certain number of bosons in certain levels and recreates them in other levels. It has the following general form (cf. [18, Chapter 1: Review of Hamiltonian Dynamics])

$$\mathbf{b}^{(n)} = \tag{6.1.1}$$

$$\frac{1}{n!}\sum_{m_1}\cdots\sum_{m_n}\sum_{m'_1}\cdots\sum_{m'_n}\langle m_1,\cdots,m_n|\hat{b}^{(n)}|m'_1,\cdots,m'_n\rangle \mathbf{a}^{\dagger}_{m_1}\cdots\mathbf{a}^{\dagger}_{m_n}\mathbf{a}_{m'_1}\cdots\mathbf{a}_{m'_n}$$

where m_i, $m'_i \in \mathbb{R}^D$, $i \in \{1,\cdots n\}$, $D \in \mathbb{N}$, are sets of quantum numbers, called levels, characterizing the state of the single particle (for example, the three components of the momentum and the spin); $\mathbf{a}_{m'_i}$, $i \in \{1,\cdots n\}$, is the annihilation operator, which describes the destruction of a particle in level m'_i; $\mathbf{a}_{m_i}$, $i \in \{1,\cdots n\}$, is the creation operator, which describes the creation of a particle in level m_i; $\hat{b}^{(n)}$ is an operator of the Hilbert space H under consideration; and $\langle m_1,\cdots,m_n|\hat{b}^{(n)}|m'_1,\cdots,m'_n\rangle$ is the matrix element of the operator $\hat{b}^{(n)}$ in the ordinary representation.

The two operators $\mathbf{a}_m$ and $\mathbf{a}^{\dagger}_m$ do not commute with each other. Consider two quantum levels m, m',

$$[\mathbf{a}_m, \mathbf{a}_{m'}] = 0, \quad [\mathbf{a}^{\dagger}_m, \mathbf{a}^{\dagger}_{m'}] = 0, \quad [\mathbf{a}_m, \mathbf{a}^{\dagger}_{m'}] = \delta_{m,m'}, \tag{6.1.2}$$

© Springer Nature Switzerland AG 2019
Y. Pomeau, M.-B. Tran, *Statistical Physics of Non Equilibrium Quantum Phenomena*,
Lecture Notes in Physics 967, https://doi.org/10.1007/978-3-030-34394-1_6

where $[X, Y] = XY - YX$.

The usual quantized wave functions can therefore be defined at the position space $\mathbf{x} \in \mathbb{R}^D$

$$\Psi(\mathbf{x}) = \sum_m \mathbf{a}_m \varphi_m(\mathbf{x}), \quad \Psi^\dagger(\mathbf{x}) = \sum_m \mathbf{a}_m^\dagger \varphi_m^*(\mathbf{x}), \qquad (6.1.3)$$

where $\{\varphi_m\}$ is the basis of the Hilbert space H under consideration. Therefore, the following holds true

$$[\Psi(\mathbf{x}), \Psi(\mathbf{x}')] = 0, \quad [\Psi^\dagger(\mathbf{x}), \Psi^\dagger(\mathbf{x}')] = 0, \quad [\Psi(\mathbf{x}), \Psi^\dagger(\mathbf{x}')] = \delta(\mathbf{x} - \mathbf{x}'). \tag{6.1.4}$$

An important particular case of the relationship between $\Psi^\dagger$ and $\mathbf{a}^\dagger$ is when $\varphi_\kappa(\mathbf{x})$ are plane waves, with the continuous wave vector $\mathbb{R}^D$

$$\varphi_\kappa = \frac{1}{(2\pi)^{3/2}} e^{i\kappa \cdot \mathbf{x}}. \tag{6.1.5}$$

In this case, the wave operators are written as

$$\Psi(\mathbf{x}) = \frac{1}{\sqrt{8\pi^3}} \int_{\mathbb{R}^D} \mathbf{a}(\hbar\kappa) e^{i\kappa \cdot \mathbf{x}} \mathrm{d}\kappa, \quad \Psi^\dagger(\mathbf{x}) = \frac{1}{\sqrt{8\pi^3}} \int_{\mathbb{R}^D} \mathbf{a}^\dagger(\hbar\kappa) e^{i\kappa \cdot \mathbf{x}} \mathrm{d}\kappa. \tag{6.1.6}$$

Then $\mathbf{b}^{(n)}$ can be expressed as

$$\mathbf{b}^{(n)} = \frac{1}{n!} \int_{\mathbb{R}^{D \times n}} \Psi^\dagger(\mathbf{x}_1) \cdots \Psi^\dagger(\mathbf{x}_n) \hat{\mathbf{b}}^{(n)} \Psi(\mathbf{x}_n) \cdots \Psi(\mathbf{x}_1) \mathrm{d}\mathbf{x}_1 \cdots \mathrm{d}\mathbf{x}_n. \tag{6.1.7}$$

Taking the sum of $\mathbf{b}^{(n)}$ over all indices n, we find a new operator, which reads

$$\mathbf{b} = \sum_{n=0}^{\infty} \mathbf{b}^{(n)}, \tag{6.1.8}$$

whose average value can be expressed in terms of the density matrix ρ as

$$\langle \mathbf{b} \rangle = \mathrm{Tr}\rho\mathbf{b} \tag{6.1.9}$$

$$= \sum_{n=0}^{\infty} \frac{1}{n!} \int_{\mathbb{R}^{D \times n}} \mathrm{Tr}\{\rho\Psi^\dagger(\mathbf{x}_1) \cdots \Psi^\dagger(\mathbf{x}_n) \hat{\mathbf{b}}^{(n)} \Psi(\mathbf{x}_n) \cdots \Psi(\mathbf{x}_1)\} \mathrm{d}\mathbf{x}_1 \cdots \mathrm{d}\mathbf{x}_n.$$

By the Weyl correspondence rule (cf. [18, Chapter 1: Review of Hamiltonian Dynamics]), the quantum operator $\hat{\mathbf{b}}^{(n)}$ can be expressed as

$$\hat{\mathbf{b}}^{(n)} = \qquad\qquad\qquad\qquad\qquad\qquad\qquad\qquad (6.1.10)$$

$$\int_{\mathbb{R}^{2D \times n}} \beta_n(\mathbf{k}_1, \mathbf{j}_1, \cdots, \mathbf{k}_n, \mathbf{j}_n) \exp\left[i \sum_{s=1}^{n} (\mathbf{k}_s \cdot \hat{\mathbf{q}}_s + \mathbf{j}_s \cdot \hat{\mathbf{p}}_s) \right] d\mathbf{k}_1 d\mathbf{j}_1 \cdots d\mathbf{k}_n d\mathbf{j}_n,$$

for some function β_n, where $\hat{\mathbf{q}}_s = \mathbf{x}_s$ and $\mathbf{j}_s = -i\hbar\frac{\partial}{\partial \mathbf{x}_s}$ ($\hbar$ is Planck's constant).

For $n = 1$, we have

$$\langle \mathbf{b}^{(1)} \rangle = \qquad\qquad\qquad\qquad\qquad\qquad\qquad\qquad (6.1.11)$$

$$\int_{\mathbb{R}^{2D}} \beta_1(\mathbf{k}, \mathbf{j}) \mathrm{Tr}\left\{ \rho \left\{ \Psi^\dagger(\mathbf{x}) \exp\left[i\mathbf{k} \cdot \mathbf{x} + i\mathbf{j} \cdot \left(-i\hbar\frac{\partial}{\partial \mathbf{x}} \right) \right] \Psi(\mathbf{x}) \right\} \right\} d\mathbf{k} d\mathbf{j}.$$

Using the identity

$$\exp\left[i\mathbf{k} \cdot \mathbf{x} + i\mathbf{j} \cdot \left(\frac{\partial}{\partial \mathbf{x}} \right) \right]$$

$$= \exp(i\mathbf{k} \cdot \mathbf{x}) \exp\left(\hbar\mathbf{j} \cdot \frac{\partial}{\partial \mathbf{x}} \right) \exp\left(-\frac{1}{2}\left[i\mathbf{k} \cdot \mathbf{x}, \hbar\mathbf{j} \cdot \frac{\partial}{\partial \mathbf{x}} \right] \right)$$

$$= \exp(i\mathbf{k} \cdot \mathbf{x}) \exp\left(\hbar\mathbf{j} \cdot \frac{\partial}{\partial \mathbf{x}} \right) \exp\left(\frac{1}{2}\hbar\mathbf{k} \cdot \mathbf{j} \right)$$

and the formula

$$\exp\left(\alpha \cdot \frac{\partial}{\partial \mathbf{x}} \right) f(\mathbf{x}) = f(\mathbf{x} + \alpha),$$

we deduce from (6.1.11) that

$$\langle \mathbf{b}^{(1)} \rangle = \int_{\mathbb{R}^{2D}} \beta_1(\mathbf{k}, \mathbf{j}) \mathrm{Tr}\left\{ \rho \left\{ \Psi^\dagger(\mathbf{x}) \exp\left[i\mathbf{k} \cdot \left(\mathbf{x} + \frac{1}{2}\hbar\mathbf{j} \right) \right] \Psi\left(\mathbf{x} + \hbar\mathbf{j} \right) \right\} \right\} d\mathbf{k} d\mathbf{j}$$

$$=: \int_{\mathbb{R}^{4D}} \beta_1(\mathbf{k}, \mathbf{j}) \exp(i\mathbf{k} \cdot \mathbf{q} + i\mathbf{j} \cdot \mathbf{p}) \mathcal{F}_1^W(\mathbf{p}, \mathbf{q}) d\mathbf{p} d\mathbf{q} d\mathbf{k} d\mathbf{j}$$

$$= \int_{\mathbb{R}^{2D}} \mathbf{b}^{(1)}(\mathbf{q}, \mathbf{p}) \mathcal{F}_1^W(\mathbf{p}, \mathbf{q}) d\mathbf{p} d\mathbf{q},$$

$$(6.1.12)$$

in which the one-particle Wigner function can be written as

$$\mathcal{F}_1^W(\mathbf{p}, \mathbf{q}) := \frac{1}{8\pi^3} \int_{\mathbb{R}^D} \exp(-i\mathbf{j} \cdot \mathbf{p}) \mathrm{Tr}\left[\rho \Psi^\dagger\left(\mathbf{q} - \frac{1}{2}\hbar\mathbf{j} \right) \Psi\left(\mathbf{q} + \frac{1}{2}\hbar\mathbf{j} \right) \right] d\mathbf{j}. \quad (6.1.13)$$

The result can easily be generalized. The n-particle Wigner function is defined as follows

$$\mathcal{F}_n^W(\mathbf{p}, \mathbf{q}) := \frac{1}{(8\pi^3)^n} \int_{\mathbb{R}^{D \times n}} \exp\left(-i \sum_{s=1}^{n} \mathbf{j}_s \cdot \mathbf{p}_s\right)$$

$$\times \mathrm{Tr}\left\{\rho \Psi^\dagger\left(\mathbf{q}_1 - \frac{1}{2}\hbar\mathbf{j}_1\right) \cdots \Psi^\dagger\left(\mathbf{q}_n - \frac{1}{2}\hbar\mathbf{j}_n\right)\right. \qquad (6.1.14)$$

$$\left. \times \Psi\left(\mathbf{q}_1 + \frac{1}{2}\hbar\mathbf{j}_1\right) \cdots \Psi\left(\mathbf{q}_n + \frac{1}{2}\hbar\mathbf{j}_n\right)\right\} \mathrm{d}\mathbf{j}_1 \cdots \mathrm{d}\mathbf{j}_n,$$

and

$$\langle \mathbf{b} \rangle = \sum_{n=0}^{\infty} \frac{1}{n!} \int_{\mathbb{R}^{2D \times n}} \mathbf{b}^{(n)}(\mathbf{q}_1, \mathbf{p}_1, \cdots, \mathbf{q}_n, \mathbf{p}_n) \qquad (6.1.15)$$

$$\times \mathcal{F}_n^W(\mathbf{q}_1, \mathbf{p}_1, \cdots, \mathbf{q}_n, \mathbf{p}_n) \mathrm{d}\mathbf{p}_1 \mathrm{d}\mathbf{q}_1 \cdots \mathrm{d}\mathbf{p}_n \mathrm{d}\mathbf{q}_n.$$

Now, let us look more closely at the formulation of the one-particle Wigner function. By substituting (6.1.3) into (6.1.13), one gets

$$\mathcal{F}_1^W(\mathbf{q}, \mathbf{p}) = \frac{1}{(8\pi^3)^2} \int_{\mathbb{R}^{2D}} \exp\left(-i\mathbf{j} \cdot \mathbf{p} - i\kappa \cdot \left(\mathbf{q} - \frac{1}{2}\hbar\mathbf{j}\right) + i\kappa' \cdot \left(\mathbf{q} + \frac{1}{2}\hbar\mathbf{j}\right)\right)$$

$$\times \mathrm{Tr}\left[\rho \mathbf{a}^\dagger(\hbar\kappa)\mathbf{a}(\hbar\kappa')\right] \mathrm{d}\mathbf{j}\mathrm{d}\kappa\mathrm{d}\kappa'$$

$$= \frac{1}{8\pi^3} \int_{\mathbb{R}^{2D}} \delta\left(-\mathbf{p} + \frac{1}{2}\hbar\kappa + \frac{1}{2}\hbar\kappa'\right) \exp(i(\kappa' - \kappa) \cdot \mathbf{q})$$

$$\times \mathrm{Tr}\left[\rho \mathbf{a}^\dagger(\hbar\kappa)\mathbf{a}(\hbar\kappa')\right] \mathrm{d}\kappa\mathrm{d}\kappa'.$$

Applying the change of variables $\kappa' - \kappa = \mathbf{k}$, $\kappa' + \kappa = 2\mathbf{K}$, whose Jacobian is just 1, one finds

$$\mathcal{F}_1^W(\mathbf{q}, \mathbf{p}) = \frac{1}{8\pi^3} \int_{\mathbb{R}^{2D}} \delta(-\mathbf{p} + \hbar\mathbf{K}) \exp(i\mathbf{k} \cdot \mathbf{q})$$

$$\times \mathrm{Tr}\left[\rho \mathbf{a}^\dagger\left(\hbar\mathbf{K} - \frac{1}{2}\hbar\mathbf{k}\right)\mathbf{a}\left(\hbar\mathbf{K} + \frac{1}{2}\hbar\mathbf{k}\right)\right] \mathrm{d}\mathbf{k}\mathrm{d}\mathbf{K}$$

$$= \frac{1}{8\pi^3\hbar^3} \int_{\mathbb{R}^D} \exp(i\mathbf{k} \cdot \mathbf{q}) \mathrm{Tr}\left[\rho \mathbf{a}^\dagger\left(\mathbf{p} - \frac{1}{2}\hbar\mathbf{k}\right)\mathbf{a}\left(\mathbf{p} + \frac{1}{2}\hbar\mathbf{k}\right)\right] \mathrm{d}\mathbf{k}.$$

Generalizing the above argument, we get the following form for the n-particle function

$$\mathcal{F}_n^W(\mathbf{q}_1, \mathbf{p}_1, \cdots, \mathbf{q}_n, \mathbf{p}_n) = \qquad (6.1.16)$$

$$\frac{1}{(8\pi^3)^n} \int_{\mathbb{R}^{D \times n}} \exp\left(-i \sum_{s=1}^{n} \mathbf{k}_s \cdot \mathbf{q}_s\right) \mathcal{F}_s^W(\mathbf{k}_1, \mathbf{p}_1, \cdots, \mathbf{k}_n, \mathbf{p}_n) \mathrm{d}\mathbf{k}_1 \cdots \mathrm{d}\mathbf{k}_n,$$

in which, with an abuse of notation,

$$\mathcal{F}_n^W(\mathbf{k}_1, \mathbf{p}_1, \cdots, \mathbf{k}_n, \mathbf{p}_n) = \hbar^{-3n} \mathrm{Tr}\left\{\rho \mathbf{a}^\dagger\left(\mathbf{p}_1 - \frac{1}{2}\hbar\mathbf{k}_1\right) \cdots\right.$$

$$\mathbf{a}^{\dagger}\Big(\mathbf{p}_n - \frac{1}{2}\hbar\mathbf{k}_n\Big)\mathbf{a}\Big(p_n + \frac{1}{2}\hbar\mathbf{k}_n\Big)\cdots\mathbf{a}\Big(\mathbf{p}_1 + \frac{1}{2}\hbar\mathbf{k}_1\Big)\Big\}.$$

Therefore

$$\mathcal{F}_n^W(\mathbf{k}_1, \mathbf{p}_1, \cdots, \mathbf{k}_n, \mathbf{p}_n) = \hbar^{-3n}\text{Tr}[\rho\mathbf{a}^{\dagger}(\mathbf{P}_1)\cdots\mathbf{a}^{\dagger}(\mathbf{P}_n)\mathbf{a}(\mathbf{P}_n')\cdots\mathbf{a}(\mathbf{P}_1')],$$
$$(6.1.17)$$

in which

$$\mathbf{k}_j = \hbar^{-1}(\mathbf{P}_j' - \mathbf{P}_j), \quad \mathbf{p}_j = \frac{1}{2}(\mathbf{P}_j' + \mathbf{P}_j).$$

Let us use the Greek symbol ξ_n to denote the pair comprising the wave vector and momentum $(\mathbf{k}_n, \mathbf{p}_n)$. The distribution vector is now defined to be the set of Wigner functions

$$\tilde{\mathcal{F}}^W = \{\mathcal{F}_n^W(\xi_1, \cdots, \xi_n); n = 0, 1, 2, \cdots\}. \quad (6.1.18)$$

Let us now discuss the symmetry properties of the Wigner functions. It is clear that

$$\text{Tr}\Big\{\rho\mathbf{a}^{\dagger}\Big(\mathbf{p}_1 - \frac{1}{2}\hbar\mathbf{k}_1\Big)\mathbf{a}^{\dagger}\Big(\mathbf{p}_2 - \frac{1}{2}\hbar\mathbf{k}_2\Big)\mathbf{a}\Big(\mathbf{p}_2 + \frac{1}{2}\hbar\mathbf{k}_2\Big)\mathbf{a}\Big(\mathbf{p}_1 + \frac{1}{2}\hbar\mathbf{k}_1\Big)\Big\}$$
$$= \text{Tr}\Big\{\rho\mathbf{a}^{\dagger}\Big(\mathbf{p}_2 - \frac{1}{2}\hbar\mathbf{k}_2\Big)\mathbf{a}^{\dagger}\Big(\mathbf{p}_1 - \frac{1}{2}\hbar\mathbf{k}_1\Big)\mathbf{a}\Big(\mathbf{p}_1 + \frac{1}{2}\hbar\mathbf{k}_1\Big)\mathbf{a}\Big(\mathbf{p}_2 + \frac{1}{2}\hbar\mathbf{k}_2\Big)\Big\}.$$

Indeed, this operation is the permutation of $\mathbf{a}^{\dagger}$ and $\mathbf{a}$. Therefore, for bosons, the operation leaves the value of the product unchanged. As a consequence

$$\mathcal{F}_2^W(\xi_1, \xi_2) = \mathcal{F}_2^W(\xi_2, \xi_1).$$

Moreover, a more subtle symmetry property also holds true for bosons, which is

$$\text{Tr}\Big\{\rho\mathbf{a}^{\dagger}\Big(\mathbf{p}_1 - \frac{1}{2}\hbar\mathbf{k}_1\Big)\mathbf{a}^{\dagger}\Big(\mathbf{p}_2 - \frac{1}{2}\hbar\mathbf{k}_2\Big)\mathbf{a}\Big(\mathbf{p}_2 + \frac{1}{2}\hbar\mathbf{k}_2\Big)\mathbf{a}\Big(\mathbf{p}_1 + \frac{1}{2}\hbar\mathbf{k}_1\Big)\Big\}$$
$$= \text{Tr}\Big\{\rho\mathbf{a}^{\dagger}\Big(\mathbf{p}_1 - \frac{1}{2}\hbar\mathbf{k}_1\Big)\mathbf{a}^{\dagger}\Big(\mathbf{p}_2 - \frac{1}{2}\hbar\mathbf{k}_2\Big)\mathbf{a}\Big(\mathbf{p}_1 + \frac{1}{2}\hbar\mathbf{k}_1\Big)\mathbf{a}\Big(\mathbf{p}_2 + \frac{1}{2}\hbar\mathbf{k}_2\Big)\Big\}.$$

Using (6.1.17), the following identity can be found

$$\mathcal{F}_2^W(\mathbf{k}_1, \mathbf{p}_1, \mathbf{k}_2, \mathbf{p}_2) =$$
$$\mathcal{F}_2^W\Big[\frac{1}{2}(\mathbf{k}_1 + \mathbf{k}_2) + \hbar^{-1}(\mathbf{p}_2 - \mathbf{p}_1), \frac{1}{2}(\mathbf{p}_1 + \mathbf{p}_2) + \frac{1}{4}\hbar(\mathbf{k}_2 - \mathbf{k}_1), \quad (6.1.19)$$
$$\frac{1}{2}(\mathbf{k}_1 + \mathbf{k}_2) + \hbar^{-1}(\mathbf{p}_1 - \mathbf{p}_2), \frac{1}{2}(\mathbf{p}_1 + \mathbf{p}_2) + \frac{1}{4}\hbar(\mathbf{k}_1 - \mathbf{k}_2)\Big].$$

This property is a very important quantum-statistical feature. For an n-particle function, this relation holds for every pair of variables that can be chosen among the n particles.

6.1.2 Step 2: The Quantum Liouville Equation

Let us consider the classical Hamiltonian

$$H = H^0 + H' + H^F. \tag{6.1.20}$$

The term H^0 represents the free motion of non-interacting particles in the absence of an external field. That is, the sum of N terms, each depending only on the momentum of a single particle

$$H^0 = \sum_{j=1}^{N} H_j^0, \tag{6.1.21}$$

with

$$H_j^0 = \frac{p_j^2}{2m}, $$

where p_j is the momentum of the j-th particle and m is the mass of each of them.

The term H' describes the interactions between particles. The main role of this term is to provide a coupling between particles. It can be written as a sum of many terms, each depending on the canonical variables of two particles

$$H' = \sum_{j<n=1}^{N} V_{jn}. \tag{6.1.22}$$

In our case, the function V_{jn} is assumed to depend only on the positions $\mathbf{q}_j$ and $\mathbf{q}_n$,

$$V_{jn} = V(|\mathbf{q}_j - \mathbf{q}_n|). \tag{6.1.23}$$

The term H^F describes the action of an external field, which is assumed to be 0 in our case. Therefore $H^F = 0$.

We then write

$$\begin{aligned} H = H^0 + H' &= \int_{\mathbb{R}^D} \Psi^\dagger(\mathbf{x}) \left[-\frac{\hbar^2}{2m} \Delta \right] \Psi(\mathbf{x}) \\ &+ \frac{1}{2} \int_{\mathbb{R}^{2D}} \Psi^\dagger(\mathbf{x}) \Psi^\dagger(\mathbf{x}') V(\mathbf{x} - \mathbf{x}') \Psi(\mathbf{x}') \Psi(\mathbf{x}) \mathrm{d}\mathbf{x} \mathrm{d}\mathbf{x}'. \end{aligned} \tag{6.1.24}$$

Developing H^0 and H', we obtain

$$H^0 = \frac{\hbar^2}{2m} \int_{\mathbb{R}^D} \kappa^2 \mathbf{a}^\dagger(\hbar\kappa) \mathbf{a}^\dagger(\hbar\kappa) \mathrm{d}\kappa, \tag{6.1.25}$$

and

$$H' = \frac{1}{2} \int_{\mathbb{R}^{3D}} \tilde{V}_1 \mathbf{a}^\dagger(\hbar\mathbf{k}_1) \mathbf{a}^\dagger(\hbar\mathbf{k}_2) \mathbf{a}^\dagger(\hbar\mathbf{k}_2 + \hbar\mathbf{l}) \mathbf{a}^\dagger(\hbar\mathbf{k}_1 - \hbar\mathbf{l}) \mathrm{d}\kappa_1 \mathrm{d}\kappa_2 \mathrm{d}\mathbf{l}, \tag{6.1.26}$$

where

$$\tilde{V}_1 \;=\; \frac{1}{8\pi^3}\int_{\mathbb{R}^D} V(\mathbf{r})e^{-i\mathbf{l}\cdot\mathbf{r}}d\mathbf{r}. \qquad (6.1.27)$$

The quantum Liouville equation for the density function ρ of the system then reads

$$i\hbar\frac{\partial}{\partial t}\rho(t) \;=\; [H,\rho(t)], \qquad (6.1.28)$$

which leads to

$$\frac{\partial}{\partial t}\mathcal{F}_2^W \;=$$

$$\hbar^{-6}\mathrm{Tr}\left\{(\partial_t\rho)\mathbf{a}^\dagger\left(\mathbf{p}_1 - \frac{1}{2}\hbar\mathbf{k}_1\right)\mathbf{a}^\dagger\left(\mathbf{p}_2 - \frac{1}{2}\hbar\mathbf{k}_2\right)\mathbf{a}\left(\mathbf{p}_2 + \frac{1}{2}\hbar\mathbf{k}_2\right)\mathbf{a}\left(\mathbf{p}_1 + \frac{1}{2}\hbar\mathbf{k}_1\right)\right\}$$

$$= \hbar^{-6}(i\hbar)^{-1}\mathrm{Tr}\left\{[H^0 + H',\rho]\mathbf{a}^\dagger\mathbf{a}^\dagger\mathbf{a}\mathbf{a}\right\}$$

$$= \hbar^{-6}(i\hbar)^{-1}\mathrm{Tr}\left\{\rho\left[\mathbf{a}^\dagger\left(\mathbf{p}_1 - \frac{1}{2}\hbar\mathbf{k}_1\right)\right.\right.$$

$$\left.\left.\times\,\mathbf{a}^\dagger\left(\mathbf{p}_2 - \frac{1}{2}\hbar\mathbf{k}_2\right)\mathbf{a}\left(\mathbf{p}_2 + \frac{1}{2}\hbar\mathbf{k}_2\right)\mathbf{a}\left(\mathbf{p}_1 + \frac{1}{2}\hbar\mathbf{k}_1\right), H^0 + H'\right]\right\},$$

$$(6.1.29)$$

where the last identity follows from the identity below concerning the trace of a product of operators

$$\mathrm{Tr}[ABC] \;=\; \mathrm{Tr}[BCA] \;=\; \mathrm{Tr}[CAB].$$

We compute

$$(i\hbar)^{-1}[\mathbf{a}^\dagger\mathbf{a}^\dagger\mathbf{a}\mathbf{a}, H^0] = (i\hbar m)^{-1}\left\{\left(\mathbf{p}_1 + \frac{1}{2}\hbar\mathbf{k}_1\right)^2 - \left(\mathbf{p}_1 - \frac{1}{2}\hbar\mathbf{k}_1\right)^2\right.$$

$$\left. + \left(\mathbf{p}_2 + \frac{1}{2}\hbar\mathbf{k}_2\right)^2 - \left(\mathbf{p}_2 - \frac{1}{2}\hbar\mathbf{k}_2\right)^2\right\}\mathbf{a}^\dagger\mathbf{a}^\dagger\mathbf{a}\mathbf{a}$$

$$= -\frac{i}{m}\left(\mathbf{k}_1\cdot\mathbf{p}_1 + \mathbf{k}_2\cdot\mathbf{p}_2\right)\mathbf{a}^\dagger\left(\mathbf{p}_1 - \frac{1}{2}\hbar\mathbf{k}_1\right)\mathbf{a}^\dagger\left(\mathbf{p}_2 - \frac{1}{2}\hbar\mathbf{k}_2\right)$$

$$\times\,\mathbf{a}\left(\mathbf{p}_2 + \frac{1}{2}\hbar\mathbf{k}_2\right)\mathbf{a}\left(\mathbf{p}_1 + \frac{1}{2}\hbar\mathbf{k}_1\right).$$

$$(6.1.30)$$

In the next step, we find

$$[\mathbf{a}^\dagger\mathbf{a}^\dagger\mathbf{a}\mathbf{a}, H'] =$$

$$
\int_{\mathbb{R}^{2D}} \tilde{V}_1\Big\{ \mathbf{a}^\dagger\Big(\mathbf{p}_1 - \tfrac{1}{2}\hbar\mathbf{k}_1\Big)\mathbf{a}^\dagger\Big(\mathbf{p}_2 - \tfrac{1}{2}\hbar\mathbf{k}_2\Big)\mathbf{a}\Big(\mathbf{p}_2 + \tfrac{1}{2}\hbar\mathbf{k}_2\Big)\mathbf{a}^\dagger\big(\hbar\mathbf{k}\big)
$$

$$
\times\, \mathbf{a}\big(\hbar\mathbf{k} + \hbar\mathbf{l}\big)\mathbf{a}\Big(\mathbf{p}_1 + \tfrac{1}{2}\hbar\mathbf{k}_1 - \hbar\mathbf{l}\Big) + \mathbf{a}^\dagger\Big(\mathbf{p}_1 - \tfrac{1}{2}\hbar\mathbf{k}_1\Big)\mathbf{a}^\dagger\Big(\mathbf{p}_2 - \tfrac{1}{2}\hbar\mathbf{k}_1\Big)
$$

$$
\times\, \mathbf{a}^\dagger\big(\hbar\mathbf{k}\big)\mathbf{a}\big(\hbar\mathbf{k} + \hbar\mathbf{l}\big)\mathbf{a}\Big(\mathbf{p}_2 + \tfrac{1}{2}\hbar\mathbf{k}_2 - \hbar\mathbf{l}\Big)\mathbf{a}\Big(\mathbf{p}_1 + \tfrac{1}{2}\hbar\mathbf{k}_1\Big)
$$

$$
-\, \mathbf{a}^\dagger\Big(\mathbf{p}_1 - \tfrac{1}{2}\hbar\mathbf{k}_1\Big)\mathbf{a}^\dagger\big(\hbar\mathbf{k}\big)\mathbf{a}^\dagger\Big(\mathbf{p}_2 - \tfrac{1}{2}\hbar\mathbf{k}_2 - \hbar\mathbf{l}\Big)\mathbf{a}\big(\hbar\mathbf{k} - \hbar\mathbf{l}\big)
$$

$$
\times\, \mathbf{a}\Big(\mathbf{p}_2 + \tfrac{1}{2}\hbar\mathbf{k}_2\Big)\mathbf{a}\Big(\mathbf{p}_1 + \tfrac{1}{2}\hbar\mathbf{k}_1\Big) - \mathbf{a}^\dagger\big(\hbar\mathbf{k}\big)\mathbf{a}^\dagger\Big(\mathbf{p}_1 - \tfrac{1}{2}\hbar\mathbf{k}_1 - \hbar\mathbf{l}\Big)
$$

$$
\times\, \mathbf{a}\big(\hbar\mathbf{k} - \hbar\mathbf{l}\big)\mathbf{a}^\dagger\Big(\mathbf{p}_2 - \tfrac{1}{2}\hbar\mathbf{k}_2\Big)\mathbf{a}\Big(\mathbf{p}_2 + \tfrac{1}{2}\hbar\mathbf{k}_2\Big)\mathbf{a}\Big(\mathbf{p}_1 + \tfrac{1}{2}\hbar\mathbf{k}_1\Big) \Big\}\mathrm{d}\mathbf{k}\mathrm{d}\mathbf{l},
$$

$$\tag{6.1.31}$$

which can be developed into

$$[\mathbf{a}^\dagger\mathbf{a}^\dagger\mathbf{a}\mathbf{a}, H']$$

$$
= \int_{\mathbb{R}^{D}} \tilde{V}_1\Big\{ \mathbf{a}^\dagger\Big(\mathbf{p}_1 - \tfrac{1}{2}\hbar\mathbf{k}_1\Big)\mathbf{a}^\dagger\Big(\mathbf{p}_2 - \tfrac{1}{2}\hbar\mathbf{k}_2\Big)
$$

$$
\times\, \mathbf{a}\Big(\mathbf{p}_2 + \tfrac{1}{2}\hbar\mathbf{k}_2 + \hbar\mathbf{l}\Big)\mathbf{a}^\dagger\Big(\mathbf{p}_1 + \tfrac{1}{2}\hbar\mathbf{k}_1 - \hbar\mathbf{l}\Big)
$$

$$
-\, \mathbf{a}^\dagger\Big(\mathbf{p}_1 - \tfrac{1}{2}\hbar\mathbf{k}_1 + \hbar\mathbf{l}\Big)\mathbf{a}^\dagger\Big(\mathbf{p}_2 - \tfrac{1}{2}\hbar\mathbf{k}_2 - \hbar\mathbf{l}\Big)
$$

$$
\times\, \mathbf{a}\Big(\mathbf{p}_2 + \tfrac{1}{2}\hbar\mathbf{k}_2\Big)\mathbf{a}\Big(\mathbf{p}_1 + \tfrac{1}{2}\hbar\mathbf{k}_1\Big) \Big\}\mathrm{d}\mathbf{l}
$$

$$
+\, \hbar^{-3}\int_{\mathbb{R}^{2D}} \tilde{V}_1\Big\{ \mathbf{a}^\dagger\Big(\mathbf{p}_1 - \tfrac{1}{2}\hbar\mathbf{k}_1\Big)\mathbf{a}^\dagger\Big(\mathbf{p}_2 - \tfrac{1}{2}\hbar\mathbf{k}_2\Big)\mathbf{a}^\dagger\big(\mathbf{p}_3\big)
$$

$$
\times\, \mathbf{a}\big(\mathbf{p}_3 + \hbar\mathbf{l}\big)\mathbf{a}\Big(\mathbf{p}_2 + \tfrac{1}{2}\hbar\mathbf{k}_2\Big)\mathbf{a}\Big(\mathbf{p}_1 + \tfrac{1}{2}\hbar\mathbf{k}_1 - \hbar\mathbf{l}\Big)
\tag{6.1.32}
$$

$$
+\, \mathbf{a}^\dagger\Big(\mathbf{p}_1 - \tfrac{1}{2}\hbar\mathbf{k}_1\Big)\mathbf{a}^\dagger\Big(\mathbf{p}_2 - \tfrac{1}{2}\hbar\mathbf{k}_2\Big)\mathbf{a}^\dagger\big(\mathbf{p}_3\big)
$$

$$
\times\, \mathbf{a}\big(\mathbf{p}_3 + \hbar\mathbf{l}\big)\mathbf{a}\Big(\mathbf{p}_2 + \tfrac{1}{2}\hbar\mathbf{k}_2 - \hbar\mathbf{l}\Big)\mathbf{a}\Big(\mathbf{p}_1 + \tfrac{1}{2}\hbar\mathbf{k}_1\Big)
$$

$$
-\, \mathbf{a}^\dagger\Big(\mathbf{p}_1 - \tfrac{1}{2}\hbar\mathbf{k}_1\Big)\mathbf{a}^\dagger\Big(\mathbf{p}_2 - \tfrac{1}{2}\hbar\mathbf{k}_2 + \hbar\mathbf{l}\Big)\mathbf{a}^\dagger\big(\mathbf{p}_3 - \hbar\mathbf{l}\big)
$$

$$
\times\, \mathbf{a}\big(\mathbf{p}_3\big)\mathbf{a}\Big(\mathbf{p}_2 + \tfrac{1}{2}\hbar\mathbf{k}_2\Big)\mathbf{a}\Big(\mathbf{p}_1 + \tfrac{1}{2}\hbar\mathbf{k}_1\Big)
$$

$$
-\, \mathbf{a}^\dagger\Big(\mathbf{p}_1 - \tfrac{1}{2}\hbar\mathbf{k}_1 + \hbar\mathbf{l}\Big)\mathbf{a}^\dagger\Big(\mathbf{p}_2 - \tfrac{1}{2}\hbar\mathbf{k}_2\Big)\mathbf{a}^\dagger\big(\mathbf{p}_3 - \hbar\mathbf{l}\big)
$$

$$
\times\, \mathbf{a}\big(\mathbf{p}_3\big)\mathbf{a}\Big(\mathbf{p}_2 + \tfrac{1}{2}\hbar\mathbf{k}_2\Big)\mathbf{a}\Big(\mathbf{p}_1 + \tfrac{1}{2}\hbar\mathbf{k}_1\Big) \Big\}\mathrm{d}\mathbf{l}\mathrm{d}\mathbf{p}_3.
$$

Plugging (6.1.31) and (6.1.32) into (6.1.29) yields

$$\frac{\partial}{\partial t} \mathcal{F}_2^W(\mathbf{k}_1, \mathbf{p}_1, \mathbf{k}_2, \mathbf{p}_2) =$$

$$- \frac{i}{m}(\mathbf{k}_1 \cdot \mathbf{p}_1 + \mathbf{k}_2 \cdot \mathbf{p}_2) \mathcal{F}_2^W(\mathbf{k}_1, \mathbf{p}_1, \mathbf{k}_2, \mathbf{p}_2)$$

$$+ (i\hbar)^{-1} \int_{\mathbb{R}^D} \tilde{V}_1 \Big\{ \mathcal{F}_2^W\Big(\mathbf{k}_1 - \mathbf{l}, \mathbf{p}_1 - \frac{1}{2}\hbar\mathbf{l}, \mathbf{k}_2 + \mathbf{l}, \mathbf{p}_2 + \frac{1}{2}\hbar\mathbf{l}\Big)$$

$$- \mathcal{F}_2^W\Big(\mathbf{k}_1 - \mathbf{l}, \mathbf{p}_1 + \frac{1}{2}\hbar\mathbf{l}, \mathbf{k}_2 + \mathbf{l}, \mathbf{p}_2 - \frac{1}{2}\hbar\mathbf{l}\Big) \Big\} d\mathbf{l}$$

$$+ (i\hbar)^{-1} \int_{\mathbb{R}^{2D}} \tilde{V}_1 \Big\{ \mathcal{F}_3^W\Big(\mathbf{k}_1 - \mathbf{l}, \mathbf{p}_1 - \frac{1}{2}\hbar\mathbf{l}, \mathbf{k}_2, \mathbf{p}_2, \mathbf{l}, \mathbf{p}_3 + \frac{1}{2}\hbar\mathbf{l}\Big)$$

$$- \mathcal{F}_3^W\Big(\mathbf{k}_1 - \mathbf{l}, \mathbf{p}_1 + \frac{1}{2}\hbar\mathbf{l}, \mathbf{k}_2, \mathbf{p}_2, \mathbf{l}, \mathbf{p}_3 - \frac{1}{2}\hbar\mathbf{l}\Big)$$

$$+ \mathcal{F}_3^W\Big(\mathbf{k}_1, \mathbf{p}_1, \mathbf{k}_2 - \mathbf{l}, \mathbf{p}_2 - \frac{1}{2}\hbar\mathbf{l}, \mathbf{l}, \mathbf{p}_3 + \frac{1}{2}\hbar\mathbf{l}\Big)$$

$$- \mathcal{F}_3^W\Big(\mathbf{k}_1, \mathbf{p}_1, \mathbf{k}_2 - \mathbf{l}, \mathbf{p}_2 + \frac{1}{2}\hbar\mathbf{l}, \mathbf{l}, \mathbf{p}_3 - \frac{1}{2}\hbar\mathbf{l}\Big) \Big\} d\mathbf{l} d\mathbf{p}_3. \tag{6.1.33}$$

Let us then define the operators

$$L_j^0 = -\frac{j}{m} \mathbf{k}_j \cdot \mathbf{p}_j,$$

$$L_{jn}' = \frac{1}{i\hbar} \int_{\mathbb{R}^D} \tilde{V}_1 \Big[\exp\Big(-\frac{1}{2}\hbar\mathbf{l} \cdot \partial_{jn}\Big) - \exp\Big(\frac{1}{2}\hbar\mathbf{l} \cdot \partial_{jn}\Big) \Big] \exp(-\mathbf{l} \cdot \delta_{jn}) d\mathbf{l}, \tag{6.1.34}$$

in which

$$\partial_j = \partial_{\mathbf{p}_j}, \quad \partial_{ij} = \partial_{\mathbf{p}_i} - \partial_{\mathbf{p}_j},$$
$$\delta_j = \partial_{\mathbf{k}_j}, \quad \delta_{ij} = \partial_{\mathbf{k}_i} - \partial_{\mathbf{k}_j}. \tag{6.1.35}$$

Then (6.1.33) can be rewritten as

$$\frac{\partial}{\partial t} \mathcal{F}_2^W(\mathbf{k}_1, \mathbf{p}_1, \mathbf{k}_2, \mathbf{p}_2) =$$

$$(L_1^0 + L_2^0) \mathcal{F}_2^W(\mathbf{k}_1, \mathbf{p}_1, \mathbf{k}_2, \mathbf{p}_2) + L_{12}' \mathcal{F}_2^W(\mathbf{k}_1, \mathbf{p}_1, \mathbf{k}_2, \mathbf{p}_2) \tag{6.1.36}$$

$$+ \int_{\mathbb{R}^{2D}} \delta(\mathbf{k}_3)(L_{13}' + L_{23}') \mathcal{F}_3^W(\mathbf{k}_1, \mathbf{p}_1, \mathbf{k}_2, \mathbf{p}_2, \mathbf{k}_3, \mathbf{p}_3) d\mathbf{p}_3 d\mathbf{k}_3.$$

The equation for $\mathcal{F}_n^W$ follows by the same process

$$\frac{\partial}{\partial t} \mathcal{F}_n^W(\xi_1, \cdots, \xi_n) = \sum_{j=1}^n L_j^0 \mathcal{F}_n^W(\xi_1, \cdots, \xi_n) + \sum_{j<s=1}^n L_{js}' \mathcal{F}_n^W(\xi_1, \cdots, \xi_n)$$

$$+ \sum_{j=1}^n \int_{\mathbb{R}^{2D}} \delta(\mathbf{k}_{n+1}) L_{j,n+1}' \mathcal{F}_{n+1}^W(\xi_1, \cdots, \xi_{n+1}).$$

$$\tag{6.1.37}$$

6.1.3 Step 3: Quantum Correlations

Let us now discuss correlation in quantum systems. Consider the two-particle Wigner function. In the case when the particles are statistically independent, the function $\mathcal{F}_2^W(\mathbf{k}_1, \mathbf{p}_1, \mathbf{k}_2, \mathbf{p}_2)$ would be factorized as:

$$\mathcal{F}_2^W(\mathbf{k}_1, \mathbf{p}_1, \mathbf{k}_2, \mathbf{p}_2) = \mathcal{F}_1^W(\mathbf{k}_1, \mathbf{p}_1)\mathcal{F}_1^W(\mathbf{k}_2, \mathbf{p}_2) + \mathcal{G}_2^W(\mathbf{k}_1, \mathbf{p}_1, \mathbf{k}_2, \mathbf{p}_2).$$
$$(6.1.38)$$

This equation does not give any hint about the form of the function $\mathcal{G}_2^W$. From (6.1.19), the function $\mathcal{F}_2^W(\xi_1, \xi_2)$ possesses the characteristic symmetry related to the boson nature of the particles. However, the first term on the right-hand side of (6.1.38) does not have this property and therefore neither does $\mathcal{G}_2^W$. Due to the symmetry of the global wave function for bosons, the constituent particles cannot be independent. In a classical system, the only source for the correlations is the interactions between particles. In a system of bosons, the correlations also come from the quantum statistical boson constraints. In order to have quantum correlations, we can add another term to the product $\mathcal{F}_1^W(\mathbf{k}_1, \mathbf{p}_1)\mathcal{F}_1^W(\mathbf{k}_2, \mathbf{p}_2)$ based on the form of (6.1.19)

$$\mathcal{F}_2^W(\mathbf{k}_1, \mathbf{p}_1, \mathbf{k}_2, \mathbf{p}_2) = \mathcal{F}_1^W(\mathbf{k}_1, \mathbf{p}_1)\mathcal{F}_1^W(\mathbf{k}_2, \mathbf{p}_2)$$
$$+ \mathcal{F}_1^W\left(\frac{1}{2}(\mathbf{k}_1 + \mathbf{k}_2) + \hbar^{-1}(\mathbf{p}_2 - \mathbf{p}_1), \frac{1}{2}(\mathbf{p}_1 + \mathbf{p}_2) + \frac{1}{4}\hbar(\mathbf{k}_2 - \mathbf{k}_1)\right)$$
$$\times \mathcal{F}_1^W\left(\frac{1}{2}(\mathbf{k}_1 + \mathbf{k}_2) + \hbar^{-1}(\mathbf{p}_1 - \mathbf{p}_2), \frac{1}{2}(\mathbf{p}_1 + \mathbf{p}_2) + \frac{1}{4}\hbar(\mathbf{k}_1 - \mathbf{k}_2)\right)$$
$$+ \mathcal{G}_2^W(\mathbf{k}_1, \mathbf{p}_1, \mathbf{k}_2, \mathbf{p}_2).$$
$$(6.1.39)$$

The sum of the first and the second terms satisfies the symmetry (6.1.19), thus so does $\mathcal{G}_2^W$. Let us rewrite this sum as

$$P(1|2)\pi_2(1|2) \qquad\qquad (6.1.40)$$

in which

$$\pi_2(1|2) = \mathcal{F}_1^W(\mathbf{k}_1, \mathbf{p}_1)\mathcal{F}_1^W(\mathbf{k}_2, \mathbf{p}_2),$$

$$P(1|2)\pi_2(1|2) = \int_{\mathbb{R}^{2D}} \langle \mathbf{k}_1\mathbf{k}_2|P(1|2)|\mathbf{k}_1'\mathbf{k}_2'\rangle \mathcal{F}_1^W(\mathbf{k}_1', \mathbf{p}_1)\mathcal{F}_1^W(\mathbf{k}_2', \mathbf{p}_2)d\mathbf{k}_1'd\mathbf{k}_2',$$

and

$$\langle \mathbf{k}_1 \mathbf{k}_2 | P(1|2) | \mathbf{k}_1' \mathbf{k}_2' \rangle$$

$$= \delta\left[- \mathbf{k}_1' + \hbar^{-1}(\mathbf{p}_1 - \mathbf{p}_1) + \frac{1}{2}(\mathbf{k}_1 + \mathbf{k}_1) \right]$$

$$\times \delta\left[- \mathbf{k}_2' + \hbar^{-1}(\mathbf{p}_2 - \mathbf{p}_2) + \frac{1}{2}(\mathbf{k}_2 + \mathbf{k}_2) \right]$$

$$+ \delta\left[- \mathbf{k}_1' + \hbar^{-1}(\mathbf{p}_2 - \mathbf{p}_1) + \frac{1}{2}(\mathbf{k}_2 + \mathbf{k}_1) \right]$$

$$\times \delta\left[- \mathbf{k}_2' + \hbar^{-1}(\mathbf{p}_1 - \mathbf{p}_2) + \frac{1}{2}(\mathbf{k}_1 + \mathbf{k}_2) \right]$$

$$\times \exp\left[\frac{1}{2}\hbar((\mathbf{k}_1' - \mathbf{k}_1) \cdot \partial_1 + (\mathbf{k}_2' - \mathbf{k}_2) \cdot \partial_2) \right]$$

$$= \sum_{(1|2)} \left\{ \delta\left[- \mathbf{k}_1' + \hbar^{-1}(\mathbf{p}_{i_1} - \mathbf{p}_1) + \frac{1}{2}(\mathbf{k}_{i_1} + \mathbf{k}_1) \right] \right.$$

$$\left. \times \delta\left[- \mathbf{k}_2' + \hbar^{-1}(\mathbf{p}_{i_2} - \mathbf{p}_2) + \frac{1}{2}(\mathbf{k}_{i_2} + \mathbf{k}_2) \right] \right\}$$

$$\times \exp\left[\frac{1}{2}\hbar((\mathbf{k}_1' - \mathbf{k}_1) \cdot \partial_1 + (\mathbf{k}_2' - \mathbf{k}_2) \cdot \partial_2) \right],$$

where the summation is over all permutations of the indices i_1, i_2 of $1, 2$.

The third term can also be written as the action of a symmetrization operator

$$P(12) = \delta(-\mathbf{k}_1' + \mathbf{k}_1)\delta(-\mathbf{k}_2' + \mathbf{k}_2)$$

$$= \sum_{(12)} \delta\left(- \mathbf{k}_1' + \hbar^{-1}(\mathbf{p}_{i_1} - \mathbf{p}_1) + \frac{1}{2}(\mathbf{k}_{i_1} + \mathbf{k}_1) \right)$$

$$\times \delta\left(- \mathbf{k}_2' + \hbar^{-1}(\mathbf{p}_{i_2} - \mathbf{p}_2) + \frac{1}{2}(\mathbf{k}_{i_2} + \mathbf{k}_2) \right) \qquad (6.1.41)$$

$$\times \exp\left\{ \frac{1}{2}\hbar\left[(\mathbf{k}_1' - \mathbf{k}_1) \cdot \partial_1 + (\mathbf{k}_2' - \mathbf{k}_2) \cdot \partial_2 \right] \right\},$$

where the summation is taken only for $i_1 = 1$ and $i_2 = 2$, on the function

$$\pi_2(12) = \mathcal{G}_2^W(\mathbf{k}_1, \mathbf{p}_1, \mathbf{k}_2, \mathbf{p}_2). \qquad (6.1.42)$$

Therefore $\mathcal{F}_2^W$ takes the form

$$\mathcal{F}_2^W(1|2) = P(1|2)\pi_2(1|2) + P(12)\pi_2(12). \qquad (6.1.43)$$

We can therefore generalize the above identity to

$$\mathcal{F}_n^W(\xi_1, \cdots, \xi_n) = \sum_{\Gamma_n} P_n(\Gamma_n)\pi_n(\xi_1, \cdots, \xi_n; [\Gamma_n]), \qquad (6.1.44)$$

where the summation runs over all partitions of the set $(1, \cdots, n)$, each partition is denoted by Γ_n, and

$$(P\pi)(\mathbf{k}_1, \mathbf{p}_1, \cdots, \mathbf{k}_n, \mathbf{p}_n; [\Gamma_n]) = \tag{6.1.45}$$

$$\int_{\mathbb{R}^{D \times n}} \langle \mathbf{k}_1, \cdots, \mathbf{k}_n | P_n(\Gamma_n) | \mathbf{k}'_1, \cdots, \mathbf{k}'_n \rangle \pi_n(\mathbf{k}'_1, \mathbf{p}_1, \cdots, \mathbf{k}'_n, \mathbf{p}_n; [\Gamma_n]) \mathrm{d}\mathbf{k}'_1 \cdots \mathrm{d}\mathbf{k}'_n,$$

with

$$P_n(\Gamma_n) = \sum_{\Gamma_n} \delta\left(-\mathbf{k}'_1 + \hbar^{-1}(\mathbf{p}_{i_1} - \mathbf{p}_1) + \frac{1}{2}(\mathbf{k}_{i_1} + \mathbf{k}_1)\right) \times \cdots$$

$$\times \delta\left(-\mathbf{k}'_n + \hbar^{-1}(\mathbf{p}_{i_n} - \mathbf{p}_n) + \frac{1}{2}(\mathbf{k}_{i_n} + \mathbf{k}_n)\right) \tag{6.1.46}$$

$$\times \exp\left\{\frac{1}{2}\hbar\left[(\mathbf{k}'_1 - \mathbf{k}_1) \cdot \partial_1 + \cdots + (\mathbf{k}'_n - \mathbf{k}_n) \cdot \partial_n\right]\right\},$$

where the sum runs over all permutations of particles within the same subset. For example, if $n = 3$, for the pattern $\Gamma_j = (1|2|3)$, all 3! permutations of the indices are considered. However, for the $\Gamma_s = (1|23)$, only $3!/2!$ terms are counted since the permutation of particles in the subset (23) are not counted.

6.1.4 Step 4: Dynamics of Quantum Correlations

Let us start with the dynamical definition of the correlation patterns. In this section, we assume $\hbar = m = 1$. The state of the system is characterized by a distribution vector $\tilde{f}$

$$\tilde{f} = \{\pi_n(\xi_1, \cdots, \xi_n; [\Gamma_n]\}. \tag{6.1.47}$$

This distribution vector is then the solution of the set of equations (6.1.37), which can be denoted as a single quantum Liouville equation

$$\frac{\partial}{\partial t}\tilde{f} = \mathcal{L}\tilde{f}. \tag{6.1.48}$$

Equation (6.1.48) can be decomposed into component equations

$$\frac{\partial}{\partial t}\pi_n([\Gamma_n]; t) = \sum_{r=0}^{\infty}\sum_{\Gamma'_r}\langle(n)[\Gamma_n]|\mathcal{L}^0 + \mathcal{L}'|(r)[\Gamma'_r]\rangle\pi_r([\Gamma'_r]; t). \tag{6.1.49}$$

In combination with (6.1.49), this gives

$$\langle 1|\mathcal{L}^0|1\rangle = L_1^0,$$
$$\langle 1|\mathcal{L}'|1\rangle = 0,$$
$$\langle 1|\mathcal{L}'|1|2\rangle = \int_{\mathbb{R}^{2D}} \delta(\mathbf{k}_2)L'_{12}P_2(1|2)\delta(\mathbf{k}_2)\mathrm{d}\xi_2, \tag{6.1.50}$$
$$\langle 1|\mathcal{L}'|12\rangle = \int_{\mathbb{R}^{2D}} \delta(\mathbf{k}_2)L'_{12}P_2(12)\delta(\mathbf{k}_2)\mathrm{d}\xi_2.$$

Following (6.1.37), one can obtain an equation for π_1

$$\left(\frac{\partial}{\partial t} - L_1^0\right)\pi_1(1) = \int_{\mathbb{R}^{2D}} \delta(\mathbf{k}_2)\{L'_{12}P_2(1|2)\pi_2(1|2) + L'_{12}P_2(12)\pi_2(12)\}\mathrm{d}\xi_2$$

(6.1.51)

and π_2, and we have

$$(\frac{\partial}{\partial t} - L_1^0 - L_2^0)[P_2(1|2)\pi_2(1|2) + \pi_2(12)]$$

$$= L'_{12}[P_2(1|2)\pi_2(1|2) + P_2(12)\pi_2(12)]$$

$$+ \int_{\mathbb{R}^{2D}} \delta(\mathbf{k}_3)(L'_{13} + L'_{23})\Big\{P_3(1|2|3)\pi_3(1|2|3)$$

$$+ P_3(1|23)\pi_3(1|23) + P_3(2|13)\pi_3(2|13)$$

$$+ P_3(3|12)\pi_3(3|12) + P_3(123)\pi_3(123)\Big\}\mathrm{d}\xi_3.$$

(6.1.52)

By using

$$\pi_2(1|2) = \pi_1(1)\pi_1(2),$$

we get

$$\frac{\partial}{\partial t}\pi_2(1|2) =$$

$$\pi_1(1)\Big\{L_2^0\pi_1(2) + \int_{\mathbb{R}^{2D}} \delta(\mathbf{k}_3)L'_{23}[P_2(2|3)\pi_2(2|3) + P_2(23)\pi_2(23)]\mathrm{d}\xi_3\Big\}$$

$$+ \pi_1(2)\Big\{L_1^0\pi_1(1) + \int_{\mathbb{R}^{2D}} \delta(\mathbf{k}_3)L'_{13}[P_2(1|3)\pi_2(1|3) + P_2(13)\pi_2(13)]\mathrm{d}\xi_3\Big\},$$

(6.1.53)

and by using

$$\pi_2(1|23) = \pi_1(1)\pi_1(23),$$
$$\pi_2(1|2|3) = \pi_1(1)\pi_1(2)\pi_1(3),$$

we obtain

$$\frac{\partial}{\partial t}\pi_2(1|2) = (L_1^0 + L_2^0)\pi_2(1|2)$$

$$+ \int_{\mathbb{R}^{2D}} \delta(\mathbf{k}_3)\Big\{L'_{13}P_2(1|3)\pi_3(1|2|3) + L'_{23}P_2(2|3)\pi_3(1|2|3)$$

$$+ L'_{13}P_2(13)\pi_2(2|13) + L'_{23}P_2(23)\pi_3(1|23)\Big\}\mathrm{d}\xi_3.$$

(6.1.54)

Multiplying all terms of (6.1.54) by $P_2(1|2)$, using the identity

$$P_2(1|2)(L_1^0 + L_2^0) = (L_1^0 + L_2^0)P_2(1|2),$$

and subtracting (6.1.52) yields

$$P_2(1|2)(L_1^0 + L_2^0)\pi_2(1|2)$$

$$= -i \int_{\mathbb{R}^{2D}} \left\{ \delta(-\mathbf{k}_1' + \mathbf{k}_1)\delta(-\mathbf{k}_2' + \mathbf{k}_2) + \delta\left(-\mathbf{k}_1' + (\mathbf{p}_2 - \mathbf{p}_1) + \frac{1}{2}(\mathbf{k}_2 + \mathbf{k}_1) \right) \right.$$

$$\left. \times \delta\left[-\mathbf{k}_2' + (\mathbf{p}_1 - \mathbf{p}_2) + \frac{1}{2}(\mathbf{k}_1 + \mathbf{k}_2) \right] \right\}$$

$$\times \exp\left[\frac{1}{2}(\mathbf{k}_1' - \mathbf{k}_1) \cdot \partial_1 + \frac{1}{2}(\mathbf{k}_2' - \mathbf{k}_2) \cdot \partial_2 \right]$$

$$\times (\mathbf{k}_1' \cdot \mathbf{p}_1 + \mathbf{k}_2' \cdot \mathbf{p}_2)\pi_1(\mathbf{k}_1', \mathbf{p}_1)\pi(\mathbf{k}_2', \mathbf{p}_2)$$

$$= (L_1^0 + L_2^0)P_2(1|2)\pi_2(1|2).$$

$$(6.1.55)$$

The symmetrizer for all the correlations commutes with the unperturbed Liouvillian, which allows us to write

$$\left(\frac{\partial}{\partial t} - L_1^0 - L_2^0 \right) P_2(1|2)\pi_{12}(1|2)$$

$$= \int_{\mathbb{R}^{2D}} \delta(\mathbf{k}_3)P_2(1|2)\{L_{13}'P_2(1|3)\pi_3(1|2|3) + L_{23}'P_2(2|3)\pi_3(1|2|3) \qquad (6.1.56)$$

$$+ L_{13}'P_2(13)\pi_3(2|13) + L_{23}'P_2(23)\pi_3(1|23)\}\mathrm{d}\xi_3.$$

Subtracting (6.1.56) and (6.1.52) gives

$$\left(\frac{\partial}{\partial t} - L_1^0 - L_2^0 \right) \pi_{12}(12)$$

$$= L_{12}'P_2(1|2)\pi_2(1|2) + L_{12}'P_2(12)\pi_2(12),$$

$$+ \int_{\mathbb{R}^{2D}} \delta(\mathbf{k}_3)\{L_{13}'P_3(1|2|3) - P_2(1|2)L_{13}'P_2(1|3)$$

$$+ L_{23}'P_3(1|2|3) - P_2(1|2)L_{23}'P_2(2|3)\pi_3(1|2|3),$$

$$+ [L_{13}'P_3(1|23) + L_{23}'P_3(1|23) - P_2(1|2)L_{23}'P_2(23)]\pi_3(1|23),$$

$$+ [L_{23}'P_3(2|13) + L_{13}'P_3(2|13) - P_2(1|2)L_{13}'P_2(13)]\pi_3(2|13),$$

$$+ (L_{13}' + L_{23}')P_3(3|12)\pi_3(3|12) + (L_{13}' + L_{23}')P_3(123)\pi_3(123)\}\mathrm{d}\xi_3.$$

$$(6.1.57)$$

Finally, we have

$$\langle 1|2|\mathcal{L}'|1|2 \rangle = \langle 1|2|\mathcal{L}'|12 \rangle = 0,$$

$$\langle 1|2|\mathcal{L}'|1|2|3 \rangle = \int_{\mathbb{R}^{2D}} \delta(\mathbf{k}_3)[L_{13}'P_2(1|3) + L_{23}'P_2(2|3)]\mathrm{d}\xi_3,$$

$$\langle 1|2|\mathcal{L}'|1|23 \rangle = \int_{\mathbb{R}^{2D}} \delta(\mathbf{k}_3)L_{23}'P_2(2|3)\mathrm{d}\xi_3, \qquad (6.1.58)$$

$$\langle 1|2|\mathcal{L}'|2|13 \rangle = \int_{\mathbb{R}^{2D}} \delta(\mathbf{k}_3)L_{13}'P_2(1|3)\mathrm{d}\xi_3,$$

$$\langle 1|2|\mathcal{L}'|3|12 \rangle = \langle 1|2|\mathcal{L}'|123 \rangle = 0$$

and

$$\langle 12|\mathcal{L}'|1|2\rangle = L'_{12}P_2(1|2),$$

$$\langle 12|\mathcal{L}'|12\rangle = L'_{12}P_2(12),$$

$$\langle 12|\mathcal{L}'|1|2|3\rangle = \int_{\mathbb{R}^{2D}} \delta(\mathbf{k}_3)[L'_{13}P_3(1|2|3) - P_2(1|2)L'_{13}P_2(1|3)$$
$$+ L'_{23}P_3(1|2|3) - P_2(1|2)L'_{23}P_2(2|3)]\mathrm{d}\xi_3,$$

$$\langle 12|\mathcal{L}'|1|23\rangle = \int_{\mathbb{R}^{2D}} \delta(\mathbf{k}_3)[L'_{13}P_3(2|13)$$
$$+ L'_{23}P_3(1|23) - P_2(1|2)L'_{23}P_2(23)]\mathrm{d}\xi_3, \qquad (6.1.59)$$

$$\langle 12|\mathcal{L}'|2|13\rangle = \int_{\mathbb{R}^{2D}} \delta(\mathbf{k}_3)[L'_{23}P_3(1|23)$$
$$+ L'_{13}P_3(2|13) - P_2(1|2)L'_{13}P_2(13)]\mathrm{d}\xi_3,$$

$$\langle 12|\mathcal{L}'|3|12\rangle = \int_{\mathbb{R}^{2D}} \delta(\mathbf{k}_3)[L'_{13} + L'_{23}]P_3(3|12)\mathrm{d}\xi_3,$$

$$\langle 12|\mathcal{L}'|123\rangle = \int_{\mathbb{R}^{2D}} \delta(\mathbf{k}_3)[L'_{13} + L'_{23}]P_3(123)\mathrm{d}\xi_3.$$

6.1.5 Step 5: The Quantum Kinetic Equation

For the sake of simplicity, we only consider spatially homogeneous systems of Bose gases. In this case

$$\tilde{f}(t,\xi) = 8\pi^3\delta(\mathbf{k})\varphi(t,\mathbf{p}). \qquad (6.1.60)$$

Since there is no free-flow term

$$L_1^0\tilde{f}(t,\xi_1) = 8\pi^3(\mathbf{k}_1 \cdot \mathbf{p}_1)\delta(\mathbf{k}_1)\varphi(t,\mathbf{p}_1). \qquad (6.1.61)$$

Notice that in the homogeneous case, the Vlasov term vanishes

$$\int_{\mathbb{R}^{2D}} \delta(\mathbf{k}_2)L'_{12}P_2(1|2)\tilde{f}(\xi_1)\tilde{f}(\xi_2)$$
$$= \int_{\mathbb{R}^{5D}} \delta(\mathbf{k}_2)\tilde{V}_1\Big[\exp(-\frac{1}{2}\mathbf{l}\cdot\partial_{12}) - \exp(\frac{1}{2}\mathbf{l}\cdot\partial_{12})\Big]\exp(-\mathbf{l}\cdot\delta_{12})$$
$$\times \Big[\delta(-\mathbf{k}'_1 + \mathbf{k}_1)\delta(-\mathbf{k}'_2 + \mathbf{k}_2) + \delta\Big(-\mathbf{k}'_1 + \mathbf{p}_2 - \mathbf{p}_1 + \frac{1}{2}\mathbf{k}_1 + \frac{1}{2}\mathbf{k}_2\Big)$$
$$\times \delta\Big(-\mathbf{k}'_2 + \mathbf{p}_1 - \mathbf{p}_2 + \frac{1}{2}\mathbf{k}_1 + \frac{1}{2}\mathbf{k}_2\Big)\Big]$$
$$\times \delta(\mathbf{k}'_1)\delta(\mathbf{k}'_2)\varphi(\mathbf{p}_1)\varphi(\mathbf{p}_2)\mathrm{d}\mathbf{k}'_1\mathrm{d}\mathbf{k}'_2\mathrm{d}\mathbf{k}_2\mathrm{d}\mathbf{p}_2\mathrm{d}\mathbf{l}$$

$$= \delta(\mathbf{k}_1) \int_{\mathbb{R}^{2D}} \tilde{V}_{\mathbf{l}} \left[\exp\left(-\frac{1}{2} \cdot \partial_{12} \right) - \exp\left(\frac{1}{2}\mathbf{l} \cdot \partial_{12} \right) \right]$$

$$\times [\delta(\mathbf{l}) + \delta(\mathbf{p}_1 - \mathbf{p}_2)]\varphi(\mathbf{p}_1)\varphi(\mathbf{p}_2)\mathrm{d}\mathbf{p}_2\mathrm{d}\mathbf{l} = 0. \qquad (6.1.62)$$

We can now therefore focus only on the collision operator. In order to do that, we need to evaluate $\langle 12|\mathcal{L}'|1|2|3\rangle$. By using (6.1.45)

$$\langle \mathbf{k}_1\mathbf{k}_2\mathbf{k}_3|(L'_{13} + L'_{23})P_3(1|2|3) - P_2(1|2)[L'_{13}P_2(1|3) + L'_{23}P_2(2|3)]|\mathbf{k}'_1\mathbf{k}'_2\mathbf{k}'_3\rangle$$

$$= \int_{\mathbb{R}^D} \tilde{V}_{\mathbf{l}} \left\{ \left[\exp\left(-\frac{1}{2}\mathbf{l} \cdot \partial_{13} \right) - \exp\left(\frac{1}{2}\mathbf{l} \cdot \partial_{13} \right) \right] \right.$$

$$\times \delta(-\mathbf{k}'_1 + \mathbf{k}_1 - \mathbf{l})\delta\left(-\mathbf{k}'_2 + \mathbf{p}_3 - \mathbf{p}_2 + \frac{1}{2}\mathbf{l} \right)\delta\left(-\mathbf{k}'_3 + \mathbf{p}_2 - \mathbf{p}_3 + \frac{1}{2}\mathbf{l} \right)$$

$$+ \exp\left(-\frac{1}{2}\mathbf{l} \cdot \partial_{13} \right)\delta\left(-\mathbf{k}'_1 + \mathbf{p}_3 - \mathbf{p}_1 + \frac{1}{2}\mathbf{k}_1 \right)$$

$$\times \delta\left(-\mathbf{k}'_2 + \mathbf{p}_1 - \mathbf{p}_2 + \frac{1}{2}\mathbf{k}_1 - \frac{1}{2}\mathbf{l} \right)\delta\left(-\mathbf{k}'_3 + \mathbf{p}_2 - \mathbf{p}_3 + \frac{1}{2}\mathbf{l} \right)$$

$$- \exp\left(\frac{1}{2}\mathbf{l} \cdot \partial_{13} \right)\delta\left(-\mathbf{k}'_1 + \mathbf{p}_2 - \mathbf{p}_1 + \frac{1}{2}\mathbf{k}_1 + \frac{1}{2}\mathbf{l} \right) \qquad (6.1.63)$$

$$\times \delta\left(-\mathbf{k}'_2 + \mathbf{p}_3 - \mathbf{p}_2 + \frac{1}{2}\mathbf{l} \right)\delta\left(-\mathbf{k}'_3 + \mathbf{p}_1 - \mathbf{p}_3 + \frac{1}{2}\mathbf{k}_1 \right) \right\} \mathrm{d}\mathbf{l}$$

$$\times \exp\left[\frac{1}{2}(\mathbf{k}'_1 - \mathbf{k}_1 + \mathbf{l}) \cdot \partial_1 + (\mathbf{k}'_2 - \mathbf{k}_2) \cdot \partial_2 + (\mathbf{k}'_3 - \mathbf{k}_3 - \mathbf{l}) \cdot \partial_3 \right] + [1 \leftrightarrow 2],$$

in which $[1 \leftrightarrow 2]$ denotes the same terms with the indices 1 and 2 permuted. The derivation of this identity is based on the observation that

$$\exp\left(-\frac{1}{2}\mathbf{l} \cdot \partial_{13} \right)\delta\left(-\mathbf{k}'_1 + \mathbf{p}_2 - \mathbf{p}_1 - \frac{3}{2}\mathbf{l} + \frac{1}{2}\mathbf{k}_1 + \frac{1}{2}\mathbf{k}_2 \right)$$

$$\times \delta\left(-\mathbf{k}'_2 + \mathbf{p}_1 - \mathbf{p}_2 + \frac{1}{2}\mathbf{l} + \frac{1}{2}\mathbf{k}_1 + \frac{1}{2}\mathbf{k}_2 \right)$$

$$= \exp\left(-\frac{1}{2}\mathbf{l} \cdot \partial_{23} \right)\delta\left(-\mathbf{k}'_1 + \mathbf{p}_2 - \mathbf{p}_1 - \frac{1}{2}\mathbf{l} + \frac{1}{2}\mathbf{k}_1 + \frac{1}{2}\mathbf{k}_2 \right)$$

$$\times \delta\left(-\mathbf{k}'_2 + \mathbf{p}_1 - \mathbf{p}_2 + \frac{1}{2}\mathbf{l} + \frac{1}{2}\mathbf{k}_1 + \frac{1}{2}\mathbf{k}_2 \right).$$

Now, putting together (6.1.45), (6.1.59), (6.1.50), (6.1.63), with straightforward computations, we obtain an equation for f. Substituting (6.1.60) into this equation, we find

$$\frac{\partial}{\partial t}\varphi(t, \mathbf{p}_1)$$

$$= (i\hbar)^{-2}8\pi^3 \int_{\mathbb{R}^{2D}} \int_{-\infty}^{\infty} \exp\left[i\left(\mathbf{l} \cdot (\mathbf{v}_1 - \mathbf{v}_2) + \frac{\hbar \mathbf{l}^2}{m} \right)\tau \right] \tilde{V}_{\mathbf{l}}[\tilde{V}_{\mathbf{l}} + \tilde{V}_{\mathbf{l}+(\mathbf{p}_1-\mathbf{p}_2)/\hbar}]$$

$$\times \{ \varphi(\mathbf{p}_1)\varphi(\mathbf{p}_2)[1 + 8\pi^3\hbar^3\varphi(\mathbf{p}_1 + \hbar\mathbf{l}) + 8\pi^3\hbar^3\varphi(\mathbf{p}_2 - \hbar\mathbf{l})]$$

$$- \varphi(\mathbf{p}_1 + \hbar\mathbf{l})\varphi(\mathbf{p}_2 - \hbar\mathbf{l})[1 + 8\pi^3\hbar^3\varphi(\mathbf{p}_1) + 8\pi^3\hbar^3\varphi(\mathbf{p}_2)] \} \mathrm{d}\tau\mathrm{d}\mathbf{p}_2\mathrm{d}\mathbf{l}.$$

$$(6.1.64)$$

Integrating in τ gives

$$\frac{\partial}{\partial t}\varphi(t, \mathbf{p}_1) = 16\pi^4\hbar^{-2} \int_{\mathbb{R}^{2D}} \delta\left[\mathbf{l}\cdot(\mathbf{v}_1 - \mathbf{v}_2) + \frac{\hbar\mathbf{l}^2}{m}\right] [\tilde{V}_{\mathbf{l}} + \tilde{V}_{\mathbf{l}+(\mathbf{p}_1-\mathbf{p}_2)/\hbar}]^2$$

$$\times \{-\varphi(\mathbf{p}_1)\varphi(\mathbf{p}_2)[1 + \pi^3\hbar^3\varphi(\mathbf{p}_1 + \hbar\mathbf{l}) + \pi^3\hbar^3\varphi(\mathbf{p}_2 - \hbar\mathbf{l})]$$

$$+ \varphi(\mathbf{p}_1 + \hbar\mathbf{l})\varphi(\mathbf{p}_2 - \hbar\mathbf{l})[1 + \pi^3\hbar^3\varphi(\mathbf{p}_1) + \pi^3\hbar^3\varphi(\mathbf{p}_2)]\}d\mathbf{p}_2 d\mathbf{l}.$$

$$(6.1.65)$$

Now, the above equation can be rewritten in the form

$$\frac{\partial}{\partial t}\varphi(t, \mathbf{p}_1) = \int_{\mathbb{R}^{2D}} W\left(\mathbf{p}_1 + \frac{\hbar}{2}\mathbf{l}, \mathbf{p}_1 - \frac{\hbar}{2}\mathbf{l}; \mathbf{l}\right)$$

$$\times \left\{ -\varphi(\mathbf{p}_1)\varphi(\mathbf{p}_2)\left[1 + \pi^3\hbar^3\varphi(\mathbf{p}_1 + \hbar\mathbf{l})\right]\left[1 + \pi^3\hbar^3\varphi(\mathbf{p}_2 - \hbar\mathbf{l})\right] \right.$$

$$\left. + \varphi(\mathbf{p}_1 + \hbar\mathbf{l})\varphi(\mathbf{p}_2 - \hbar\mathbf{l})\left[1 + \pi^3\hbar^3\varphi(\mathbf{p}_1)\right]\left[1 + \pi^3\hbar^3\varphi(\mathbf{p}_2)\right] \right\}d\mathbf{p}_2 d\mathbf{l},$$

$$(6.1.66)$$

where

$$W\left(\mathbf{p}_1 + \frac{\hbar}{2}\mathbf{l}, \mathbf{p}_1 - \frac{\hbar}{2}\mathbf{l}; \mathbf{l}\right) =$$

$$16\pi^4\hbar^{-2}\delta\left[\mathbf{l}\cdot(\mathbf{p}_1 - \mathbf{p}_2) + \frac{\hbar\mathbf{l}^2}{m}\right][\tilde{V}_{\mathbf{l}} + \tilde{V}_{\mathbf{l}+(\mathbf{p}_1-\mathbf{p}_2)/\hbar}]^2.$$

Now, approximating $\delta\left[\mathbf{l}\cdot(\mathbf{p}_1 - \mathbf{p}_2) + \frac{\hbar\mathbf{l}^2}{m}\right]$ by $\delta\left[\mathbf{l}\cdot(\mathbf{p}_1 - \mathbf{p}_2)\right]$, and defining $\mathbf{p}_3 = \mathbf{p}_1 + \frac{\hbar}{2}\mathbf{l}$, $\mathbf{p}_4 = \mathbf{p}_2 - \frac{\hbar}{2}\mathbf{l}$, we get

$$\mathbf{p}_3 + \mathbf{p}_4 = \mathbf{p}_1 + \mathbf{p}_2$$

and

$$\mathbf{p}_3^2 + \mathbf{p}_4^2 = \mathbf{p}_1^2 + \mathbf{p}_2^2.$$

By reducing the matrix elements of V to their low energy limit, proportional to the scattering length, we obtain (5.0.1) from (6.1.66).

6.2 The Coupling BEC and Thermal Cloud System (5.0.2)–(5.0.3)

The theory of dilute quantum gases has two different time scales. The average time between collisions, or mean free flight time, is defined as

$$t_{mfp} = \frac{1}{\rho g^2}\left(\frac{m}{k_B T}\right)^{\frac{1}{2}}, \qquad (6.2.1)$$

formed by the small dimensionless parameter measuring the density, in which m is the mass of the particles, k_B is the Boltzmann constant, T is the temperature of the gas, ρ is the number density, and g is the scattering length.

On the other hand, the theory of Vlasov (cf. [18]) concerns short time scales of order

$$t_{Vl} = \frac{m}{g\hbar\rho},$$
(6.2.2)

which is of order g^{-1}. This means the corresponding collision operator is of order g.

In order to obtain the kinetic equation, we follow [118] by starting with the quantum BBGKY hierarchy, relying on the density matrices of one and two bodies $R(1; 1')$ and $R(1, 1'; 2, 2')$ $(1, 1', 2, 2'...$ abbreviate the positions $r_1, r_1', r_2, r_2'...)$

$$ih\frac{\partial R(1; 1')}{\partial t} = -\frac{\hbar^2}{2m}(\Delta_1 - \Delta_2)R(1; 1')$$
(6.2.3)
$$+ \frac{g\hbar^2}{m}(R_2(1, 1; 1', 1') - R_2(1, 1'; 1', 1')),$$

where Δ_1 and Δ_2 denote the Laplace operators with respect to the two variables r_1 and r_2. This equation is written under the assumption that the interacting potential is a delta function, which allows us to integrate the coordinate of the second particle in the interacting term. This approximation is justified since the diffusion is of long wavelength (cf. [106]).

Next, we write the matrix R_1 in terms of the wavefunction of the condensate Φ and the contribution of the thermal excitation $\tilde{R}$

$$R(1; 1') = \Phi(1)\Phi(1') + \tilde{R}(1; 1'),$$
(6.2.4)

in which $\tilde{R} \to 0$ as $|r_1 - r_1'| \to \infty$.

Following the Vlasov approximation, one can write R_2 as

$$
\begin{aligned}
R_2(1, 2, ; 1', 2') = \ &\Phi(1)\Phi(2)\bar{\Phi}(1')\bar{\Phi}(2') + \Phi(1)\Phi(1')\tilde{R}(2; 2') \\
&+ \Phi(2)\bar{\Phi}(2')\tilde{R}(1; 1') + \Phi(1)\Phi(2')\tilde{R}(1; 1') \\
&+ \Phi(1)\bar{\Phi}(2')\tilde{R}(2; 1') + \Phi(2)\Phi(1')\tilde{R}(1; 2') \\
&+ \tilde{R}(1; 1')\tilde{R}(2; 2') + \tilde{R}(1; 2')\tilde{R}(2; 1').
\end{aligned}
$$
(6.2.5)

Substituting (6.2.5) into (6.2.3) gives

$$ih\frac{\partial \Phi(1)}{\partial t} = -\frac{\hbar^2}{2m}\Delta_1\Phi(1) + \frac{g\hbar^2}{m}(\rho(1) + \tilde{R}(1; 1))\Phi(1),$$
(6.2.6)

and

$$ih\frac{\partial \tilde{R}(1; 1')}{\partial t} = -\frac{\hbar^2}{2m}(\Delta_1 - \Delta_{1'})\tilde{R}(1; 1')$$
$$+ \frac{g\hbar^2}{m}\tilde{R}(1; 1')[\rho(1) + \tilde{R}(1; 1) - \rho(1') - \tilde{R}(1'; 1')],$$
(6.2.7)

where

$$\rho(1) \;=\; R(1;1) \;=\; |\Phi(1)|^2 \;+\; \tilde{R}(1;1)$$

is the local numerical density. At equilibrium, $|\Phi|^2 = \rho_0$, which is the condensate density, while $\tilde{R}(1;1')$ is a function of $r_1 - r_1'$ and the temperature. The Fourier transform of $\tilde{R}(1;1')$ is then the Bose factor

$$\tilde{R}_{eq}(1;1') \;=\; \frac{1}{(2\pi\hbar)^3} \int_{\mathbb{R}^2} e^{i\frac{p}{\hbar}(r_1-r_1')} \varphi_{eq}(p)dp,$$

where

$$\varphi_{eq}(p) \;=\; \frac{1}{e^{\frac{p^2}{2mk_BT}} - 1}.$$

From this coupling equation, one can derive the dispersion relation, or spectrum, in the terminology of Landau (cf. [102]), from an energy equation for the momentum of small fluctuations in a neighborhood around the equilibrium

$$1 \;=\; \frac{\mathcal{L}(p)}{1-\mathcal{L}(p)} \;-\; \frac{2g\rho_s\hbar^2}{m} \frac{\frac{p^2}{2m}}{\left(\frac{p^2}{2m}\right)^2 - \omega^2(p)}, \qquad (6.2.8)$$

with

$$\mathcal{L}(p) \;=\; \frac{2g\hbar^2}{m(2\pi\hbar)^2} \int_{\mathbb{R}^3} \varphi_{eq}(p_1) \frac{\frac{p^2-2pp_2}{2m}}{\left(\frac{p^2-2pp_2}{2m}\right)^2 - \omega^2(p)} dp.$$

When there is no thermal particle ($T = \varphi_{eq} = 0$), we find the Bogoliubov spectrum (cf. [28])

$$\omega(p) = \sqrt{\kappa_1|p|^2 + \kappa_2|p|^4}, \qquad \kappa_1 = \frac{g\rho_0}{m} > 0, \quad \kappa_2 = \frac{1}{4m^2} > 0. \quad (6.2.9)$$

Because of the integration with respect to p_2 in (6.2.8), there is a Landau damping effect caused by the resonance between waves and particles of the same velocities.

At the Vlasov order, the masses of the condensate $\int_{\mathbb{R}^3} |\Phi(r_1)|^2 dr_1$ and of the normal fluid $\int_{\mathbb{R}^3} |\tilde{R}(1,1)|^2 dr_1$ are conserved separately, while an exchange of energy between the condensate and normal gas may happen, constrained by the conservation of total energy equal to

$$\int_{\mathbb{R}^3} \left[\frac{\hbar^2}{2m} \left(-\nabla_1\nabla_{1'}\tilde{R}(1;1')|_{1=1'} + |\nabla_1\Phi(1)|^2 \right) + \frac{g\hbar^2}{2m} \left(\rho^2 - \frac{\rho_0^2}{2} \right) \right] dr_1.$$

This expression of energy is correct to first order with respect to g (cf. [104]).

In the classical kinetic theory of gases, the next order in developing the collision operator, which is g^2, gives an irreversible evolution operator. In the absence of the condensate, this collision operator is the Boltzmann–Nordheim operator, which can be deduced to be exactly of order g^2. In this case, the dispersion relation is the classical one

$$\omega(p) = \frac{1}{2}|p|^2. \tag{6.2.10}$$

Let us first consider the spatially homogeneous Boltzmann–Nordheim equation

$$\frac{\partial}{\partial t} f_{p_1} = C_{22}[f_{p_1}], \tag{6.2.11}$$

where we recall that

$$C_{22}[f_{p_1}] = \int_{\mathbb{R}^9} W_{p_1,p_2;p_3,p_4}[f_{p_3}f_{p_4}(f_{p_1}+1)(f_{p_2}+1)$$
$$- f_{p_1}f_{p_2}(f_{p_3}+1)(f_{p_4}+1)]\mathrm{d}p_2\mathrm{d}p_3\mathrm{d}p_4.$$

In this homogeneous case, $R(1;1')$ is related to f_p by

$$R(1;1') = \frac{1}{(2\pi\hbar)^3}\int_{\mathbb{R}^3} e^{\frac{p(r_1-r'_r)}{\hbar}} f_p\mathrm{d}p.$$

The quantity W is defined by

$$W_{p_1,p_2;p_3,p_4} = \frac{2g^2}{m\hbar^2}\delta(p_1+p_2-p_3-p_4)\delta(p_1^2+p_2^2-p_3^2-p_4^2).$$

For the sake of simplicity, in the sequel, we take $\hbar = m = 1$. In our case, with the presence of the Bose–Einstein condensate, the function f_ω can be decomposed as

$$f_\omega(t) = \frac{\rho_0(t)}{\sqrt{\omega}}\delta(\omega) + \tilde{f}_\omega(t), \tag{6.2.12}$$

in which ρ_0 is the density of the condensate and $\tilde{f}_\omega$ the part continued from the continuous part of the Wigner distribution.

Plugging this ansatz into (6.2.11), we then obtain the following coupling system for ρ_0 and $\tilde{f}$:

$$\frac{\partial}{\partial t}\rho_0 = \rho_0\tilde{Q}_{12}[\tilde{f}], \tag{6.2.13}$$

$$\frac{\partial}{\partial t}\tilde{f} = C_{22}[\tilde{f}] + \rho_0 Q_{12}[\tilde{f}], \tag{6.2.14}$$

where

$$\tilde{Q}_{12}[\tilde{f}] = 2g^2\int_{\mathbb{R}^9}\delta(p_2-p_3-p_4)\delta(p_2^2-p_3^2-p_4^2)$$
$$\times [\tilde{f}_{p_3}\tilde{f}_{p_4} - \tilde{f}_{p_2}(\tilde{f}_{p_3}+\tilde{f}_{p_4}+1)]\mathrm{d}p_2\mathrm{d}p_3\mathrm{d}p_4,$$

and

$$Q_{12}[\tilde{f}] = \frac{g^2}{4\pi^3}\int_{\mathbb{R}^9}\delta(p_1-p_3-p_4)\delta(p_1^2-p_3^2-p_4^2)$$
$$\times [\tilde{f}_{p_3}\tilde{f}_{p_4} - \tilde{f}_{p_1}(\tilde{f}_{p_3}+\tilde{f}_{p_4}+1)]\mathrm{d}p_2\mathrm{d}p_3\mathrm{d}p_4$$
$$+ \frac{g^2}{2\pi^3}\int_{\mathbb{R}^9}\delta(p_1+p_2-p_4)\delta(p_1^2+p_2^2-p_4^2)$$

$$\times\ [\tilde{f}_{p_4}(\tilde{f}_{p_1} + \tilde{f}_{p_2} + 1)\ -\ \tilde{f}_{p_1}\tilde{f}_{p_2}]\mathrm{d}p_2\mathrm{d}p_3\mathrm{d}p_4.$$

We will now generalize the spatially homogeneous system (6.2.13)–(6.2.14) to the inhomogeneous case in several steps.

Firstly, one needs to transform the equation for ρ_0 into an equation for Φ. The equation for ρ_0 is then divided into two equations, one for Φ and one for the conjugate function

$$\frac{\partial \Phi}{\partial t}\ =\ ig\left[\rho(1)\ +\ \tilde{R}(1;1)\right]\Phi\ -\ 2\Phi\tilde{C}_{12}[\tilde{f}]. \qquad (6.2.15)$$

The form of the evolution equation of $\tilde{f}$ does not change. The function $\rho_0(1)$ is taken to be $|\Phi|^2$. The right-hand side of the equation (6.2.15) contains two parts. The first part is unitary and is written following the Vlasov order $O(g)$. The second part does not preserve the norm and is of order $O(g^2)$. This part is non-unitary and represents the exchange of mass between the condensate and the thermal cloud, under the condition that the gas is out of equilibrium. Notice that for a Bose gas whose distribution has a null potential, $\tilde{Q}_{12}$ is 0. This exchange of mass depends on the same type of effect that contributes to the entropy production. Since the system (6.2.14)–(6.2.15) conserves the total mass, it does not conserve the energy due to the condition of omitting the interaction part of this energy. The Dirac function $\delta(p_2^2 - p_3^2 - p_4^2)$ expresses the conservation of energy in $\tilde{Q}_{12}$ for the processes creating the condensate. This function is written under the assumption that the energy of the condensate is null. This assumption is of course only true in the absence of interactions. With the interactions, one needs to add the frequency $\frac{d\theta}{dt}$, in which $\theta(t)$ is the phase of the wavefunction Φ. This means one needs to replace the Dirac function for the conservation of energy by

$$\delta\left(2\frac{d\theta}{dt}\ +\ p_2^2\ -\ p_3^2\ -\ p_4^2\right).$$

Now, we need to write the coupling system for the inhomogeneous case. In this case, $\tilde{f}_p$ depends also on $r = \frac{r_1 + r_1'}{2}$, and the kinetic equation needs to include the normal kinetic term $\frac{p}{m}\nabla_r\tilde{f}(t, r)$.

Another effect of the inhomogeneity of the system is that Ψ now becomes a function of r, which means that the condensate carries a local wavenumber of value $\hbar\nabla_r\theta(t, r)$, where $\theta(t, r)$ is the phase of Φ. This number of waves must be inserted in the argument of the Dirac function expressing the conservation of total vibration. The function $\delta(p_2 - p_3 - p_4)$ in $\tilde{Q}_{12}$ must be replaced by $\delta(\nabla_r\theta + p_2 - p_3 - p_4)$ in the inhomogeneous case. The conservation of energy, like in the homogeneous case, leads to a new modification of the Dirac function for the energy in the collision operator. We have already seen how to take into account a non-null frequency. This frequency can arise from the interaction with the frequency $g\rho$. At the same order in the interaction, the thermal particle energy is also modified. To have a coupling system that preserves the energy, one needs to replace the energy of particles, from $\frac{p^2}{2}$

to $\frac{p^2}{2} + g\left(\rho - \frac{\rho_0}{2}\right)$. This replacement does not change the expression of the Dirac function. However, in $\tilde{Q}_{12}$ and Q_{12}, one needs to write the constraint of conservation of energy by putting the following Dirac function into the integrand of the kinetic operator:

$$\delta\left(2\frac{\partial\theta}{\partial t} - 2g\left(\rho - \frac{\rho_0}{2}\right) + p_2^2 - p_3^2 - p_4^2\right).$$

The coupling system can then be written as, after dropping the tilde sign on f,

$$\frac{\partial}{\partial t}f + p\cdot\nabla_r f + g\nabla_r(2\rho - \rho_0)\cdot\nabla_p f = C_{22}[f] + \rho_0 C_{12}[f], \quad (6.2.16)$$

and

$$\frac{\partial}{\partial t}\Phi = -\frac{1}{2}\Delta_r\Phi + g\left(\rho(1) + \tilde{R}(1;1)\right)\Phi + 2i\Phi\tilde{C}_{12}[f], \quad (6.2.17)$$

where

$$\tilde{C}_{12}[f] =$$
$$2g^2\int_{\mathbb{R}^9}\delta(\nabla_r\theta + p_2 - p_3 - p_4)\delta\left(2\frac{\partial\theta}{\partial t} - 2g\left(\rho - \frac{\rho_0}{2}\right) + p_2^2 - p_3^2 - p_4^2\right)$$
$$\times[f_{p_3}f_{p_4} - f_{p_2}(f_{p_3} + f_{p_4} + 1)]dp_2 dp_3 dp_4,$$

and

$$C_{12}[f] =$$
$$\frac{g^2}{4\pi^3}\int_{\mathbb{R}^9}\delta(\nabla_r\theta + p_1 - p_3 - p_4)\delta\left(2\frac{\partial\theta}{\partial t} - 2g\left(\rho - \frac{\rho_0}{2}\right) + p_1^2 - p_3^2 - p_4^2\right)$$
$$\times[f_{p_3}f_{p_4} - f_{p_1}(f_{p_3} + f_{p_4} + 1)]dp_2 dp_3 dp_4$$
$$+ \frac{g^2}{2\pi^3}\int_{\mathbb{R}^9}\delta(\nabla_r\theta p_1 + p_2 - p_4)\delta\left(2\frac{\partial\theta}{\partial t} - 2g\left(\rho - \frac{\rho_0}{2}\right) + p_1^2 + p_2^2 - p_4^2\right)$$
$$\times[f_{p_4}(f_{p_1} + f_{p_2} + 1) - f_{p_1}f_{p_2}]dp_2 dp_3 dp_4.$$

The system (6.2.16)–(6.2.17) constitutes an extension of the Boltzmann–Nordheim equation to the case in which the condensate is present, including a first-order correction in the interaction that allows the conservation of the exact energy

$$\int_{\mathbb{R}^3}\int_{\mathbb{R}^3}f(t,r,p)\frac{p^2}{2}drdp + \int_{\mathbb{R}^3}\left[\frac{|\nabla_r\Phi|^2}{2} + \frac{g}{2}\left(\rho^2 - \frac{\rho_0^2}{2}\right)\right]dr.$$

This equilibrium distribution of momenta takes the form:

$$\tilde{f}_{eq,u} = \frac{1}{e^{\frac{p^2 - 2pu + 2g\rho_0}{k_B T}} - 1}, \quad (6.2.18)$$

which represents a normal fluid of velocity u with respect to the superfluid rest frame. This velocity u needs to satisfy the condition $u^2 < 4g\rho_0$ so that $\tilde{f}_{eq,u}$ stays normalizable, following Landau (cf. [102]).

Let us emphasize that, due to the randomness of the collisions between bosons in the condensate and in the thermal gas, the ground state also has a random phase. It is then reasonable not to consider the condensate as an eigenstate in the traditional sense, namely a state with a phase exactly proportional to time (cf. [132, 148]).

6.3 Yet Another Coupling System (5.0.20)–(5.0.21)–(5.0.22)–(5.0.23) – The Ultra Low Temperature Regime

6.3.1 The Coupling Quantum Boltzmann and Hugenholtz–Pines System – A Modification of the Coupling Quantum Boltzmann and Gross–Pitaevskii System

Let us recall from the previous chapters that particles in the condensate, when it exists, occupy the lowest quantum state, which make possible various macroscopic quantum phenomena. Furthermore, the condensate excited atoms occupy higher quantum states. There are two types of interactions:

- excited atoms - excited atoms: C_{22} (Boltzmann–Nordheim or Uehling–Uhlenbeck term),
- condensate atoms - excited atoms: C_{12} (condensate growth term).

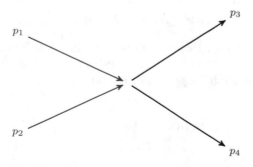

Fig. 6.1 The C_{22} collision. In the picture, two atoms with momenta p_1 and p_2 collide, and they change momenta into p_3 and p_4.

In order to get an intuition of what represent the various collision operators, let us start with the collision operator for classical particles. Suppose we are studying the density of a spatially homogeneous gas, whose distribution function is $f(t, p)$, which indicates the probability of finding a particle with momentum p at time t. Suppose two particles with momenta p_1 and p_2 collide and change their momenta into p_3 and p_4. That means we lose two particles with momenta p_1 and p_2, and gain two new particles with momenta p_3 and p_4. The probability of losing two particles with momenta p_1 and p_2 is $-f(t, p_1)f(t, p_2)$ and of gaining two particles with momenta p_3 and p_4 is $f(t, p_3)f(t, p_4)$. As a result, the collision operator has to contain

$$f(t, p_3)f(t, p_4) - f(t, p_1)f(t, p_2).$$

Since collisions conserve mass and momentum, we also have $p_1 + p_2 = p_3 + p_4$ and $\omega_{p_1} + \omega_{p_2} = \omega_{p_3} + \omega_{p_4}$, in which ω_p is the energy of the particle with momentum p. As a consequence, the collision operator should be proportional to

$$\delta(p_1 + p_2 - p_3 - p_4)\delta(\omega_{p_1} + \omega_{p_2} - \omega_{p_3} - \omega_{p_4})[f(t, p_3)f(t, p_4) - f(t, p_1)f(t, p_2)].$$

The classical Boltzmann equation then reads

$$\frac{\partial}{\partial t}f(t, p_1) = \int_{\mathbb{R}^9} \delta(p_1 + p_2 - p_3 - p_4)\delta(\omega_{p_1} + \omega_{p_2} - \omega_{p_3} - \omega_{p_4})$$
$$\times [f(t, p_3)f(t, p_4) - f(t, p_1)f(t, p_2)]\mathrm{d}p_2\mathrm{d}p_3\mathrm{d}p_4,$$

where the collision kernel is omitted for the sake of simplicity.

Let us apply the above argument for a gas of fermions. For fermions, we cannot consider the collisions of each quantum particle as in the case of a classical gas. We then modify the argument as follows: Suppose that two momentum states with momenta p_1 and p_2 collide and after the collision some particles will move to the new momentum states p_3 and p_4. As a result, we will have a term proportional to

$$f(t, p_3)f(t, p_4) - f(t, p_1)f(t, p_2)$$

in the collision operator. However, fermions follow the Pauli Exclusion Principle, which says that if the momentum states p_3 and p_4 are already full, then particles cannot move from p_1 and p_2 to p_3 and p_4. We therefore modify the collision as

$$f(t, p_3)f(t, p_4)(1 - f(t, p_1))(1 - f(t, p_2))$$
$$- f(t, p_1)f(t, p_2)(1 - f(t, p_3))(1 - f(t, p_4)),$$

where the factor $1 - f(t, p_i)$ indicates that momentum state p_i is occupied.

Rigorously speaking, the difference between the boson case and the fermion case is due to the fact that the annihilation/creation operators for fermions behave differently. Instead of (6.1.2), the annihilation/creation operators for fermions satisfy

$$\{\mathbf{a}_i, \mathbf{a}_j^\dagger\} \equiv \mathbf{a}_i \mathbf{a}_j^\dagger + \mathbf{a}_j^\dagger \mathbf{a}_i = \delta_{ij},$$
$$\{\mathbf{a}_i^\dagger, \mathbf{a}_j^\dagger\} = \{\mathbf{a}_i, \mathbf{a}_j\} = 0.$$

$$(6.3.1)$$

As a consequence, exchanging disjoint (i.e. $i \neq j$) operators in a product of annihilation or creation operators will reverse the sign in a system of fermions, but not in a system of bosons.

The quantum Boltzmann equation for fermions can be written as follows

$$\frac{\partial}{\partial t} f(t, p_1) = \int_{\mathbb{R}^9} \delta(p_1 + p_2 - p_3 - p_4) \delta(\omega_{p_1} + \omega_{p_2} - \omega_{p_3} - \omega_{p_4})$$
$$\times [f(t, p_3) f(t, p_4)(1 - f(t, p_1))$$
$$\times (1 - f(t, p_2)) - f(t, p_1) f(t, p_2)(1 - f(t, p_3))(1 - f(t, p_4))] dp_2 dp_3 dp_4.$$

The question now is how to adapt the above arguments for fermions to the case of the operator C_{12} for bosons. Suppose again that two momentum states with momenta p_1 and p_4 collide but the momentum state p_4 is hidden inside the condensate: we have a collision between p_1 and the condensate. After the collision some particles will move to the new momentum states p_2 and p_3. Similar to the above, in the collision operator, we will have a term proportional to $f(t, p_2) f(t, p_3) - f(t, p_1)$. If we apply the same argument as above, the collision operator is proportional to

$$f(t, p_2) f(t, p_3)(1 - f(t, p_1)) - f(t, p_1)(1 - f(t, p_2))(1 - f(t, p_3)).$$

However, unlike fermions, bosons do not follow the Pauli Exclusion Principle. Therefore, instead of multiplying by $1 - f(t, p_i)$, we use the factor $1 + f(t, p_i)$ and modify the collision as

$$f(t, p_2) f(t, p_3)(1 + f(t, p_1)) - f(t, p_1)(1 + f(t, p_2))(1 + f(t, p_3)).$$

The conservation of momentum and energy in this case are $p_1 = p_2 + p_3$ and $\omega_{p_1} = \omega_{p_2} + \omega_{p_3}$. The collision operator should contain

$$\delta(p_1 - p_2 - p_3)\delta(\omega_{p_1} - \omega_{p_2} - \omega_{p_3})$$
$$\times [f(t, p_2) f(t, p_3)(1 + f(t, p_1)) - f(t, p_1)(1 + f(t, p_2))(1 + f(t, p_3))].$$

$$(6.3.2)$$

Note that a more precise derivation of $1 + f(t, p_i)$ has been presented in Section 6.1.

Denoting this case as $p_1 \rightarrow p_2 + p_3$, we observe that there are still two other cases $p_2 \rightarrow p_1 + p_3$ and $p_3 \rightarrow p_1 + p_2$. Due to the symmetry of p_2 and p_3, we can combine them into

$$- 2\delta(p_2 - p_1 - p_3)\delta(\omega_{p_2} - \omega_{p_1} - \omega_{p_3})$$
$$\times [f(t, p_1) f(t, p_3)(1 + f(t, p_2)) - f(t, p_2)(1 + f(t, p_1))(1 + f(t, p_3))].$$

$$(6.3.3)$$

Putting (6.3.2) and (6.3.3) together, the form of C_{12} then follows. The interaction is of the type $1 \leftrightarrow 2$.

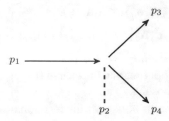

Fig. 6.2 The C_{12} collision. In the picture, an atom with momentum p_1 collides with a "condensate atom", and they change momenta into p_3 and p_4. The dashed line represents the "condensate atom".

The operator C_{22} describes the process that two momentum states with momenta p_1 and p_2 collide and after the collision some bosons will move to the new momentum states p_3 and p_4. The same argument as above also gives the form of C_{22}. The C_{22} operator describes the $2 \leftrightarrow 2$ interaction.

The above model has been remarkably successful in describing several Bose–Einstein Condensate (BEC) dynamical problems, in both the hydrodynamic and collisionless regimes, with extremely broad temperature ranges, and even the description of the condensate growth. It also provides a clear picture of what physical processes are taking place in a partially condensed Bose gas and gives information about the superfluid properties of the system. As discussed in [76], the description of the thermal component of the BEC is simplified in this system due to the use of an approximate particle-like Hartree–Fock excitation spectrum, which makes predictions for transport properties at ultra low temperature inaccurate, for instance, the hydrodynamic modes at $T = 0.01T_{BE}$. As a consequence, in [77, 78, 79, 80, 142, 143], the Reichl et al. model has been introduced as a modified version of the Pomeau et al./Zaremba et al. model, in which a collision operator is added to the model.

Let us look carefully into the structure of C_{22}: We call this a four-excitation process due to the fact that there are four particles in each collision. In a four-excitation process, besides the $2 \leftrightarrow 2$ interaction, there should be another $3 \leftrightarrow 1$ interaction. The missing collision operator is, therefore, of the type C_{31}, which describes the process of three momentum states p_1, p_2 and p_3 colliding and particles moving into a new momentum state p_4.

In the modified model, the collision operator $C_{12} + C_{22}$ is then replaced by

$$\mathbf{G} \; = \; \mathbf{C}_{12} + \mathbf{C}_{22} + \mathbf{C}_{31}.$$

As a consequence, at low temperatures, not only are the $2 \leftrightarrow 2$ collisions as strong as the $1 \leftrightarrow 2$ collisions, but also the $1 \leftrightarrow 3$ collisions come into play.

Notice that n_c appears in the expressions of both $\mathbf{C}_{12}$ and $\mathbf{C}_{31}$. When $n_c = 0$, they both vanish and the system becomes the standard Boltzmann–Nordheim equation. This leads us to think that n_c plays a very important role in the long time dynamics of the full system. However, it is pointed

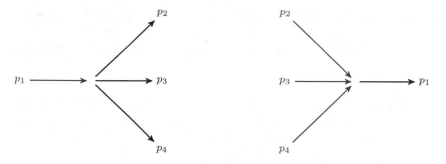

Fig. 6.3 The C_{31} collision. This represents the $3 \leftrightarrow 1$ process.

out in the above-mentioned works that after linearizing the kinetic equations around the equilibrium, treating n_c as a dynamical variable has no effect on the evolution of the linearized system. The reason for this is that when we linearize, the equation for n_c becomes an equation for $\tilde{n}_c$, the perturbative parameter of N_c. In this system, only $\mathbf{C}_{12}$ contributes to the evolution of $\tilde{n}_c$, and, to the first-order approximation, we will end up with $\tilde{n}_c$ multiplied by $\mathbf{C}_{12}$ applied to the equilibrium distributions $\mathcal{N}_i^{\mathrm{eq}}$, which eventually gives zero. As a consequence, in the perturbative regime, the contribution of $\tilde{n}_c$ to the long time dynamics of the kinetic equation is quite small. In Chapter 8, we will see that, in order to compute the hydrodynamic modes, we calculate the eigenvalues and eigenfunctions of the bogolon collision operator $\mathcal{C}(p_1, p_2)$, whose formulation depends on the parameters given in (8.2.6). In these parameters, n_c is approximated by N_c.

This phenomenon can also be explained by applying a perturbative argument to the following spatially homogeneous and simplified model of the system (5.0.20)–(5.0.21)–(5.0.22)–(5.0.23)

$$\frac{\partial}{\partial t} f = \mathbf{C}_{12}[f] + \mathbf{C}_{22}[f] + \theta_{3,1}\mathbf{C}_{31}[f], \quad f(0,p) = f_0(p), \forall (t,p) \in \mathbb{R}_+ \times \mathbb{R}^3,$$

$$\frac{\partial}{\partial t} n_c = -\int_{\mathbb{R}^3} \mathbf{C}_{12}[f]\mathrm{d}p, \quad n_c(0) = n_{c,0}, \forall t \in \mathbb{R}_+,$$

$$(6.3.4)$$

where we have introduced the $3 \leftrightarrow 1$ collisional parameter $\theta_{3,1}$. In the low temperature regime $\theta_{3,1} = 0$ and in the ultra-low temperature $(T < 0.3T_{BE})$ regime $\theta_{3,1} = 1$. We perform the linearizations $f(t,p) = \mathcal{N}_p^{\mathrm{eq}} + \mathcal{N}_p^{\mathrm{eq}}(1 + \mathcal{N}_p^{\mathrm{eq}})h(t,p)$ and $n_c = N_c + \tilde{n}_c$, in which N_c is the number of condensate atoms at equilibrium. Up to first order, the linearized system takes the form

$$\frac{\partial}{\partial t} h = \mathbf{L}_{12}[h] + \mathbf{L}_{22}[h] + \theta_{3,1}\mathbf{L}_{31}[h], \quad f(0,p) = f_0(p), \forall (t,p) \in \mathbb{R}_+ \times \mathbb{R}^3,$$

$$\frac{\partial}{\partial t} \tilde{n}_c = -\int_{\mathbb{R}^3} \mathcal{N}_p^{\mathrm{eq}}(1 + \mathcal{N}_p^{\mathrm{eq}})\mathbf{L}_{12}[h]\mathrm{d}p, \quad \tilde{n}_c(0) = \tilde{n}_{c,0}, \forall t \in \mathbb{R}_+,$$

$$(6.3.5)$$

where $\mathbf{L}_{12}, \mathbf{L}_{22}, \mathbf{L}_{31}$ are the linearized operators of $\mathbf{C}_{12}, \mathbf{C}_{22}, \mathbf{C}_{31}$, with n_c being replaced by N_c. In this system, the total mass is conserved

$$\frac{\partial}{\partial t} \left(\int_{\mathbb{R}^3} \mathcal{N}_p^{\mathrm{eq}} (1 + \mathcal{N}_p^{\mathrm{eq}}) h \mathrm{d}p \; + \; \tilde{n}_c \right) \; = \; 0,$$

but the evolution of $\tilde{n}_c$ is completely decoupled from the kinetic equation. Therefore, in the perturbative regime, the contribution of n_c as a dynamical parameter to the long time dynamics of the kinetic equation is negligible.

We thank Linda Reichl for our fruitful discussions about the models. One of the authors would also like to thank J. Robert Dorfman for the meeting at the University of Maryland, College Park in 2016 and the many email exchanges explaining the differences between the models.

6.3.2 Derivation of the Coupling Quantum Boltzmann and Hugenholtz–Pines Equations

Let us now give a brief derivation of the systems (5.0.20)–(5.0.23). The computations of the hydrodynamic modes will be given in Chapter 8. We partially follow the presentation of the paper [143] by Reichl and Tran.

Let $\hat{\Psi}^\dagger(\mathbf{x})$ and $\hat{\Psi}(\mathbf{x})$ be the quantum field operators that create and annihilate a particle at position $\mathbf{x}$. Following Section 6.1.1, the classical Hamiltonian for N bosons of mass m, in a cubic box with volume Ω, normally takes the form

$$\hat{H} = \int_B \hat{\Psi}^\dagger(\mathbf{x}) \left(-\frac{\hbar^2}{2m} \Delta_{\mathbf{x}} \right) \hat{\Psi}(\mathbf{x}) \mathrm{d}\mathbf{x} \qquad (6.3.6)$$
$$+ \frac{1}{2} \iint_{\Omega \times \Omega} \mathrm{U}(|\mathbf{x}_1 - \mathbf{x}_2|) \hat{\Psi}^\dagger(\mathbf{x}_1) \hat{\Psi}^\dagger(\mathbf{x}_2) \hat{\Psi}(\mathbf{x}_2) \hat{\Psi}(\mathbf{x}_1) \mathrm{d}\mathbf{x}_1 \mathrm{d}\mathbf{x}_2.$$

Recall that these operators satisfy the boson commutation relations

$$[\hat{\Psi}(\mathbf{x}_1), \hat{\Psi}^\dagger(\mathbf{x}_2)] = \delta(\mathbf{x}_1 - \mathbf{x}_2).$$

We make the assumption that the interaction between particles is given by the contact potential $\mathrm{U}(|\mathbf{x}_1 - \mathbf{x}_2|) = g\delta(\mathbf{x}_1 - \mathbf{x}_2)$. Let $\hat{\sigma}$ be the probability density operator for this system. Its evolution follows the quantum Liouville equation (see also Section 6.1.2)

$$\frac{\partial \hat{\sigma}(t)}{\partial t} = -\frac{i}{\hbar} [\hat{H}, \hat{\sigma}(t)]. \qquad (6.3.7)$$

Since the temperature of our system is below the Bose–Einstein condensation transition temperature, T_{BE}, the gauge symmetry of the fluid is broken. This is the main difference between the derivation presented in this section and the derivation of the Boltzmann–Nordheim equation discussed in Section 6.1. The broken symmetry needs to be incorporated into the dynamics of the BEC, by using the one-body reduced density operator

$$\hat{\bar{\Theta}}(\mathbf{x}_1, \mathbf{x}_2) = \begin{pmatrix} \hat{\Phi}^\dagger(\mathbf{x}_1)\hat{\Phi}(\mathbf{x}_2) & \hat{\Phi}^\dagger(\mathbf{x}_1)\hat{\Phi}^\dagger(\mathbf{x}_2) \\ \hat{\Phi}(\mathbf{x}_1)\hat{\Phi}(\mathbf{x}_2) & \hat{\Phi}(\mathbf{x}_1)\hat{\Phi}^\dagger(\mathbf{x}_2) \end{pmatrix}. \tag{6.3.8}$$

We now proceed using techniques from quantum field theory. The one-body reduced density matrix then takes the form

$$\bar{\mathcal{G}}(t, \mathbf{x}_1, \mathbf{x}_2) = \mathrm{Tr}\left[\hat{\sigma}(t)\hat{\bar{\Theta}}(\mathbf{x}_1, \mathbf{x}_2)\right] = \begin{pmatrix} \langle\hat{\Phi}^\dagger(\mathbf{x}_1)\hat{\Phi}(\mathbf{x}_2)\rangle & \langle\hat{\Phi}^\dagger(\mathbf{x}_1)\hat{\Phi}^\dagger(\mathbf{x}_2)\rangle \\ \langle\hat{\Phi}(\mathbf{x}_1)\hat{\Phi}(\mathbf{x}_2)\rangle & \langle\hat{\Phi}(\mathbf{x}_1)\hat{\Phi}^\dagger(\mathbf{x}_2)\rangle \end{pmatrix}, \tag{6.3.9}$$

which follows the time evolution equation

$$-i\hbar\frac{\partial\bar{\mathcal{G}}(t, \mathbf{x}_1, \mathbf{x}_2)}{\partial t} = \mathrm{Tr}[\hat{\sigma}(t)\ [\hat{H}, \hat{\bar{\Theta}}(\mathbf{x}_1, \mathbf{x}_2)]]. \tag{6.3.10}$$

Actually, the density operator $\hat{\rho}(t)$ becomes a functional of the single particle reduced density operator $\bar{\mathcal{G}}(t, \mathbf{x}_1, \mathbf{x}_2)$, after a very short time, following the Bogoliubov assumption. Therefore, we can write the density operator, with an abuse of notation, as follows

$$\hat{\sigma}(t) = \hat{\sigma}'(\{\bar{\mathcal{G}}\}), \tag{6.3.11}$$

where $\{\bar{\mathcal{G}}\}$ is the notation for the vector containing $\bar{\mathcal{G}}(t, \mathbf{x}_1, \mathbf{x}_2)$ for all values of $(\mathbf{x}_1, \mathbf{x}_2)$. The quantity $\bar{\mathcal{G}}(t, \mathbf{x}_1, \mathbf{x}_2)$ is defined such that

$$\bar{\mathcal{G}}(t, \mathbf{x}_1, \mathbf{x}_2) = \mathrm{Tr}[\hat{\sigma}'(\{\bar{\mathcal{G}}\})\hat{\bar{\Theta}}(\mathbf{x}_1, \mathbf{x}_2)]. \tag{6.3.12}$$

By dividing the total Hamiltonian into a mean field contribution $\hat{H}_0$ and a deviation from the mean field $\hat{H}_1$, the existence of the broken symmetry can now be made explicit. After this procedure, the total Hamiltonian takes the form $\hat{H} = \hat{H}_0 + \hat{H}_1$, in which the mean field Hamiltonian is written as

$$\hat{H}_0 = \int_B \hat{\Psi}^\dagger(\mathbf{x})\left(-\frac{\hbar^2}{2m}\Delta_\mathbf{x} - \mu\right)\hat{\Psi}(\mathbf{x})\mathrm{d}\mathbf{x} + \hat{H}_3, \tag{6.3.13}$$

where

$$\hat{H}_3 = \frac{1}{2}\int_B [\nu(\mathbf{x}_1)\hat{\Psi}^\dagger(\mathbf{x}_1)\hat{\Psi}(\mathbf{x}_1) + \nu(\mathbf{x}_1)\hat{\Psi}(\mathbf{x}_1)\hat{\Psi}^\dagger(\mathbf{x}_1)]\mathrm{d}\mathbf{x}_1 \tag{6.3.14}$$

$$+ \frac{1}{2}\int_B d\mathbf{\Lambda}^\dagger(\mathbf{x}_1)\hat{\Psi}(\mathbf{x}_1)\hat{\Psi}(\mathbf{x}_1)\mathrm{d}\mathbf{x}_1 + \frac{1}{2}\int_B \mathbf{\Lambda}(\mathbf{x}_1)\hat{\Psi}^\dagger(\mathbf{x}_1)\hat{\Psi}^\dagger(\mathbf{x}_1)\mathrm{d}\mathbf{x}_1,$$

and $\hat{H}_1$ has deviations from the mean field Hamiltonian

$$\hat{H}_1 = \frac{1}{2}\int_B\int_B U(|\mathbf{x}_1 - \mathbf{x}_2|)\hat{\Psi}^\dagger(\mathbf{x}_1)\hat{\Psi}^\dagger(\mathbf{x}_2)\hat{\Psi}(\mathbf{x}_2)\hat{\Psi}(\mathbf{x}_1)\mathrm{d}\mathbf{x}_1\mathrm{d}\mathbf{x}_2 - \hat{H}_3. \tag{6.3.15}$$

In the above set of equations,

$$\nu(\mathbf{x}_1) = 2g \, \langle \hat{\Psi}^\dagger(\mathbf{x}_1)\hat{\Psi}(\mathbf{x}_1)\rangle,$$
$$\Lambda(\mathbf{x}_1) = g \, \langle \hat{\Psi}(\mathbf{x}_1)\hat{\Psi}(\mathbf{x}_1)\rangle,$$
$$\Lambda^\dagger(\mathbf{x}_1) = g \, \langle \hat{\Psi}^\dagger(\mathbf{x}_1)\hat{\Psi}^\dagger(\mathbf{x}_1)\rangle,$$

and μ is the equilibrium chemical potential.

From the above Hamiltonians, we can write the kinetic equation describing the dynamic evolution of the one-body density matrix as follows

$$-i\hbar \frac{\partial \bar{\mathcal{G}}(t,\mathbf{x}_1,\mathbf{x}_2)}{\partial t} = \mathrm{Tr}\{\hat{\sigma}'(\{\bar{\mathcal{G}}\}), [\hat{H}_0, \hat{\bar{\Theta}}(\mathbf{x}_1,\mathbf{x}_2)]\} + \mathrm{Tr}\{\hat{\sigma}'(\{\bar{\mathcal{G}}\}), [\hat{H}_1, \hat{\bar{\Theta}}(\mathbf{x}_1,\mathbf{x}_2)]\}$$
$$+\frac{i}{\hbar}\int_{-\infty}^0 \mathrm{Tr}\{\hat{\sigma}'(\{\bar{\mathcal{G}}\}), [\hat{H}_1, \hat{V}^{0,\dagger}(0,s)[\hat{\bar{\Theta}}(\mathbf{x}_1,\mathbf{x}_2), \hat{H}_1]\hat{V}^0(0,s)\}\mathrm{d}s, \quad (6.3.16)$$

in which $\hat{V}^0$ denotes the semigroup operator

$$\hat{V}^0(s_1,s_2) = \mathrm{e}^{-\hat{H}_0(s_1-s_2)/\hbar}, \qquad (6.3.17)$$

and $\hat{V}^{0,\dagger}$ denotes the adjoint operator of $\hat{V}^0$.

In order to remove secular effects in the evolution of the one-body density matrix, the mean field Hamiltonian $\hat{H}_0$, defined in (6.3.13), should follow the equation

$$\mathrm{Tr}\{\hat{\sigma}'(\{\bar{\mathcal{G}}\})[\hat{H}_1, \hat{\Theta}(\mathbf{x}_1,\mathbf{x}_2)]\} = 0. \qquad (6.3.18)$$

Now, the unitary transformation to the reference frame moving with the superfluid (superfluid rest frame) can be introduced

$$\hat{V}(t) = \exp\left[-i\int_B \phi(t,\mathbf{x})\hat{\Psi}^\dagger(\mathbf{x})\hat{\Psi}(\mathbf{x})\mathrm{d}\mathbf{x}\right], \qquad (6.3.19)$$

recalling that $\phi(t,\mathbf{x})$ is the macroscopic phase of the condensate wave function. Let $\hat{\psi}^\dagger(\mathbf{x})$ and $\hat{\psi}(\mathbf{x})$ be the particle creation and annihilation operators in the superfluid rest frame. It follows that

$$\hat{V}^\dagger(t)\hat{\Psi}(\mathbf{x})\hat{V}(t) = \mathrm{e}^{-i\phi(t,\mathbf{x})}\hat{\psi}(\mathbf{x}) = \hat{\varrho}(\mathbf{x})$$

yielding

$$-i\hbar\frac{\partial}{\partial t}\langle\hat{\varrho}_1^\dagger\hat{\varrho}_2\rangle = (\mathcal{K}_1^{(+)} - \mathcal{K}_2^{(-)})\langle\hat{\varrho}_1^\dagger\hat{\varrho}_2\rangle - \Lambda_2\langle\hat{\varrho}_1^\dagger\hat{\varrho}_2^\dagger\rangle + \Lambda_1^\dagger\langle\hat{\varrho}_1\hat{\varrho}_2\rangle + \mathcal{I}_{11},$$
$$-i\hbar\frac{\partial}{\partial t}\langle\hat{\varrho}_1^\dagger\hat{\varrho}_2^\dagger\rangle = (\mathcal{K}_2^{(+)} + \mathcal{K}_1^{(+)})\langle\hat{\varrho}_1^\dagger\hat{\varrho}_2^\dagger\rangle + \Lambda_2^\dagger\langle\hat{\varrho}_1^\dagger\hat{\varrho}_2\rangle + \Lambda_1^\dagger\langle\hat{\varrho}_1\hat{\varrho}_2^\dagger\rangle + \mathcal{I}_{12},$$
$$-i\hbar\frac{\partial}{\partial t}\langle\hat{\varrho}_1\hat{\varrho}_2\rangle = -(\mathcal{K}_2^{(-)} + \mathcal{K}_1^{(-)})\langle\hat{\varrho}_1\hat{\varrho}_2\rangle - \Lambda_1\langle\hat{\psi}_1^\dagger\hat{\varrho}_2\rangle - \Lambda_2\langle\hat{\varrho}_1\hat{\varrho}_2^\dagger\rangle + \mathcal{I}_{21},$$
$$-i\hbar\frac{\partial}{\partial t}\langle\hat{\varrho}_1\hat{\varrho}_2^\dagger\rangle = (\mathcal{K}_2^{(+)} - \mathcal{K}_1^{(-)})\langle\hat{\varrho}_1\hat{\varrho}_2^\dagger\rangle - \Lambda_1\langle\hat{\varrho}_1^\dagger\hat{\varrho}_2^\dagger\rangle + \Lambda_2^\dagger\langle\hat{\varrho}_1\hat{\varrho}_2\rangle + \mathcal{I}_{22},$$
$$(6.3.20)$$

where

$$\hat{\varrho}_j = \hat{\varrho}(\mathbf{x}_j),$$

and

$$\mathcal{K}_j^{(\pm)} = K(\mathbf{x}_j) \pm i\frac{\hbar}{2}(\nabla_{\mathbf{x}_j}\cdot v_c(t,\mathbf{x}_j)) + \frac{m}{2}v_c^2(\mathbf{x}_j) \pm i\hbar v_c(\mathbf{x}_j)\cdot\nabla_{\mathbf{x}_j} + \hbar\frac{\partial\phi(t,\mathbf{x}_j)}{\partial t},$$
$$(6.3.21)$$

with

$$K(\mathbf{x}_j) = -\frac{\hbar^2}{2m}\Delta_{\mathbf{x}_j} + \nu(\mathbf{x}_j) - \mu,$$
$$\nu(\mathbf{x}_j) = 2g\langle\hat{\varrho}^\dagger(\mathbf{x}_j)\hat{\varrho}(\mathbf{x}_j)\rangle,$$
$$\Lambda_j = \Lambda(\mathbf{x}_j) = g\langle\hat{\varrho}(\mathbf{x}_j)\hat{\varrho}(\mathbf{x}_j)\rangle,$$
$$\Lambda_j^\dagger = \Lambda^\dagger(\mathbf{x}_j) = g\langle\hat{\varrho}^\dagger(\mathbf{x}_j)\hat{\varrho}^\dagger(\mathbf{x}_j)\rangle,$$
$$(6.3.22)$$

for $j = 1,2$. The quantity $\mathbf{v}_s(\mathbf{x}_j) = \frac{\hbar}{m}\nabla_{\mathbf{x}_j}\phi(\mathbf{x}_j)$ is the superfluid velocity.

With the notation,

$$\hat{\hat{\theta}}(\mathbf{x}_1,\mathbf{x}_2) = \begin{pmatrix} \hat{\varrho}^\dagger(\mathbf{x}_1)\hat{\varrho}(\mathbf{x}_2) & \hat{\varrho}^\dagger(\mathbf{x}_1)\hat{\varrho}^\dagger(\mathbf{x}_2) \\ \hat{\varrho}(\mathbf{x}_1)\hat{\varrho}(\mathbf{x}_2) & \hat{\varrho}(\mathbf{x}_1)\hat{\varrho}^\dagger(\mathbf{x}_2) \end{pmatrix},$$
$$(6.3.23)$$

the quantities

$$\begin{pmatrix} \mathcal{I}_{1,1} & \mathcal{I}_{1,2} \\ \mathcal{I}_{2,1} & \mathcal{I}_{2,2} \end{pmatrix} = \frac{i}{\hbar}\int_{-\infty}^0 \text{Tr}\{\hat{\sigma}'(\{\bar{\mathcal{G}}\})[\hat{H}_1, \hat{V}^{0,\dagger}(0,s)[\hat{\hat{\theta}}(\mathbf{x}_1,\mathbf{x}_2), \hat{H}_1]\hat{V}^0(0,s)]\}ds$$
$$(6.3.24)$$

are indeed the collision integrals governing relaxation processes in the BEC gas.

The coupled equations (6.3.20) contain the full quantum dynamics of the BEC gas. By transforming these kinetic equations to equations for the Wigner functions, we can write the kinetic equations in the hydrodynamic regime in which all macroscopic quantities are slowly varying in space and time.

Let us recall that the field operators $\hat{\varrho}_1^\dagger$ and $\hat{\varrho}_1$ are related to operators $\hat{a}_{p_1}^\dagger$ and $\hat{a}_{p_1}$, which create and annihilate, respectively, a particle with momentum $\hbar p_1$, via the Fourier transforms

$$\hat{\varrho}_1^\dagger = \frac{1}{\sqrt{\Omega}}\sum_{p_1}e^{-ip_1\cdot\mathbf{r}_1}\hat{a}_{p_1}^\dagger \quad\text{and}\quad \hat{\varrho}_1 = \frac{1}{\sqrt{\Omega}}\sum_{p_1}e^{p_1\cdot\mathbf{r}_1}\hat{a}_{p_1}.$$
$$(6.3.25)$$

Therefore, the configuration space distributions can be related to the momentum space distributions, via the Fourier transformation

$$\begin{pmatrix} \langle\hat{\varrho}_1^\dagger\hat{\varrho}_2\rangle & \langle\hat{\varrho}_1^\dagger\hat{\varrho}_2^\dagger\rangle \\ \langle\hat{\varrho}_1\hat{\varrho}_2\rangle & \langle\hat{\varrho}_1\hat{\varrho}_2^\dagger\rangle \end{pmatrix} = \frac{1}{\Omega}\sum_{p_1,p_2}e^{-ip_1\cdot\mathbf{r}_1}e^{+ip_2\cdot\mathbf{r}_2}\begin{pmatrix} \langle\hat{a}_{p_1}^\dagger\hat{a}_{p_2}\rangle & \langle\hat{a}_{p_1}^\dagger\hat{a}_{-p_2}^\dagger\rangle \\ \langle\hat{a}_{-p_1}\hat{a}_{p_2}\rangle & \langle\hat{a}_{-p_1}\hat{a}_{-p_2}^\dagger\rangle \end{pmatrix}.$$
$$(6.3.26)$$

We now introduce the center of mass $R = \frac{1}{2}(\mathbf{x}_1 + \mathbf{x}_2)$ and relative coordinates $r = \mathbf{x}_1 - \mathbf{x}_2$, and the center of mass and relative wavevectors $p = \frac{1}{2}(p_1 + p_2)$ and $q = p_1 - p_2$, respectively. Being distribution functions in phase space for quantum systems [162], Wigner functions are particularly useful when dealing with transport processes, due to the fact that, in the classical limit, these functions reduce to classical probability distributions in phase space. The Wigner functions for the BEC, whose spatial disturbance has wavevector q, are written as

$$\begin{pmatrix} G_{11}(p,q) \ G_{12}(p,q) \\ G_{21}(p,q) \ G_{22}(p,q) \end{pmatrix} = \iint_{\mathbb{R}^3 \times \mathbb{R}^3} e^{ik \cdot r} e^{iq \cdot R} \begin{pmatrix} \langle \hat{\varrho}_1^\dagger \hat{\varrho}_2 \rangle \ \langle \hat{\varrho}_1^\dagger \hat{\varrho}_2^\dagger \rangle \\ \langle \hat{\varrho}_1 \hat{\varrho}_2 \rangle \ \langle \hat{\varrho}_1 \hat{\varrho}_2^\dagger \rangle \end{pmatrix} dr dR,$$

$$(6.3.27)$$

where $\hbar p$ is the momentum of particles. The particle number density whose spatial variation has wavevector q can be computed as $N(q) = \sum_p G_{11}(p,q)$. Let us compute the number of particles $N(p)$ with momentum $\hbar p$

$$N(p) = \int_{\mathbb{R}^3} G_{11}(p, R) dR = \langle \hat{a}_p^\dagger \hat{a}_p \rangle.$$

The expression of the component of the order parameters, whose spatial variation has wavevector q, is as follows

$$\Delta^\dagger(q) = g \sum_p G_{12}(k, q)$$

and

$$\Delta(q) = g \sum_p G_{21}(p, q).$$

Since in the hydrodynamic regime that we are interested in, all macroscopic quantities are slowly varying in space, it is possible for us to keep only the lowest order derivatives with respect to R in the kinetic equations. This is equivalent to keeping only the lowest order contributions from the wave vector q (up to order q^2). The expressions for transport coefficients can be computed from the linearized kinetic equations, being linearized around absolute equilibrium. The hydrodynamic variables in terms of their equilibrium values plus small perturbations from their equilibrium values can now be written as

$$G_{i,j}(q,p) = G_{i,j}^{eq}(p) + \Delta G_{i,j}(q,p), \quad v_c(q) = v_c^0 + \Delta v_c(q),$$

$$\Lambda(q) = \Lambda_0 + \Delta\Lambda(q), \quad \Lambda^\dagger(q) = \Lambda_0 + \Delta\Lambda^\dagger(q). \quad (6.3.28)$$

where $G_{i,j}^{eq}(p)$, v_c^0 and Λ_0 are the equilibrium values of the various quantities. We are interested in the behavior of the Bose gas at temperatures below $0.6T_{BE}$, in which the Popov approximation has been shown to give good agreement with experiments and $G_{11}^{eq}(0) \approx G_{12}^{eq}(0) \approx G_{21}^{eq}(0) \approx G_{22}^{eq}(0) \approx N_c$, with N_c being the number density of particles in the condensate at equilibrium.

Since the kinetic equations are linearized, each wavevector component evolves independently. Let us define

$$e_{p,q}^{(\pm)} = \frac{\hbar^2}{2m} |p \pm \frac{1}{2} q|^2 + \nu^0 - \mu. \tag{6.3.29}$$

The resulting linearized kinetic equations can be written in the following matrix form,

$$-i\hbar \frac{\partial \Delta G}{\partial t} = \{\epsilon_{p,q}^{(+)} \Delta G - \Delta G \, \epsilon_{p,q}^{(-)}\} + \hbar q {\cdot} v_c(q) G^{eq} - \hbar p {\cdot} v_c(q) \, q {\cdot} \nabla_p G^{eq}$$
$$+ \{B \, G^{eq} - G^{eq} \, B'\} + q {\cdot} \nabla_p \{D \, G^{eq} - G^{eq} \, D'\} + \Delta \mathcal{I}, \tag{6.3.30}$$

where

$$\Delta G = \begin{pmatrix} \Delta G_{11}(t,q,p) & \Delta G_{12}(t,q,p) \\ \Delta G_{21}(t,q,p) & \Delta G_{22}(t,q,p) \end{pmatrix}, \quad G^{eq} = \begin{pmatrix} G_{11}^{eq}(p) & G_{12}^{eq}(p) \\ G_{21}^{eq}(p) & G_{22}^{eq}(p) \end{pmatrix}, \tag{6.3.31}$$

$$\epsilon_{p,q}^{(+)} = \begin{pmatrix} e_{p,q}^{(+)} & \Lambda \\ -\Lambda & -e_{p,q}^{(+)} \end{pmatrix}, \quad \epsilon_{p,q}^{(-)} = \begin{pmatrix} e_{p,q}^{(-)} & -\Lambda \\ \Lambda & -e_{p,q}^{(-)} \end{pmatrix}, \tag{6.3.32}$$

$$B = \begin{pmatrix} \Phi(q) & \Delta \Lambda^\dagger(q) \\ -\Delta \Lambda(q) & -\check{\Phi}(q) \end{pmatrix}, \quad B' = \begin{pmatrix} \Phi(q) & -\Delta \Lambda^\dagger(q) \\ \Delta \Lambda(q) & -\Phi(q) \end{pmatrix}, \tag{6.3.33}$$

$$D = \begin{pmatrix} -\frac{1}{2}\Phi(q) & -\frac{1}{2}\Delta \Lambda^\dagger(q) \\ \frac{1}{2}\Delta \Lambda(q) & \frac{1}{2}\Phi(q) \end{pmatrix}, \quad D' = \begin{pmatrix} \frac{1}{2}\Phi(q) & -\frac{1}{2}\Delta \Lambda^\dagger(q) \\ \frac{1}{2}\Delta \check{\Lambda}(q) & -\frac{1}{2}\check{\Phi}(q) \end{pmatrix}, \tag{6.3.34}$$

$\Phi(q) = \hbar \frac{\partial \phi(q)}{\partial t} + \delta \nu(q)$ and

$$\Delta \mathcal{I} = \begin{pmatrix} \Delta \mathcal{I}_{11}(t,q,p) & \Delta \mathcal{I}_{12}(t,q,p) \\ \Delta \mathcal{I}_{21}(t,q,p) & \Delta \mathcal{I}_{22}(t,q,p) \end{pmatrix} \tag{6.3.35}$$

are the linearized collision integrals for the particle kinetic equations. For simplicity and without loss of generality, we set $v_c^0 = 0$ (superfluid velocity at equilibrium).

Let us compute the total particle number density in the interval $q \to q + dq$ at time t

$$\Delta N(q,t) = \frac{1}{\Omega} \sum_P \Delta G_{11}(P,q,t). \tag{6.3.36}$$

It follows from (6.3.30) that

$$-i\hbar \frac{\partial \Delta G_{11}(t,P,q,t)}{\partial t} = \left(\tilde{\epsilon}_{P,q}^{(+)} - \tilde{\epsilon}_{P,q}^{(-)} \right) \Delta G_{11}(t,P,q) + \hbar q {\cdot} v_c(t,q) N_P^{eq}$$
$$+ \hbar q {\cdot} v_c(t,q) \, q {\cdot} \nabla_P N_P^{eq} - \delta \Delta G_{12}(t,P,q) - \Delta \delta(t,q) G_{12}^{eq}(P)$$
$$+ \delta \Delta G_{21}(t,P,q) + \Delta \delta^\dagger(t,q) G_{21}^{eq}(P) + \Delta \mathcal{I}_{11}(t,P,q). \tag{6.3.37}$$

Observe that

$$\delta = \frac{g}{\Omega}\sum_P G_{12}^{eq}(P) = \frac{g}{\Omega}\sum_P G_{12}^{eq}(P), \tag{6.3.38}$$

$$\Delta\delta(t,q) = \frac{g}{\Omega}\sum_P \Delta G_{21}(t,P,q), \quad \Delta\delta^\dagger(t,q) = \frac{g}{\Omega}\sum_P \Delta G_{12}(t,P,q).$$

Summing over all momentum states in (6.3.37), the terms that depend on δ cancel. One can also check that the third term on the right-hand side of (6.3.37) is negligible in comparison to the second term, when one integrates over P. Equation (6.3.37) then reduces to

$$-i\hbar\frac{\partial\Delta N(t,q)}{\partial t} = \frac{\hbar^2}{m}\frac{1}{\Omega}\sum_P P{\cdot}q\Delta G_{11}(t,P,q) + \hbar q{\cdot}v_c(t,q)N^{eq}, \tag{6.3.39}$$

which is the continuity equation for the total particle number density.

When the BEC is in equilibrium, the mean field Hamiltonian (in the superfluid rest frame) can be written as

$$\hat{H}_0 = \sum_i \left[(\epsilon_i - \Lambda_0)\hat{a}_i^\dagger\hat{a}_i + \frac{\Lambda_0}{2}\left(\hat{a}_i^\dagger\hat{a}_i + \hat{a}_i^\dagger\hat{a}_i\right)\right] = \frac{g}{2}n_c^2 + \sum_i E_i\hat{b}_i^\dagger\hat{b}_i, \tag{6.3.40}$$

where

$$E_1 = \sqrt{e_1^2 - \Lambda_0{}^2} \quad \text{with} \quad e_1 = \frac{\hbar^2 p_1^2}{2m} + \nu^0 - \mu = \frac{\hbar^2 p_1^2}{2m} + \Lambda_0, \tag{6.3.41}$$

with the Hugenholtz–Pines relation being used: $\mu = \nu^0 - \Lambda_0$ [88]. In terms of these equilibrium quantities, the Bogoliubov transformation parameters can be written as follows

$$u_1 = \frac{1}{\sqrt{2}}\sqrt{1 + \frac{e_1}{E_1}}, \quad v_1 = \frac{1}{\sqrt{2}}\sqrt{\frac{e_1}{E_1} - 1}. \tag{6.3.42}$$

Notice that

$$u_1^2 - v_1^2 = 1, \quad \Lambda_0(u_1^2 + v_1^2) - 2e_1 u_1 v_1 = 0, \quad e_1(u_1^2 + v_1^2) - 2\Lambda_0 u_1 v_1 = E_1 \tag{6.3.43}$$

and

$$V_1^{-1}{\cdot}\begin{pmatrix} e_1 & \Lambda_0 \\ -\Lambda_0 & -e_1 \end{pmatrix}{\cdot}V_1 = V_1{\cdot}\begin{pmatrix} e_1 & -\Lambda_0 \\ \Lambda_0 & -e_1 \end{pmatrix}{\cdot}V_1^{-1} = \begin{pmatrix} E_1 & 0 \\ 0 & -E_1 \end{pmatrix}. \tag{6.3.44}$$

The dynamics of the bogolons governs the hydrodynamic relaxation of the BEC. As a consequence, to determine the hydrodynamic behavior of the BEC, we must transform particle kinetic equations into bogolon kinetic equations. This can be done via the Bogoliubov transformation V_j transforming particle creation and annihilation operators, $\hat{a}_j^\dagger = \hat{a}_{p_j}^\dagger$ and $\hat{a}_j = \hat{a}_{p_j}$, respectively, into bogolon creation and annihilation operators, $\hat{b}_j^\dagger = \hat{b}_{p_j}^\dagger$ and $\hat{b}_j = \hat{b}_{p_j}$, respectively. We write

$$\begin{pmatrix} \langle \hat{b}_1^\dagger \hat{b}_2 \rangle & 0 \\ 0 & \langle \hat{b}_{-1} \hat{b}_{-2}^\dagger \rangle \end{pmatrix} = V_1^{-1} \cdot \begin{pmatrix} \langle \hat{a}_1^\dagger \hat{a}_2 \rangle & \langle \hat{a}_1^\dagger \hat{a}_{-2}^\dagger \rangle \\ \langle \hat{a}_{-1} \hat{a}_2 \rangle & \langle \hat{a}_{-1} \hat{a}_{-2}^\dagger \rangle \end{pmatrix} \cdot V_2^{-1}, \qquad (6.3.45)$$

with

$$V_i = \begin{pmatrix} u_i & -v_i \\ -v_i & u_i \end{pmatrix} \quad \text{and} \quad V_i^{-1} = \begin{pmatrix} u_i & v_i \\ v_i & u_i \end{pmatrix}, \quad i = 1, 2, \qquad (6.3.46)$$

where $u_1 = u_{p_1}$, $u_2 = u_{p_2}$. Recall that the bogolons do not form a condensate, so we need $\langle \hat{b}_1^\dagger \hat{b}_{-1}^\dagger \rangle = 0$ and $\langle \hat{b}_{-1} \hat{b}_2 \rangle = 0$. Since the kinetic equations are linearized about absolute equilibrium, the parameters u_1 and v_1 can be expressed in terms of equilibrium quantities. We find

$$\begin{pmatrix} \mathcal{N}^{eq}(p) & 0 \\ 0 & \mathcal{N}^{eq}(p) + 1 \end{pmatrix} = \begin{pmatrix} \langle \hat{b}_p^\dagger \hat{b}_p \rangle_{eq} & 0 \\ 0 & \langle \hat{b}_{-p} \hat{b}_{-p}^\dagger \rangle_{eq} \end{pmatrix} = V_p^{-1} \cdot G^{eq}(p) \cdot V_p^{-1},$$
$$(6.3.47)$$

in which $\mathcal{N}^{eq}(p) = [\exp(\beta E(p)) - 1]^{-1}$ is the Bose–Einstein distribution for bogolons.

Now, expanding the particle number distribution in terms of bogolon distributions yields

$$\Delta G_{11}(t, q, p) = u_p^2 f(t, q, p) + v_p^2 f(t, q, p), \qquad (6.3.48)$$

and expanding the particle current in terms of bogolon currents implies

$$\sum_p p \Delta G_{11}(t, q, p) = \sum_p p \left[u_p^2 f(q, p, t) + v_p^2 f(t, q, -p) \right]$$
$$= \sum_p p \Delta f(t, q, p), \qquad (6.3.49)$$

since $u_p^2 - v_p^2 = 1$. In other words, the bogolon momentum density and the particle momentum density are indeed equal.

We now derive the Hugenholtz–Pines (H-P) equation for the BEC [88]. Observing that the time derivative of the macroscopic phase $\phi(\mathbf{X}, t)$ is proportional to the chemical potential $\mu = \hbar \frac{\partial \phi}{\partial t}$, in the hydrodynamic regime, we can assume that the system is locally in equilibrium, and write

$$\hbar \frac{\partial \phi(t, q)}{\partial t} + \Delta \nu(t, q) - \Delta \tilde{\Lambda}(t, \mathbf{q}) = 0, \qquad (6.3.50)$$

where

$$\Delta \nu(t, q) = 2g \sum_P \Delta F_{11}(q, p) = 2g \Delta N(t, q), \qquad (6.3.51)$$

and

$$\Delta \tilde{\Lambda}(t, q) = \frac{g}{2} \sum_p (\Delta F_{12}(t, q, p) + \Delta F_{21}(t, q, p)). \qquad (6.3.52)$$

Making a "Bogoliubov-like" approximation for the non-equilibrium order parameter, $\Delta\tilde{\Lambda}(\mathbf{q}, t) = g\Delta N(\mathbf{q}, t)$, we deduce the Hugenholtz–Pines equation

$$\hbar\frac{\partial\phi}{\partial t} + g\Delta N = 0. \qquad (6.3.53)$$

This approximation is limited to very dilute gases, and it is expected that the accuracy of the results for the longitudinal modes is limited to the temperature range $0{\leq}T{\leq}0.3T_{BE}$. Equation (6.3.53) gives a closure condition for the hydrodynamic equations, yielding the system (5.0.20)–(5.0.21)–(5.0.22)–(5.0.23).

Chapter 7
Formation of Singularities

This chapter is devoted to a rather detailed study of the dynamical transition of the state of a gas of bosons without condensate to a state with condensate, or equivalently to the dynamical formation of a condensate. This transition manifests itself as a finite time singularity of the solution of the Boltzmann–Nordheim kinetic equation at zero wavenumber. This singularity is described by a self-similar solution of this equation of the second kind in the terminology of Zel'dovich [167]. In particular, it depends on an exponent which cannot be derived from simple scaling laws and which appears as a nonlinear eigenvalue in a problem involving an integro-differential equation with definite boundary conditions. Moreover, a careful study of the coupled equations between the condensate and the thermal parcels yields the beginning of the growth of the condensate after the time of the singularity, since at singularity the condensate has no mass yet.

7.1 Finite Time Singularities in the Boltzmann–Nordheim Equation

Suppose that the density function f is spatially homogeneous and radially symmetric in p: $f(t, r, p) = f(t, |p|) = f(t, \omega)$. Applying the change of variable $p \to \omega$, so that f_p becomes f_ω, we get

$$
\begin{aligned}
\frac{\partial}{\partial t} f_{\omega_1} &= C_{22}[f_{\omega_1}] \\
&\equiv \frac{1}{\sqrt{\omega_1}} \int_{\mathcal{D}} \tilde{W}_{\omega_1,\omega_2,\omega_3,\omega_4}[f_{\omega_3} f_{\omega_4}(1 + f_{\omega_1})(1 + f_{\omega_2}) \\
&\quad - f_{\omega_1} f_{\omega_2}(1 + f_{\omega_3})(1 + f_{\omega_4})]\mathrm{d}\omega_3 \mathrm{d}\omega_4,
\end{aligned}
\tag{7.1.1}
$$

where

$$
\tilde{W}_{\omega_1,\omega_2,\omega_3,\omega_4} = \frac{g^2}{m\hbar^3} \min\{\sqrt{\omega_1}, \sqrt{\omega_2}, \sqrt{\omega_3}, \sqrt{\omega_4}\},
$$

© Springer Nature Switzerland AG 2019
Y. Pomeau, M.-B. Tran, *Statistical Physics of Non Equilibrium Quantum Phenomena*,
Lecture Notes in Physics 967, https://doi.org/10.1007/978-3-030-34394-1_7

and

$$\omega_2 = \omega_3 + \omega_4 - \omega_1.$$

Since $\omega_2 = \omega_3 + \omega_4 - \omega_1$ must be positive, one integrates over a domain D such as $\omega_3 + \omega_4 > \omega_1$. This is an equation of evolution toward positive times of the distribution f_ω. The solution f_ω conserves mass

$$J = \int_0^\infty f_\omega(t)\sqrt{\omega}\,d\omega = \int_0^\infty f_\omega(t)\sqrt{\omega}\,d\omega \qquad (7.1.2)$$

and energy

$$P = \int_0^\infty f_\omega(t)\omega^{3/2}\,d\omega = \int_0^\infty f_\omega(t), \omega^{3/2}\,d\omega. \qquad (7.1.3)$$

The H-Theorem for the entropy

$$S[f] = \int_0^\infty [(1+f_\omega)\log(1+f_\omega) - f_\omega \log f_\omega]\sqrt{\omega}\,d\omega \qquad (7.1.4)$$

implies that the solution of (7.1.1) converges to

$$f_\omega^{eq} = \frac{1}{e^{(\omega-\mu)T} - 1}$$

as t tends to infinity, where T is the absolute temperature in energy units and f_ω^{eq} also needs to satisfy the conservation of mass and energy.

Suppose that the initial condition is of the form

$$f_\omega(t=0) = A\left(1 + a\omega + \frac{(a\omega)^2}{2}\right)e^{-a\omega}, \qquad (7.1.5)$$

in which A and a are related to the initial mass and energy. One can get a relation between A and the dimensionless chemical potential $\frac{\mu}{T}$ characterizing the asymptotic state

$$A = \frac{216\left(\zeta_{3/2}(e^{\mu/T})\right)^{5/2}}{175\sqrt{5}\left(\zeta_{5/2}(e^{\mu/T})\right)^{3/2}}, \qquad (7.1.6)$$

where

$$\zeta_s(z) = \sum_{n=1}^\infty \frac{z^n}{n^s}$$

is the incomplete Riemann ζ-function.

When A is small, the density is low and μ is negative. This is a similar situation to that of an ideal classical gas. As A increases, μ also increases. When μ reaches the critical value $\mu = 0$, which happens for $A = A_c$, we have

$$A_c = \frac{216\left(\zeta_{3/2}(1)\right)^{5/2}}{175\sqrt{5}\left(\zeta_{5/2}(1)\right)^{3/2}} = 3.91868\ldots$$

This can be seen from Figure 7.1.

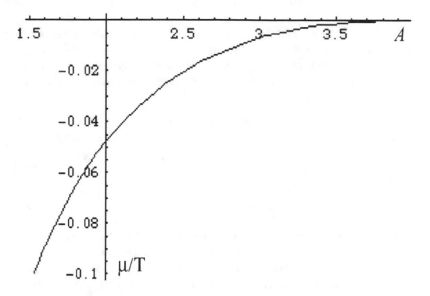

Fig. 7.1 The behavior of the amplitude of the initial condition $\mathcal{A}$ as a function of μ/T. The asymptotics reaches the critical value $\mathcal{A}_c$.

If $\mathcal{A} > \mathcal{A}_c$, equation (7.1.6) cannot be satisfied and a Bose–Einstein condensate is formed. We expect a condensation to zero energy in the isotropic case under consideration. In other words, there is a spontaneous occurrence of a singularity of the solution of equation (7.1.1) at $\omega = 0$. This should transform the solution into a singular delta function at $\omega = 0$ plus a smooth function, as in:

$$f_\omega \;=\; \frac{\rho_0}{\sqrt{\omega}}\delta(\omega) \;+\; \tilde{f}_\omega,$$

$\tilde{f}_\omega$ being the smooth function. Therefore we expect that just before the singularity, the occupation number of small energies becomes very large: $f_\omega \gg 1$ for ω small. When this happens one can neglect for ω small the quadratic term in (7.1.1), since it is small with respect to the cubic term. This yields a simpler form of (7.1.1), which is nothing but a 4-wave turbulence kinetic equation

$$\frac{\partial}{\partial t} f_{\omega_1} \;=\; C_{4wave}[f_{\omega_1}]$$

$$\equiv \frac{1}{\sqrt{\omega_1}} \int_D \tilde{W}_{\omega_1,\omega_2,\omega_3,\omega_4}[f_{\omega_3} f_{\omega_4}(f_{\omega_1} + f_{\omega_2}) - f_{\omega_1} f_{\omega_2}(f_{\omega_3} + f_{\omega_4})]\mathrm{d}\omega_3 \mathrm{d}\omega_4.$$

$$(7.1.7)$$

The steady state solution $\frac{T}{\omega-\mu}$ of this equation follows from the maximization of the entropy

$$S[f_\omega] = \int_0^\infty \log(f_\omega(t))\sqrt{\omega}d\omega. \tag{7.1.8}$$

This steady state solution does not yield converging expressions for both the energy and the total mass. With finite mass and energy, the solution of the wave kinetic equation (7.1.8) spreads forever in energy space (cf. [131]). Such a spreading is stopped in the full Boltzmann–Nordheim equation by the quadratic part of the collision term in (7.1.1).

Since (7.1.7) is a wave turbulence kinetic equation, the method of Zakharov can be applied (cf. [163]) and two other stationary solutions can be found,

$$f_\omega^{Sta1} = P^{1/3}\omega^{-3/2} \tag{7.1.9}$$

and

$$f_\omega^{Sta1} = J^{1/3}\omega^{-7/6}, \tag{7.1.10}$$

where (P/J) is the (energy/mass) flux in ω-space. These solutions make the collision integral in the Boltzmann–Nordheim equation vanish. However, it is impossible to use this kind of solution for the current situation since we expect the collapse to be a dynamical process and the corresponding solution to depend on time in a non-trivial way. Stationary solutions may be useful to understand the transfer of mass and energy through the spectrum. We will show later that the actual exponents for the self-similar solution do not follow from classical scaling estimates.

Let us notice that since the right-hand side of (7.1.1) is cubic homogeneous in f_ω, equation (7.1.1) admits a self-similar dynamical solution which tends to accumulate particles at zero energy, although at the time when the condensation occurs, there is still no mass stacked at zero energy. We consider self-similar solutions of the form

$$f_\omega = \frac{1}{\tau(t)^\nu}\phi\left(\frac{\omega}{\tau(t)}\right), \tag{7.1.11}$$

where $\tau(t)$ tends to 0 as $t \to t_*$, the condensation time, and where ν is a positive exponent to be found as well as the numerical function ϕ.

This type of singularity does not have a meaning outside the range of validity of the Boltzmann–Nordheim equation. That is, for time scales for changes of order or smaller than the mean-free flight time (6.2.1): $|t - t_*| \gg t_{mfp}$.

Now, plugging (7.1.11) into (7.1.1), and imposing the separation of pure temporal $\tau(t)$ and rescaled $\varpi = \frac{\omega}{\tau(t)}$ variables, yield

$$\frac{C_{4wave}[\phi(\varpi)]}{\nu\phi(\varpi) + \varpi\phi'(\varpi)} = -\tau(t)^{2\nu-3}\frac{d\tau(t)}{dt} \equiv 1. \tag{7.1.12}$$

For $\nu = 1$, we have that $\tau(t)$ decreases to zero in infinite time as $\tau(t) \backsim e^{-t}$. When $\nu > 1$,

$$\tau(t) = (2(\nu - 1)(t_* - t))^{\frac{1}{2(\nu-1)}}. \tag{7.1.13}$$

Moreover, $\phi(\varpi)$ satisfies the following integro-differential equation

$$\nu\phi(\varpi) + \varpi\phi'(\varpi) = C_{4wave}[\phi(\varpi)] \equiv \frac{1}{\sqrt{\varpi}} \int_D \tilde{W}_{\varpi_1,\varpi_2,\varpi_3,\varpi_4}$$
$$\times \phi_{\varpi_1}\phi_{\varpi_2}\phi_{\varpi_3}\phi_{\varpi_4} \left(\frac{1}{\phi_{\varpi_1}} + \frac{1}{\phi_{\varpi_2}} - \frac{1}{\phi_{\varpi_3}} - \frac{1}{\phi_{\varpi_4}} \right) d\varpi_3 d\varpi_4 \tag{7.1.14}$$

together with the following boundary conditions

$$\phi(\varpi) \to \phi_0 \quad \text{as} \quad \varpi \to 0 \tag{7.1.15}$$

and

$$\phi(\varpi) \to \frac{1}{\varpi^\nu} \quad \text{as} \quad \varpi \to \infty. \tag{7.1.16}$$

The behavior of $\phi(\varpi)$ for large ϖ is such that, as $\tau(t)$ goes to 0, the function f_ϖ, given in (7.1.11), does not depend on t when $\varpi \gg \tau(t)$. In the system (7.1.14)–(7.1.15)–(7.1.16), ν and ϕ_0 are the undefined parameters.

7.1.1 Bounds for ν

In [155], several self-similar solutions of the type (7.1.11) were considered, with various choices of ν and a solution with an exponent ξ which "apparently,... cannot be determined from general considerations". In [95, 155], Svistunov and coauthors considered that the relevant value for ν is 7/6. Later, Bijlsma, Zaremba, and Stoof (cf. [26]) tried to derive a law of growth of the condensate and also considered a rational exponent for this growth. In [145, 146], the authors discovered a difference between the Kolmogorov–Zakharov exponent $7/6 = 1.1666...$ and their observed numerical value $\nu \approx 1.24$. The link between the exponents for blow-up and for the growth of the condensate has been found in [94, 101] by Pomeau and collaborators. In this work, it is shown that the exponent ν is a nonlinear eigenvalue of (7.1.14), making it possible to satisfy the boundary conditions (7.1.15)–(7.1.16). Moreover, $\phi(\varpi) \backsim \varpi^{-\nu}$ as $\varpi \to \infty$ with $\nu \approx 1.234$ (see Figure 4.2). The work of Spohn [153] also confirmed the findings of Pomeau et al. (see Section 7.3).

We have already seen that $\nu > 1$ and it is not hard to show that $\nu < 3/2$. Let ω_* be a constant and define the "stretched" peak energy $\omega_*(t) = \omega_*\tau(t)$, then

$$N_{peak} = \int_0^{\omega_*(t)} \tau(t)^{-\nu}\phi\left(\frac{\omega}{\tau(t)}\right)\sqrt{\omega}d\omega = \tau(t)^{3/2-\nu}\int_0^{\omega_*}\phi(\varpi)\sqrt{\varpi}d\varpi. \tag{7.1.17}$$

Since the integral on the right-hand side cannot diverge, one needs $\nu < 3/2$.

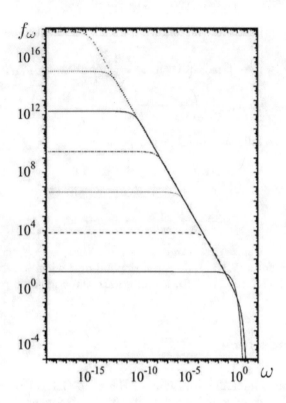

Fig. 7.2 Self-similar evolution of $f_\omega(t)$ at chosen times for successive increase of $f_{\omega=0}(t)$ by a factor of 5. From the different time slots, we can see a clear self-similar evolution. There is a build-up of the power law distribution -1.234 from large energies to small ones, as well as the $\omega^{-\nu}$ time-dependent behavior of the solution in a range expanding toward small ω, as t approaches the blow-up time.

From a physical point of view, we also expect an upper bound for ν: $\nu < 7/6$, for the following reason: Let us consider the flux

$$J_{peak} = \int_0^{\omega_*(t)} C_{4wave}[\phi_\omega]\sqrt{\omega}\,d\omega = \tau(t)^{7/2-3\nu}\int_0^{\omega_*} C_{4wave}[\phi(\varpi)]\sqrt{\varpi}\,d\varpi,$$

which vanishes as $t \to t_*$. However, for $\nu > 7/6$, the above integral diverges.

Let us also consider the entropy of the peak

$$S_{peak} = \int_0^{\omega_*(t)} \log[f_\omega]\sqrt{\omega}\,d\omega = \tau(t)^{3/2}\int_0^{\omega_*} [\log(\phi(\varpi)) - \nu\log\tau(t)]\sqrt{\varpi}\,d\varpi.$$

It vanishes as $t \to t_*$ as well as the rate of entropy production at the peak.

The following quantity

$$\mathcal{R}_{peak} = \int_0^{\omega_*(t)} \frac{C_{4wave}[f_\omega]}{f_\omega}\sqrt{\omega}d\omega = \tau^{-2\nu+7/2}\int_0^{\omega_*}\frac{C_{4wave}[f_\varpi]}{f_\varpi}\sqrt{\varpi}d\varpi$$

also vanishes as $t \to t_*$.

To conclude, we have the following bounds for ν

$$\frac{7}{6} < \nu < \frac{3}{2}. \tag{7.1.18}$$

7.1.2 The Nonlinear Eigenvalue Problem (7.1.14)

Formally integrating (7.1.14), we obtain

$$\phi(\varpi) = \frac{1}{\varpi^\nu} - \frac{1}{\varpi^\nu}\int_\varpi^\infty \varpi_1^{\nu-1}C_{4wave}[\phi(\varpi_1)]d\varpi. \tag{7.1.19}$$

Now, plugging $\phi(\varpi) = \varpi^{-\nu}$ into the right-hand side, we obtain the next order terms in the Laurent expansion for large ϖ:

$$\phi(\varpi) = \frac{1}{\varpi^\nu} - \frac{C_\nu}{2(\nu-1)\varpi^{3\nu-2}} - \frac{C_\nu D_\nu}{8(\nu-1)^2}\frac{1}{\varpi^{5\nu-4}} + \mathcal{O}\left(\frac{1}{\varpi^{7\nu-6}}\right). \tag{7.1.20}$$

The function C_ν is defined to satisfy

$$C_{4wave}[\varpi^{-\nu}] \equiv C_\nu\varpi^{-3\nu+2},$$

that is

$$C_\nu = \int_I \sqrt{z}x^{-\nu}y^{-\nu}z^{-nu}\left(1 + z^\nu - x^\nu - y^\nu\right)$$
$$\times \left(1 + z^{3\nu-7/2} - x^{3\nu-7/2} - y^{3\nu-7/2}\right)dxdy,$$

where $z = x + y - 1$ and

$$I := \{x, y \mid x + y \geq 1 \text{ and } 0 \leq x, y \leq 1\}.$$

Although the convergence of individual integrals is for $\nu < 5/4$, and D_ν vanishes at $\nu = 11/10$ and $\nu = 13/10$, the full result converges in a wider range of ν, as shown in Figure 7.3.

The convergence of C_ν implies that the interactions are mostly between particles with similar energies. In other words, the interactions in the energy space are local. This assumption was indeed used in (7.1.20).

Indeed, C_ν plays a crucial role in physical quantities. For instance, the flux of matter

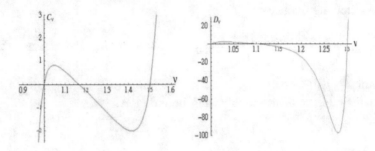

Fig. 7.3 Left: Numerical evaluation for C_ν, which vanishes at $\nu = 1$ (Rayleigh–Jean distribution), $\nu = 7/6$ (wave action inverse cascade) and $\nu = 3/2$ (energy cascade). Right: Numerical evaluation for C_ν, which vanishes at $\nu = 11/10$ and $\nu = 13/10$.

$$J = \int_0^\varpi C_{4wave}[f_{\varpi_1}]\sqrt{\varpi_1}\,d\varpi_1 = \frac{C_\nu}{3(7/6 - \nu)}\varpi^{3(7/6-\nu)},$$

the flux of energy

$$P = \int_0^\varpi C_{4wave}[f_{\varpi_1}]\sqrt{\varpi_1^3}\,d\varpi_1 = \frac{C_\nu}{3(7/6 - \nu)}\varpi^{3(3/2-\nu)},$$

and the entropy function rate

$$R = \int_0^\varpi \frac{C_{4wave}[f_{\varpi_1}]}{f_{\varpi_1}}\sqrt{\varpi_1}\,d\varpi_1 = \frac{C_\nu}{2(7/4 - \nu)}\varpi^{2(7/4-\nu)}.$$

These quantities are plotted in Figure 4.4. It is known (cf. [121, 163]) that the limits $\nu \to 7/6$ for J and $\nu \to 3/2$ for P are well defined. In [44], it is shown that the entropy production R is negative, due to the fact that $\nu \in [7/6, 3/2]$, in agreement with the qualitative idea of the condensation.

Since C_ν vanishes at $\nu = 7/6$ and $\nu = 3/2$, it is impossible to get $\nu = 7/6$ and $\nu = 3/2$, as should follow from (7.1.9) and (7.1.10). The reason is that the next order and higher order correction (7.1.20) vanishes since C_ν is 0 in both cases, and the Laurent expansion for large values of ϖ stops.

Indeed, (7.1.20) is already singular at zero energy. However, we would like to understand the evolution of a solution which remains finite at zero energy. That means $\phi(\varpi = 0) = \phi_0$ is finite and positive.

The asymptotic near $\varpi = 0$ requires

$$\phi(\varpi) \approx \phi_0 + \delta\phi,$$

hence, we need

$$\lim_{\varpi \to 0} C_{4wave}[f_\varpi] = \nu\phi_0.$$

The limit $\varpi \to 0$ of the collision integral (7.1.14) can be written

$$\lim_{\varpi \to 0} C_{4wave}[f_\varpi] = \phi_\varpi \int_0^\infty [\phi_{\varpi_3}\phi_{\varpi_4} - \phi_{\varpi_3}\phi_{\varpi_3+\varpi_4} - \phi_{\varpi_4}\phi_{\varpi_3+\varpi_4}]d\varpi_3 d\varpi_4$$

$$+ \int_0^\infty \phi_{\varpi_3}\phi_{\varpi_4}\phi_{\varpi_3+\varpi_4} d\varpi_3 d\varpi_4 + \frac{2}{3}\varpi(\phi_\varpi - \phi_0)\int_0^\infty \phi_{\varpi_4}^2 d\varpi_4$$

$$+ \varpi \int_0^\infty [\phi_\varpi(\phi_{\varpi_3} + \phi_{\varpi_4}) - \phi_{\varpi_3}\phi_{\varpi_4}]\phi'_{\varpi_3+\varpi_4}d\varpi_3 d\varpi_4 + \mathcal{O}(\varpi^2).$$

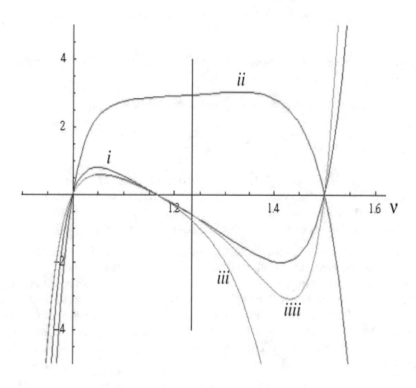

Fig. 7.4 The plot of C_ν, (i) the flux of matter $\frac{C_\nu}{3(7/6-\nu)}$, (ii) the flux of energy $\frac{C_\nu}{3(3/2-\nu)}$, (iii) entropy production $\frac{C_\nu}{2(7/4-\nu)}$.

Therefore, to zeroth-order with respect to ϖ

$$\nu\phi_0 = \int_0^\infty [\phi_0\phi_{\varpi_3}\phi_{\varpi_4} + (\phi_{\varpi_3}\phi_{\varpi_4} - \phi_0\phi_{\varpi_3} - \phi_{\varpi_0}\phi_{\varpi_4})]d\varpi_3 d\varpi_4, \quad (7.1.21)$$

and up to first order

$$\delta\phi' = \eta_1\delta\phi - \frac{\eta_2}{\varpi}\delta\phi + \eta_3, \quad (7.1.22)$$

in which

$$\eta_1 = \frac{2}{3} \int_0^\infty \phi_{\varpi_4}^2 d\varpi_4 + \int_0^\infty (\phi_{\varpi_3} + \phi_{\varpi_4}) \phi'_{\varpi_3+\varpi_4} d\varpi_3 d\varpi_4,$$

$$\eta_2 = \frac{1}{\phi_0} \int_0^\infty \int_0^\infty \phi_{\varpi_3} \phi_{\varpi_4} \phi_{\varpi_3+\varpi_4} d\varpi_3 d\varpi_4,$$

$$\eta_3 = \int_0^\infty \int_0^\infty [\phi_0 \phi_{\varpi_3+\varpi_4} - \phi_{\varpi_3} \phi_{\varpi_4}] d\varpi_3 d\varpi_4.$$

Integrating (7.1.22) gives

$$\delta\phi = \varpi^{-\eta_2} e^{\eta_1 \varpi} \left(C - \eta_3 \int_\varpi^\infty s^{\eta_2} e^{-eta_1 s} ds \right), \tag{7.1.23}$$

where $C = \frac{\eta_3 \Gamma(\eta_2+1)}{\eta_1^{\eta_2+1}}$ is chosen to avoid the $\varpi^{-\eta_2}$ singular behavior near $\varpi = 0$.

This implies the following asymptotics near 0 of ϕ

$$\phi(\varpi) = \phi_0 + \frac{\eta_3}{\eta_2+1}\varpi + \mathcal{O}(\varpi^2). \tag{7.1.24}$$

Indeed, a series such as (7.1.20) seems to imply that (7.1.14) with $\nu = 7/6$ satisfying the boundary conditions (7.1.15)–(7.1.16) has no solution since C_ν and D_ν vanish. However, a singular behavior for large ϖ is possible. Thus, $\nu = 7/6$ could be an exact and very singular solution of (7.1.14).

As is shown in [38], for suitable initial conditions, the solution of the classical Boltzmann equation for hard spheres is bounded for $t > 0$. As a consequence, a self-similar solution of the type (7.1.11) to show singularities is impossible for classical hard spheres. Notice that the classical Boltzmann equation has Kolmogorov–Zakharov spectrum $J^{1/2}/\omega^{7/4}$ for flux of matter to zero energy. We also expect that the self-similar function ϕ has no mass at the peak near the zero energy. The inequality (7.1.18) becomes $7/4 < \nu < 3/2$, which cannot be satisfied.

The transition rate S in (7.1.1) and (7.1.14) can be scaled as $S \backsim e^{3d/2}$ in d space dimensions. The Kolmogorov–Zakharov spectrum is $J^{1/3}/\omega^{d/2-1/3}$ for the particle constant flux J and $P^{1/3}/e^{d/2}$ for the energy flux P. The nonlinear eigenvalue ν satisfies $d/2 - 1/3 < \nu < d/2$. In an infinite two-dimensional space, the chemical potential μ cannot vanish at equilibrium. As a consequence, Bose–Einstein condensation does not arise in two space dimensions. However, there is no objection to the existence of a solution for the nonlinear eigenvalue problem in two dimensions. It may be possible that a singularity arises while the future evolution does not allow the condensate to be fed with particles (cf. [45, 93, 94, 131, 133, 135]).

Suppose that the dispersion relation is of the form $\omega_p = |p|^\alpha$ in a space of dimension d and suppose that the scaling of the cross section of the interactions scales as $a^2(\omega) \backsim \omega^{-2/l}$. The Kolmogorov solutions become $J^{1/3}/\omega^{d/\alpha-1/3-2/3l}$ for the particle flux and $P^{1/3}/\omega^{d/\alpha-2/3l}$ for the energy.

We obtain the inequality

$$\frac{d}{\alpha} - \frac{1}{3} - \frac{2}{3l} < \nu < \frac{d}{\alpha}.$$

7.2 Validity of the Boltzmann–Nordheim Equation Near Blow-up

Let us discuss the breaking of the validity of the Boltzmann–Nordheim kinetic theory near the blow-up. Let t_* be the blow-up time and $\mathcal{T} = t_* - t$ be the time left until blow-up. Again F is the total number density, and g the scattering length. The mean-free flight time for the core of the energy spectrum is

$$t_{mfp} = \frac{1}{Fg^2}(m/\omega_{Th})^{1/2},$$

where $\omega_{Th} = \frac{p^2}{2m}$ is the average kinetic energy per particle (a constant). Suppose that this energy is of the order of magnitude of the energy at the Boltzmann–Nordheim transition, then we obtain

$$t_{mfp} = \frac{\hbar^2 F^{\frac{2}{3}}}{mk_B}.$$

The Boltzmann–Nordheim kinetic theory applies when $\mathcal{T}$ is still much larger than $\hbar/\omega_0(\mathcal{T})$, where $\omega_0(\mathcal{T})$ is the average energy of particles taking part in this blow-up. This energy goes to 0 as $\mathcal{T}$ goes to 0. We have $\omega_0(\mathcal{T}) \sim \omega_{Th}(\mathcal{T}/t_{mfp})^\beta$ ($\beta > 0$). Therefore, the Boltzmann–Nordheim kinetic theory applies for $\mathcal{T} > \mathcal{T}_{cr} \equiv \hbar/\omega_0(\mathcal{T}_{cr})$.

From the estimates above, the inequality $\mathcal{T}_{cr} < t_{mfp}$ is equivalent to $gf^{1/3} \ll 1$, precisely the condition for a dilute gas. Therefore, in this dilute gas limit, the Boltzmann–Nordheim kinetic equation remains physically sound in the time interval $[t_{mfp}, \mathcal{T}_{cr}]$ before blow-up.

7.3 Self-Similar Dynamics for the Condensate-Thermal Cloud System After the Formation of the Singularity

Let us consider the system (6.2.13)–(6.2.14), which describes the dynamics of the growth of the condensate after the formation of the singularity, and suppose that the solution $\tilde{f}$ is radial and $\omega_p = p^2$. Then

$$\tilde{f}(p) = \tilde{f}(|p|) = \tilde{f}(\omega).$$

Dropping the tilde sign and the constants in the collision operators, we get the system

$$\frac{\partial}{\partial t}\rho_0 \ = \ \rho_0\tilde{Q}_{12}[f], \tag{7.3.1}$$

$$\frac{\partial}{\partial t}f \ = \ C_{22}[f] \ + \ \rho_0 Q_{12}[f], \tag{7.3.2}$$

where

$$\tilde{Q}_{12}[f] \ = \ \int_{\mathbb{R}_+^2} [f_{\omega_3}f_{\omega_4} \ - \ f_{\omega_3+\omega_4}(f_{\omega_3} + f_{\omega_4} + 1)]\mathrm{d}\omega_3\mathrm{d}\omega_4,$$

and

$$Q_{12}[f] \ = \ \frac{1}{\sqrt{\omega_1}}\int_{\mathbb{R}_+^2} \delta(\omega_1 - \omega_3 - \omega_4)[f_{\omega_3}f_{\omega_4} \ - \ f_{\omega_1}(f_{\omega_3} + f_{\omega_4} + 1)]\mathrm{d}\omega_3\mathrm{d}\omega_4$$

$$+ \ \frac{2}{\sqrt{\omega_1}}\int_{\mathbb{R}_+^2} \delta(\omega_1 + \omega_2 - \omega_4)[f_{\omega_4}(f_{\omega_1} + f_{\omega_2} + 1) \ - \ f_{\omega_1}f_{\omega_2}]\mathrm{d}\omega_3\mathrm{d}\omega_4.$$

These coupled equations conserve mass and energy and the H-theorem holds. For a very short time $0 < t - t_* << t_B$, it is possible to write a self-similar solution of the form

$$f_\omega(t) \ = \ \tilde{\tau}(t)^{-\nu'}\varphi\left(\frac{\omega}{\tilde{\tau}(t)}\right). \tag{7.3.3}$$

Then plugging (7.3.3) into (7.3.1)–(7.3.2), keeping the most singular term in the collision integrals of (7.3.1)–(7.3.2) and imposing the separation of temporal and re-scaled variables, we have

$$\frac{1}{\rho_0(t)}\frac{\partial\rho_0(t)}{\partial t}\tilde{\tau}^{2\nu'-2} \ = \ K_1 \ = \ \tilde{\mathcal{Q}}_{12}[\varphi], \tag{7.3.4}$$

$$\frac{\partial\tilde{\tau}(t)}{\partial t}\tilde{\tau}(t)^{2\nu'-3} \ = \ K_2, \tag{7.3.5}$$

$$\rho_0(t)\tilde{\tau}(t)^{\nu'-3/2} \ = \ K_3, \tag{7.3.6}$$

$$-K_2(\nu'\varphi(\varpi) + \varpi\varphi'(\varpi)) \ = \ C_{4wave}[\varphi] \ + \ K_3\mathcal{Q}_{12}[\varphi], \tag{7.3.7}$$

where

$$\tilde{\mathcal{Q}}_{12}[\varphi] \ = \ \int_{\mathbb{R}_+^2} [\varphi_{\omega_3}\varphi_{\omega_4} \ - \ \varphi_{\omega_3+\omega_4}(\varphi_{\omega_3} + \varphi_{\omega_4})]\mathrm{d}\omega_3\mathrm{d}\omega_4,$$

and

$$\mathcal{Q}_{12}[\varphi] \ = \ \frac{1}{\sqrt{\omega_1}}\int_{\mathbb{R}_+^2} \delta(\omega_1 - \omega_3 - \omega_4)[\varphi_{\omega_3}\varphi_{\omega_4} \ - \ \varphi_{\omega_1}(\varphi_{\omega_3} + \varphi_{\omega_4})]\mathrm{d}\omega_3\mathrm{d}\omega_4$$

$$+ \ \frac{2}{\sqrt{\omega_1}}\int_{\mathbb{R}_+^2} \delta(\omega_1 + \omega_2 - \omega_4)[\varphi_{\omega_4}(\varphi_{\omega_1} + \varphi_{\omega_2}) \ - \ \varphi_{\omega_1}\varphi_{\omega_2}]\mathrm{d}\omega_3\mathrm{d}\omega_4.$$

Integrating (7.3.5) yields

$$\tilde{\tau}(t) \; = \; (2(\nu' - 1)K_2(t - t_*))^{\frac{1}{2(\nu'-1)}}. \qquad (7.3.8)$$

Using (7.3.6),

$$\rho_0(t) = K_3 \tilde{\tau}(t)^{3/2 - \nu'},$$

which in combination with (7.3.4) implies

$$K_1 = \left(\frac{3}{2} - \nu' \right) K_2.$$

For times just after t_*, we expect that φ will be very close to the function before the condensation "far" from zero energy, since the function changes infinitely fast near the origin only. As a consequence, by continuity this implies that φ and ϕ has the same behavior for large ϖ before and after singularity. This means ν' is the same as before $\nu' \equiv \nu$. Therefore, ν is not the eigenvalue but the ratio: $\frac{K_3}{\sqrt{K_2}}$. This simple reasoning relates the growth of the "condensate" fraction after the condensation to the nonlinear eigenvalue before the condensation. We refer to [101] for more discussions on this issue.

Chapter 8
Hydrodynamic Approximations

In this chapter, the dynamics of BECs under both regimes of ultra-low and medium temperatures will be studied. At the regime of ultra-low temperatures, the microscopic hydrodynamic modes of BECs will be discussed. The Euler and Navier–Stokes approximations of the system, when the temperature is smaller but very close to the Bose–Einstein critical temperate T_{BE}, will be presented.

As discussed in the preceding chapters, the dynamics of finite temperature trapped Bose gases can be described by two coupled components – the condensate and noncondensate atoms – acting very differently. Hydrodynamic approximations of the Zaremba–Nikuni–Griffin/Pomeau–Brachet–Métens–Rica model has been a very active field of research for the last 20 years (cf. [76, 97, 98, 164, 165, 166]). The model has been used to derive two-fluid hydrodynamic equations, in which the normal fluid density is equal to the thermal cloud density and the superfluid density is equal to the condensate density. In addition, the normal superfluid and fluid densities appearing in the standard Landau two-fluid theory [29, 106] can be identified with the corresponding condensate and noncondensate densities, within the context of the Zaremba et al./Pomeau et al. model. Moreover, the transport coefficients and thermodynamic functions in these equations are based on Hartree–Fock excitations. Although this treatment of the thermal cloud is only valid at higher temperatures, in which the collective (Bogoliubov) nature of the thermal excitations can be neglected, this microscopic model for the coupled dynamics in a finite temperature trapped Bose gas includes most of the essential physics needed to describe non-equilibrium behaviors. While being quite simple, in comparison with the quantum Boltzmann–Hugenholt–Pines system (5.0.20)–(5.0.21)–(5.0.22)–(5.0.23), the Zaremba et al./Pomeau et al. model and the derived Landau two-fluid theory give considerable insight into many subtle aspects of the dynamics of a finite temperature Bose superfluid. On trying to establish a more systematic approach to the mathematical study of the hydrodynamic approximations of finite temperature trapped Bose gases, there have been some attempts, for instance [2]. In Section 8.1, we revisit the question of deriving Euler and Navier–Stokes approximations of the Zaremba

© Springer Nature Switzerland AG 2019

Y. Pomeau, M.-B. Tran, *Statistical Physics of Non Equilibrium Quantum Phenomena*,

Lecture Notes in Physics 967, https://doi.org/10.1007/978-3-030-34394-1_8

et al./Pomeau et al. model. On one hand, the derived Euler limit agrees with the Landau two-fluid system [76, 106]. On the other hand, the Navier–Stokes approximation has not been observed in the literature and is an interesting object for mathematical study.

A monatomic classical gas has one non-propagating thermal mode, one pair of propagating sound modes, and two non-propagating viscous modes. On the other hand, a monatomic BEC has six hydrodynamic modes, four of which are longitudinal modes and describe sound mode propagation in the BEC and the other two are transverse modes and describe viscous properties of the BEC. A dilute BEC has two pairs of sound modes, each being a mixture of density and temperature waves. In a BEC, the only non-propagating modes are the viscous ones. As discussed earlier in Section 6.3, the Reichl et al. model (cf. [78, 79, 80, 142, 143]) is appropriated to describe a BEC system at ultra-low temperature $(0 < T < 0.3 T_{BE})$. We will discuss in Section 8.2 how to compute the hydrodynamic modes using this theory. In particular, the decay rates of the sound modes can be computed as a function of equilibrium temperature, density, particle mass and interaction strength. Expressions for the decay rates of sound modes can be found to agree with the results of the Steinhauer experiment on a BEC of ^{87}Rb atoms [147]. The original idea for this approach to BEC hydrodynamics comes from previous work of Reichl on hydrodynamics of Fermi gases [139, 140]. The two-fluid equations, whose derivation is given later in Section 8.1 based on the Zaremba et al./Pomeau et al. approximation, treats the thermal cloud in a high temperature Hartree–Fock approximation. On the other hand, the behavior of these sound speeds, predicted by the Reichl et al. model is consistent with the theory of Lee and Yang [103]. Therefore, the two models have different ranges of validity.

The discussion of this chapter partially follows the presentations of the papers [92, 143] by Jin, Reichl and Tran.

8.1 Hydrodynamic Approximations – A Regime of Medium Temperatures

In this section, we focus on the regime where the temperature T of the system is smaller but very close to the Bose–Einstein critical temperate T_{BE}. The hydrodynamic approximations will be based on the key assumption that the collision operator C_{22} is much stronger than the collision operator C_{12}. The physics underlying this assumption is that, at this temperature regime, the collisions between excited atoms are rapid to establish a local equilibrium within the non-condensate component. As discussed above, when the temperature T is very close to $0K$, the model (5.0.2)–(5.0.3) needs to be modified to describe the collective phonon-like excitations/bogolons. In the regime under consideration, the particlelike Hartree–Fock excitation spectrum is used. We can now define the static equilibrium of the system

$$\mathfrak{E}_0(p) \;=\; \frac{1}{e^{\beta_0[(p-mv_{n0})^2/(2m)+U_0-\mu_0]} - 1},\qquad (8.1.1)$$

where v_{n0} is the static fluid velocity, μ_0 is the static chemical potential, β_0 is the static temperature parameter, and U_0 is the static mean field. The static density is set to be

$$n_{n0} \;=\; \int_{\mathbb{R}^3} \mathfrak{E}_0(p)\,\frac{dp}{(2\pi\hbar)^3}.\qquad (8.1.2)$$

The average speed of the particles is now

$$\bar{c} \;=\; \frac{1}{n_{n0}m}\int_{\mathbb{R}^3} \sqrt{p^2}\,\mathfrak{E}_0(p)\,\frac{dp}{(2\pi\hbar)^3}.\qquad (8.1.3)$$

Let us define the collision frequency with respect to C_{12}

$$
\begin{aligned}
\nu_{12}(p_1) \;=\; &\frac{\lambda_1 n_c}{m^2}\iint_{\mathbb{R}^3\times\mathbb{R}^3}\delta(mv_c+p_1-p_2-p_3)\delta(\omega_c+\omega_{p_1}-\omega_{p_2}-\omega_{p_3})\\
&\times[\mathfrak{E}_0(p_2)+\mathfrak{E}_0(p_3)+1]dp_2dp_3\\
&+2\frac{\lambda_1 n_c}{m^2}\iiint_{\mathbb{R}^3\times\mathbb{R}^3}\delta(mv_c+p_2-p_1-p_3)\\
&\times\delta(\omega_c+\omega_{p_2}-\omega_{p_1}-\omega_{p_3})\mathfrak{E}_0(p_3)dp_2dp_3,
\end{aligned}
$$

$$ (8.1.4) $$

and the mean collision frequency

$$\bar{\nu}_{12} \;=\; \frac{1}{n_{n0}m}\int_{\mathbb{R}^3}\nu_{12}(p)\mathfrak{E}_0(p)\,\frac{dp}{(2\pi\hbar)^3}.\qquad (8.1.5)$$

By reversing the collision and the mean collision frequencies, we find the free-flight time $\tau_{12}(p)$ and its average value $\bar{\tau}_{12}$, respectively

$$\tau_{12}(p) \;=\; \frac{1}{\nu_{12}(p)},\qquad \bar{\tau}_{12} \;=\; \frac{1}{\bar{\nu}_{12}}.\qquad (8.1.6)$$

Using the average speed (8.1.3), we can compute the mean-free path as $l_{12} = \bar{c}\bar{\tau}_{12}$. Similar quantities can also be defined for the second collision operator C_{22}. We start with the collision and the mean collision frequencies

$$
\begin{aligned}
\nu_{22}(p_1) \;=\; &\frac{\lambda_2}{n_{n0}m}\iiint_{\mathbb{R}^3\times\mathbb{R}^3\times\mathbb{R}^3}\delta(p_1+p_2-p_3-p_4)\delta(\omega_{p_1}+\omega_{p_2}-\omega_{p_3}-\omega_{p_4})\\
&\times\mathfrak{E}_0(p_2)(1+\mathfrak{E}_0(p_3))(1+\mathfrak{E}_0(p_4))dp_2dp_3dp_4
\end{aligned}
$$

$$ (8.1.7) $$

and

$$\bar{\nu}_{22} \;=\; \frac{1}{n_{n0}m}\int_{\mathbb{R}^3}\nu_{22}(p)\mathfrak{E}_0(p)\,\frac{dp}{(2\pi\hbar)^3}.\qquad (8.1.8)$$

Now, define the free-flight time $\tau_{22}(p)$, the mean field time $\bar{\tau}_{22}$ and the mean free path l_{22}

$$\tau_{22}(p) \;=\; \frac{1}{\nu_{22}(p)}, \qquad \bar{\tau}_{22} \;=\; \frac{1}{\bar{\nu}_{22}}, \qquad l_{22} \;=\; \bar{c}\bar{\tau}_{22}. \tag{8.1.9}$$

The derivation of the Euler and Navier–Stokes approximations will be based on the Chapman–Enskog expansion, in which L and θ are set to be the reference length and time, respectively. Since the particlelike Hartree–Fock excitation spectrum is used in the current temperature range, the rescaled variables of classical particles can be adopted (cf. [151]): $\tilde{r} = \frac{r}{L}$, $\tilde{t} = \frac{t}{\theta}$, $\tilde{p} = \frac{p}{P}$, $P = m\bar{c}$, $\tilde{v}_c = \frac{v_c}{\bar{c}}$.

Using the above new variables, we rescale mean-free paths and the mean free-flight times as $\tilde{l}_{22} = \frac{l_{22}}{P^5}$, $\tilde{\tau}_{22} = \frac{\bar{\tau}_{22}}{P^5}$ and $\tilde{l}_{12} = \frac{l_{12}}{P^2}$, $\tilde{\tau}_{12} = \frac{\bar{\tau}_{12}}{P^2}$. In addition, the potential field U also needs to be rescaled as $\tilde{U} = U/U_0$, where U_0 is the reference potential field.

Using the rescaled variables, we write

$$\tilde{C}_{12}[f](t, r, \tilde{p}_1) := \tag{8.1.10}$$

$$\tilde{\lambda}_1 n_c(t, r) \iint_{\mathbb{R}^3 \times \mathbb{R}^3} \delta(\tilde{v}_c + \tilde{p}_1 - \tilde{p}_2 - \tilde{p}_3)\delta(\omega_c + \omega_{\tilde{p}_1} - \omega_{\tilde{p}_2} - \omega_{\tilde{p}_3})$$

$$\times [(1 + f_1)f_2 f_3 - f_1(1 + f_2)(1 + f_3)] \mathrm{d}\tilde{p}_2 \mathrm{d}\tilde{p}_3$$

$$- 2\tilde{\lambda}_1 n_c(t, r) \iint_{\mathbb{R}^3 \times \mathbb{R}^3} \delta(\tilde{v}_c + \tilde{p}_2 - \tilde{p}_1 - \tilde{p}_3)\delta(\omega_c + \omega_{\tilde{p}_2} - \omega_{\tilde{p}_1} - \omega_{\tilde{p}_3})$$

$$\times [(1 + f_2)f_1 f_3 - f_2(1 + f_1)(1 + f_3)] \mathrm{d}\tilde{p}_2 \mathrm{d}\tilde{p}_3,$$

$$\tilde{C}_{22}[f](t, r, \tilde{p}_1) := \tilde{\lambda}_2 \iiint_{\mathbb{R}^3 \times \mathbb{R}^3 \times \mathbb{R}^3} \delta(\tilde{p}_1 + \tilde{p}_2 - \tilde{p}_3 - \tilde{p}_4) \tag{8.1.11}$$

$$\times \delta(\omega_{\tilde{p}_1} + \omega_{\tilde{p}_2} - \omega_{\tilde{p}_3} - \omega_{\tilde{p}_4})$$

$$\times [(1 + f_1)(1 + f_2)f_3 f_4 - f_1 f_2(1 + f_3)(1 + f_4)] \mathrm{d}\tilde{p}_2 \mathrm{d}\tilde{p}_3 \mathrm{d}\tilde{p}_4,$$

where $\tilde{\lambda}_1 = P^2 \lambda_1/\bar{c}$ and $\tilde{\lambda}_2 = P^5 \lambda_2/\bar{c}$. By putting

$$\hat{C}_{12}[f] := \sqrt{\tilde{l}_{12}\tilde{l}_{22}}\,\tilde{C}_{12}[f], \quad \hat{C}_{22}[f] := \sqrt{\tilde{l}_{12}\tilde{l}_{22}}\,\tilde{C}_{22}[f], \tag{8.1.12}$$

the following rescaled version of (5.0.3) then follows:

$$\frac{\sqrt{\tilde{l}_{12}\tilde{l}_{22}}}{\theta\bar{c}}\partial_{\tilde{t}}f \;+\; \frac{\sqrt{\tilde{l}_{12}\tilde{l}_{22}}}{L}\frac{P}{m\bar{c}}\tilde{p}\cdot\nabla_{\tilde{r}}f \;-\; \frac{\sqrt{\tilde{l}_{12}\tilde{l}_{22}}}{L}\frac{U_0}{P\bar{c}}\nabla_{\tilde{r}}\tilde{U}\cdot\nabla_{\tilde{p}}f$$

$$= \sqrt{\frac{\tilde{l}_{22}}{\tilde{l}_{12}}}\hat{C}_{12}[f] \;+\; \sqrt{\frac{\tilde{l}_{12}}{\tilde{l}_{22}}}\hat{C}_{22}[f]. \tag{8.1.13}$$

To simplify the complexity of the above expression, the constants $\frac{\sqrt{\tilde{l}_{12}\tilde{l}_{22}}}{\theta\bar{c}}$, $\frac{\sqrt{\tilde{l}_{12}\tilde{l}_{22}}}{L}$ can be set to be 1 by rescaling again the space and time variables $\tilde{t} \to \frac{\sqrt{\tilde{l}_{12}\tilde{l}_{22}}}{\theta\bar{c}}\tilde{t}$, $\tilde{r} \to \frac{\sqrt{\tilde{l}_{12}\tilde{l}_{22}}}{L}\tilde{r}$.

Notice that $\frac{\tilde{\tau}_{22}}{\tilde{\tau}_{12}} = \frac{\tilde{l}_{22}}{\tilde{l}_{12}}$ is a dimensionless parameter and is proportional to $\frac{\tilde{\lambda}_1}{\tilde{\lambda}_2}$. We denote this constant by ρ^2. The Euler approximation is valid under the physical assumption that the collisions between excited atoms are fast enough to establish a local equilibrium within the non-condensate component; the quantity $\tilde{\tau}_{22}$ is smaller than $\tilde{\tau}_{12}$ but the ratio between $\tilde{\tau}_{22}$ and $\tilde{\tau}_{12}$ is not necessarily very small. In the Euler approximation, ρ is just a parameter. On the other hand, the Navier–Stokes approximation is valid under the physical assumption that the collisions between excited atoms are fast enough to establish a local equilibrium within the non-condensate component and $\tilde{\tau}_{22} << \tilde{\tau}_{12}$. In the Navier–Stokes approximation, one needs to impose the assumption that ρ is small and it will be used as the small parameter in the usual Chapman–Enskog expansion process.

Using the momentum equation $\frac{P}{m\bar{c}} = 1$, we obtain the following equation from (8.1.13)

$$\frac{\partial}{\partial \tilde{t}}f + \tilde{p} \cdot \nabla_{\tilde{r}}f - \frac{U_0}{m\bar{c}^2}\nabla_{\tilde{r}}\tilde{U} \cdot \nabla_{\tilde{p}}f = \rho\hat{C}_{12}[f] + \frac{1}{\rho}\hat{C}_{22}[f]. \qquad (8.1.14)$$

Having rescaled the kinetic equation, we now rescale the Gross–Pitaevskii equation

$$i\frac{\hbar}{\theta}\frac{\partial}{\partial \tilde{t}}\Phi(t,r) = \left(-\frac{\hbar^2\Delta_{\tilde{r}}}{2mL^2} + g[n_c(t,r) + 2n_n(t,r)] - \frac{i\rho}{\tilde{\tau}_{12}}\tilde{\Lambda}_{12}[f](t,r) \right)\Phi(t,r),$$
$$(8.1.15)$$

where $\tilde{\Lambda}_{12}[f] = \frac{\hbar}{2n_c}\int_{\mathbb{R}^3}\hat{C}_{12}[f]\frac{dp}{(2\pi\hbar)^3}$. By the same argument as above,

$$i\frac{\sqrt{\tilde{l}_{12}\tilde{l}_{22}}}{\theta\bar{c}}\frac{\partial}{\partial \tilde{t}}\Phi(t,r) = \qquad (8.1.16)$$

$$\left(-\frac{\hbar}{mL}\frac{\sqrt{\tilde{l}_{12}\tilde{l}_{22}}\Delta_{\tilde{r}}}{2L\bar{c}} + \frac{\sqrt{\tilde{l}_{12}\tilde{l}_{22}}U_0}{\hbar\bar{c}}\tilde{U}_*(t,r) - i\rho\tilde{\Lambda}_{12}[f](t,r) \right)\Phi(t,r),$$

where $\tilde{U}_* = U_*/U_0$ and $U_*(t,r) = g[n_c(t,r) + 2n_n(t,r)]$, since $U_*(t,r)$ has the dimension of $U(t,r)$.

Since $\frac{\hbar}{m\bar{c}}$ is the Compton wavelength, it has the dimension of a length. The two quantities $\frac{\hbar}{mL\bar{c}}$ and $\frac{\hbar}{mL}\frac{\sqrt{\tilde{l}_{12}\tilde{l}_{22}}}{2L\bar{c}}$ are indeed dimensionless. In addition, $\frac{\sqrt{\tilde{l}_{12}\tilde{l}_{22}}U_0}{\hbar\bar{c}}$ is also dimensionless since it is the product of the three dimensionless parameters $\frac{\sqrt{\tilde{l}_{12}\tilde{l}_{22}}U_0}{\hbar\bar{c}} = \frac{\sqrt{\tilde{l}_{12}\tilde{l}_{22}}}{L}\frac{mL\bar{c}}{\hbar}\frac{U_0}{m\bar{c}^2} = \frac{\sqrt{\tilde{l}_{12}\tilde{l}_{22}}}{L}\frac{mL\bar{c}}{\hbar}\tilde{g}$. Setting all of the dimensionless parameters in (8.1.16) to be 1 by the same rescaling argument used for (8.1.14) and dropping the tilde and hat signs, we arrive at

$$i\frac{\partial}{\partial t}\Phi = \left(-\frac{\Delta_r}{2} + gU_* - i\rho\Lambda_{12}[f] \right)\Phi, \qquad (8.1.17)$$

where g stands for the dimensionless parameter $\tilde{g}$.

The rescaled system then follows

$$\frac{\partial}{\partial t}f + p \cdot \nabla_r f - g\nabla_r U \cdot \nabla_p f = \rho C_{12}[f] + \frac{1}{\rho}C_{22}[f],$$

$$i\frac{\partial}{\partial t}\Phi = \left(-\frac{\Delta_r}{2} + g[n_c + 2n_n] - i\rho\Lambda_{12}[g]\right)\Phi,$$

(8.1.18)

where C_{12}, C_{22} and Λ_{12} take the forms

$$C_{12}[f](t,r,p_1) = \tag{8.1.19}$$

$$n_c(t,r)\iint_{\mathbb{R}^3 \times \mathbb{R}^3} \delta(v_c + p_1 - p_2 - p_3)\delta(\omega_c + \omega_{p_1} - \omega_{p_2} - \omega_{p_3})$$

$$\times\, [(1+f_1)f_2f_3 - f_1(1+f_2)(1+f_3)]\mathrm{d}p_2\mathrm{d}p_3$$

$$-\, 2n_c(t,r)\iint_{\mathbb{R}^3 \times \mathbb{R}^3} \delta(v_c + p_2 - p_1 - p_3)\delta(\omega_c + \omega_{p_2} - \omega_{p_1} - \omega_{p_3})$$

$$\times\, [(1+f_2)f_1f_3 - f_2(1+f_1)(1+f_3)]\mathrm{d}p_2\mathrm{d}p_3,$$

$$C_{22}[f](t,r,p_1) = \tag{8.1.20}$$

$$\iiint_{\mathbb{R}^3 \times \mathbb{R}^3 \times \mathbb{R}^3} \delta(p_1 + p_2 - p_3 - p_4)\delta(\omega_{p_1} + \omega_{p_2} - \omega_{p_3} - \omega_{p_4})$$

$$\times\, [(1+f_1)(1+f_2)f_3f_4 - f_1f_2(1+f_3)(1+f_4)]\mathrm{d}p_2\mathrm{d}p_3\mathrm{d}p_4,$$

$$\Lambda_{12}[f](t,r) = \frac{1}{n_c(t,r)}\int_{\mathbb{R}^3} C_{12}[f](t,r,p)\mathrm{d}p.$$

We also define the differential operators

$$\mathcal{D}f = \partial_t f + p \cdot \nabla_r f - g\nabla_r U \cdot \nabla_p f - \rho C_{12}[f], \tag{8.1.21}$$

$$\mathbb{D}f = \partial_t f + p \cdot \nabla_r f - g\nabla_r U \cdot \nabla_p f, \tag{8.1.22}$$

and

$$\Pi f = \partial_t f + p \cdot \nabla_r f. \tag{8.1.23}$$

Since we only consider the case $U = 0$, the following system for the super-fluid of the condensate can be deduced

$$\frac{\partial}{\partial t}n_c + \nabla_r \cdot (n_c v_c) = -\rho\Gamma_{12}[f],$$

$$\frac{\partial}{\partial t}v_c + \frac{\nabla_r v_c^2}{2} = -\nabla_r \mu_c.$$

(8.1.24)

8.1.1 Basic Properties of C_{22} and the Boltzmann–Nordheim Equation

In this section, we recall the basic properties of C_{22} and the Boltzmann–Nordheim equation.

8.1.1.1 Collision Invariants and Equilibrium of C_{22}

We first observe that

$$\int_{\mathbb{R}^3} \Upsilon_i(p) C_{22}[f](p) \mathrm{d}p \; = \; 0, \quad i = 0, 1, 2, 3, 4, \tag{8.1.25}$$

where $\Upsilon_0(p) = 1$, $\Upsilon_i(p) = p^i$, $(i = 1, 2, 3)$, $\Upsilon_4(p) = |p|^2$ are the collision invariants and p^i is the i-th Cartesian component of the vector $p = (p^1, p^2, p^3)$.

The operator C_{22} also has a local equilibrium of the form

$$\mathfrak{E}(t, r, p) \; = \; \frac{1}{e^{\beta[(p-v_n)^2/2 + U - \mu]} - 1}, \tag{8.1.26}$$

where $\beta(t, r)$ is the temperature parameter, $v_n(t, r)$ is the local fluid velocity, $\mu(t, r)$ is the local chemical potential (which is different from the condensate chemical potential $\mu_c(t, r)$), and $U(t, r)$ is the mean field. Let us now define the following local Gaussian $\mathcal{G}(t, r, p) = \gamma(t, r) e^{-\frac{|p-u(t,r)|^2}{2\tau(t,r)}}$, where $\gamma(t, r) = e^{\beta(U(t,r) - \mu(t,r))}$, $u(t, r) = v_n(t, r)$, $\tau(t, r) = \frac{1}{\beta(t,r)}$. The local equilibrium $\mathfrak{E}$ can be expressed in terms of the local Gaussian $\mathcal{G}$ as $\mathfrak{E}(t, r, p) = \frac{\mathcal{G}(t,r,p)}{1-\mathcal{G}(t,r,p)}$.

8.1.1.2 The Linearized Operator of C_{22}

Let $(\cdot, \cdot)_{L^2}$ be the inner product of $L^2(\mathbb{R}^3)$, the space of real, measurable functions which are square integrable on $\mathbb{R}^3$, with the norm $\|\cdot\|_{L^2}$. We consider the linearized operator of C_{22} around a fixed equilibrium $\mathfrak{E}(t, r, p)$, defined as

$$\mathcal{L}_{22} \; := \; 2B_1(\mathfrak{E}, \cdot) \; + \; 3B_2(\mathfrak{E}, \mathfrak{E}, \cdot), \tag{8.1.27}$$

or equivalently

$$\mathcal{L}_{22}(\mathfrak{E}f)(t, r, p_1) = \int_{\mathbb{R}^3 \times \mathbb{R}^3 \times \mathbb{R}^3} \delta(p_1 + p_2 - p_3 - p_4)\delta(\omega_{p_1} + \omega_{p_2} - \omega_{p_3} - \omega_{p_4})$$

$$\times \frac{\mathcal{G}_1 \mathcal{G}_2}{(1 - \mathcal{G}_1)(1 - \mathcal{G}_2)(1 - \mathcal{G}_3)(1 - \mathcal{G}_4)}$$

$$\times \Big[(1 - \mathcal{G}_3)f(p_3) + (1 - \mathcal{G}_4)f(p_4) - (1 - \mathcal{G}_2)f(p_2)$$

$$- (1 - \mathcal{G}_1)f(p_1)\Big] dp_2 dp_3 dp_4, \qquad (8.1.28)$$

for some function $f(p)$ and fixed values $(t, r) \in \mathbb{R}_+ \times \mathbb{R}^3$ and with $\mathcal{G}_i = \mathcal{G}(t, r, p_i)$, $i = 1, 2, 3, 4$ being now the particle index.

The kernel of the linearized collision operator $\mathcal{L}_{22}$ of C_{22} can be computed explicitly as

$$\mathcal{N}_{22} := \ker\mathcal{L}_{22} = \operatorname{span}\left\{\frac{\mathfrak{E}^2}{\mathcal{G}}\Upsilon_i : i = 0, \cdots, 4\right\},$$

which allows us to define the kernel orthogonal space

$$\mathcal{R}_{22} := \mathcal{N}_{22}^\perp = \left\{G \in L^2(\mathbb{R}^3) : \left(G, \frac{\mathfrak{E}^2}{\mathcal{G}}\Upsilon_i\right)_{L^2} = 0, \ i = 0, \cdots, 4\right\}.$$

Setting the $L^2(\mathbb{R}^3)$-orthogonal projection operators onto $\mathcal{N}_{22}$ and $\mathcal{R}_{22}$ to be $\mathbb{P}$ and $\mathbb{P}^\perp = 1 - \mathbb{P}$ and normalizing $\{\Upsilon\}_{i=0,\cdots,4}$, we obtain the following orthonormal basis of $\mathcal{N}_{22}$

$$\left\{\frac{\Theta_i}{\sqrt{\Xi_i}}\frac{\mathfrak{E}^2}{\mathcal{G}} : \quad i = 0, \cdots, 4\right\}, \qquad (8.1.29)$$

with

$$\Theta_0 = 1; \quad \Theta_i = p^i - u_i, \ i = 1, 2, 3; \quad \Theta_4 = |p - u|^2 - 6\tau\frac{\Omega_1(\gamma)}{\Omega_0(\gamma)};$$

$$\Xi_0 = \int_{\mathbb{R}^3} \frac{\mathfrak{E}^2}{\mathcal{G}} dp = 2^{3/2}\pi\tau^{3/2}z\Omega_0(z);$$

$$\Xi_i = \int_{\mathbb{R}^3} \frac{\mathfrak{E}^2}{\mathcal{G}}|\Upsilon_i(p)|^2 dp = 2^{5/2}\pi\tau^{5/2}z\Omega_1(z), \quad i = 1, 2, 3;$$

$$\Xi_4 = \int_{\mathbb{R}^3} \frac{\mathfrak{E}^2}{\mathcal{G}}|\Upsilon_4(p)|^2 dp = 2^{7/2}\pi\tau^{7/2}z\Sigma(\Omega_0(z), \Omega_1(z), \Omega_2(z));$$

where

$$\Sigma(x, y, z) = \frac{5xz - 9y^2}{x},$$

and

$$\Omega_k(z) = \int_0^\infty \frac{y^{k-1/2}}{e^y + z}dy, \quad k > -1/2. \qquad (8.1.30)$$

Let us define

$$\mathfrak{A}(p) = p \otimes p - \frac{1}{3}|p|^2 \mathrm{Id}, \qquad \mathfrak{B}(p) = \frac{1}{2}p(|p|^2 - 5). \qquad (8.1.31)$$

Clearly,

$$\mathfrak{A}_{jk} \perp \ker \mathcal{L}_{22}, \qquad \mathfrak{B}_l \perp \ker \mathcal{L}_{22}, \qquad \mathfrak{B}_l \perp \mathfrak{A}_{jk}, \qquad j, k, l = 1, 2, 3. \qquad (8.1.32)$$

Let $\alpha_0(|p|)$, $\alpha_1(|p|)$, $\beta_0(|p|)$ and $\beta_1(|p|)$ be scalar-valued functions satisfying

$$
\begin{aligned}
\mathcal{L}_{22}^{-1}\left(\frac{\mathfrak{E}^2(p)}{\mathcal{G}(p)}\mathfrak{A}(p)\right) &= \alpha_0(|p|)\frac{\mathfrak{E}^2(p)}{\mathcal{G}(p)}\mathfrak{A}(p), \\
\mathcal{L}_{22}^{-1}\left(\frac{\mathfrak{E}^2(p)}{\mathcal{G}(p)}\mathfrak{B}(p)\right) &= \alpha_1(|p|)\frac{\mathfrak{E}^2(p)}{\mathcal{G}(p)}\mathfrak{B}(p)
\end{aligned} \qquad (8.1.33)
$$

and

$$
\begin{aligned}
\mathcal{L}_{22}^{-1}\left(\frac{\mathfrak{E}^2(p)}{\mathcal{G}(p)}p^i p^j\right) &= \beta_0(|p|)\frac{\mathfrak{E}^2(p)}{\mathcal{G}(p)}\mathfrak{A}_{ij}(p), \\
\mathcal{L}_{22}^{-1}\left(\frac{\mathfrak{E}^2(p)}{\mathcal{G}(p)}\left(|p|^2 - \frac{10\tau\Omega_2(\gamma)}{\Omega_1(\gamma)}\right)p^i\right) &= \beta_1(|p|)\frac{\mathfrak{E}^2(p)}{\mathcal{G}(p)}\mathfrak{B}_i(p),
\end{aligned} \qquad (8.1.34)
$$

where p^i and $\mathfrak{B}_i(p)$ are the i-th components of the vectors p and $\mathfrak{B}(p)$, respectively. In addition, $\mathfrak{A}_{ij}(p)$ is the (i, j)-th element of the matrix $\mathfrak{A}(p)$. We then define

$$\mathfrak{C}_{ij}(p) := \beta_0(|p|)\frac{\mathfrak{E}^2(p)}{\mathcal{G}(p)}\mathfrak{A}_{ij}(p), \qquad \mathfrak{C}_i(p) := \beta_1(|p|)\frac{\mathfrak{E}^2(p)}{\mathcal{G}(p)}\mathfrak{B}_i(p). \qquad (8.1.35)$$

8.1.1.3 Hydrodynamic Quantities

We define the moments of the boson density distribution $f(t, r, p)$

$$n_n[f](t, r) = \int_{\mathbb{R}^3} f(t, r, p)\mathrm{d}p, \qquad (8.1.36)$$

$$u[f](t, r) = v_n(t, r)[f](t, r) = \frac{1}{n_n[f](t, r)}\int_{\mathbb{R}^3} pf(t, r, p)\mathrm{d}p, \qquad (8.1.37)$$

$$\mathbb{E}_n[f](t, r) = e_n[f]n_n[f] = \frac{1}{2}\int_{\mathbb{R}^3} f(t, r, p)|p - v_n[f](t, r)|^2\mathrm{d}p, \qquad (8.1.38)$$

$$\tilde{\mathbb{E}}_n[f](t, r) = \frac{2\mathbb{E}_n[f](t, r)}{3}, \qquad e_n[f](t, r) = \frac{\tilde{\mathbb{E}}_n[f](t, r)}{n_n[f](t, r)}. \qquad (8.1.39)$$

Replacing f by $\mathfrak{E}$, we obtain

$$n_n[\mathfrak{E}] = 2^{5/2}\pi\tau^{3/2}\Omega_1(\gamma), \qquad \mathbb{E}_n[\mathfrak{E}] = 2^{5/2}\pi\tau^{5/2}\gamma\Omega_2(\gamma). \qquad (8.1.40)$$

where Ω_1, Ω_2 are defined in (8.1.30). For the sake of simplicity, we denote $n_n[\mathfrak{E}]$, $v_n[\mathfrak{E}]$, $u[\mathfrak{E}]$, $\mathbb{E}_n[\mathfrak{E}]$, $\tilde{\mathbb{E}}_n[\mathfrak{E}]$, $e_n[\mathfrak{E}]$ by $n_n, v_n, u, \mathbb{E}_n, \tilde{\mathbb{E}}_n$, and e_n.

We can compute γ and τ as

$$\gamma = \left(\frac{\mathrm{Id}\Omega_2}{\Omega_1^{5/3}}\right)^{-1} \left(\frac{2^{5/3}\pi^{2/3}\mathbb{E}_n}{n_n{}^{5/3}}\right) \tag{8.1.41}$$

and

$$\tau = \left(\frac{n_n}{2^{5/2}\pi\Omega_1\left(\left(\frac{\mathrm{Id}\Omega_2}{\Omega_1^{5/3}}\right)^{-1}\left(\frac{2^{5/3}\pi^{2/3}\mathbb{E}_n}{n_n{}^{5/3}}\right)\right)}\right)^{2/3}. \tag{8.1.42}$$

8.1.1.4 The Euler Limit

For the sake of completeness, in this section, we will derive the Euler quantum hydrodynamic limit of the Boltzmann–Nordheim equation

$$\frac{\partial}{\partial t}f + p \cdot \nabla_r f = C_{22}[f]. \tag{8.1.43}$$

Let ϵ be any small, positive parameter. Following the Enskog–Hilbert expansion, we write (cf. [36])

$$f = \sum_{i=0}^{n} \epsilon^i f^{(i)} + \epsilon^l \varsigma, \tag{8.1.44}$$

where n, l are positive integers. In (8.1.43), we replace f by its Hilbert expansion (8.1.44) to get a system of linear equations and a weakly nonlinear equation for the remainder ς. The system of linear equations takes the form

$$B_1(f^{(0)}, f^{(0)}) + B_2(f^{(0)}, f^{(0)}, f^{(0)}) = 0, \tag{8.1.45}$$

$$2B_1(f^{(0)}, f^{(1)}) + 3B_2(f^{(0)}, f^{(0)}, f^{(1)}) = \frac{\partial}{\partial t}f^{(0)} + p \cdot \nabla_r f^{(0)}, \tag{8.1.46}$$

$$2B_1(f^{(0)}, f^{(i)}) + 3B_2(f^{(0)}, f^{(0)}, f^{(i)}) = \frac{\partial}{\partial t}f^{(i-1)} + p \cdot \nabla_r f^{(i-1)} \tag{8.1.47}$$

$$- \sum_{j=1}^{i-1} B_1(f^{(i)}, f^{(i-j)}) - \sum_{j,k=1, 0<j+k<i}^{i-1} B_2(f^{(i)}, f^{(k)}, f^{(i-j-k)}),$$

for $i = 2, 3, \cdots, n$.

The equation for the remainder ς is as follows

$$\frac{\partial}{\partial t}\varsigma + p \cdot \nabla_r \varsigma = \frac{1}{\epsilon}\mathcal{L}_{22}[\varsigma] + 2\sum_{i=1}^{n}\epsilon^{i-1}B_1(f^{(i)},\varsigma) + \epsilon^{l-1}B_1(\varsigma,\varsigma)$$

$$+ 3\sum_{i=1}^{n}B_2(\mathfrak{E},f^{(i)},\varsigma) + 3\sum_{i,j=1}^{n}\epsilon^{i+j-1}B_2(f^{(i)},f^{(j)},\varsigma) + 3\epsilon^{(l-1)}B_2(\mathfrak{E},\varsigma,\varsigma)$$

$$+ 3\epsilon^{l-1}\sum_{i=1}^{n}\epsilon^{i}B_2(f^{(i)},\varsigma,\varsigma) + \epsilon^{2l-1}B_2(\varsigma,\varsigma,\varsigma) + \epsilon^{n-1}\mathfrak{Q},$$

$$(8.1.48)$$

where $\mathfrak{Q}$ is an operator of $\mathfrak{E}, f^{(1)}, \cdots, f^{(n)}$.

It follows from the first equation (8.1.45), that $f^{(0)}$ has to be a Bose–Einstein distribution $f^{(0)} = \mathfrak{E}$. Let us now rewrite (8.1.46) in the following equivalent form

$$\mathcal{L}_{22}[f^{(1)}] = \frac{\partial}{\partial t}\mathfrak{E} + p \cdot \nabla_r \mathfrak{E}. \qquad (8.1.49)$$

Since

$$\mathbb{P}\left[\mathcal{L}_{22}[f^{(1)}]\right] = 0,$$

we find that

$$\mathbb{P}\left[\frac{\partial}{\partial t}\mathfrak{E} + p \cdot \nabla_r \mathfrak{E}\right] = 0. \qquad (8.1.50)$$

We recall that $\mathbb{P}$ and $\mathbb{P}^{\perp} = 1 - \mathbb{P}$ are the orthogonal projection operators onto $\mathcal{N}_{22}$ and $\mathcal{R}_{22}$ in $L^2(\mathbb{R}^3)$. Equation (8.1.50) then leads to the following Euler limit

$$\frac{\partial}{\partial t}n_n + \nabla_r \cdot (n_n v_n) = 0,$$

$$n_n\frac{\partial}{\partial t}v_{nj} + n_n v_n \cdot \nabla v_{nj} + \frac{2}{3}\frac{\partial}{\partial r_j}(e_n n_n) = 0, \qquad (8.1.51)$$

$$\frac{\partial}{\partial t}e_n + v_n \cdot \nabla_r e_n + \frac{2}{3}e_n\nabla_r \cdot v_n = 0.$$

8.1.1.5 The Navier–Stokes Approximation

Now, we will discuss the derivation of the Navier–Stokes approximation, using the Chapman–Enskog expansion. Arguing similarly as in the previous section, we also have the expansion

$$f = \sum_{i=0}^{n}\epsilon^{i}f^{(i)} + \epsilon^{l}\varsigma, \qquad (8.1.52)$$

in which n and l are positive integers.

Again, $f^{(0)}$ is also a Bose–Einstein distribution $f^{(0)} = \mathfrak{E}$. We now decompose $f^{(i)}$ into two parts $f^{(i)} = h^{(i)} + k^{(i)}$, in which $h^{(i)} \in \mathcal{R}_{22}$, $k^{(i)} \in \mathcal{N}_{22}$.

It follows from (8.1.46) that

$$h^{(1)} = \mathcal{L}_{22}^{-1} \left[\frac{\partial}{\partial t} \mathfrak{E} + p \cdot \nabla_r \mathfrak{E} \right]. \tag{8.1.53}$$

By the same techniques used in [17], we now split $h^{(1)}$ into the sum of h' and h'', where h' and h'' satisfy the following system of equations

$$\mathcal{L}_{22} h' = \mathbb{P}^\perp \left[\frac{\partial}{\partial t} \mathfrak{E} + p \cdot \nabla_r \mathfrak{E} \right], \tag{8.1.54}$$

$$\mathbb{P} \left[\frac{\partial}{\partial t} \mathfrak{E} + p \cdot \nabla_r \mathfrak{E} \right] = -\epsilon \mathbb{P} \left[\frac{\partial}{\partial t} h' + p \cdot \nabla_r h' \right], \tag{8.1.55}$$

$$\mathcal{L}_{22} h'' = \epsilon \mathbb{P}^\perp \left[\frac{\partial}{\partial t} k^{(1)} + p \cdot \nabla_r k^{(1)} \right], \tag{8.1.56}$$

$$\mathbb{P} \left[\frac{\partial}{\partial t} k^{(1)} + p \cdot \nabla_r k^{(1)} \right] = -\mathbb{P} \left[\frac{\partial}{\partial t} h'' + p \cdot \nabla_r h'' \right], \tag{8.1.57}$$

and

$$\mathcal{L}_{22} h^{(i)} = \epsilon \mathbb{P}^\perp \left[\frac{\partial}{\partial t} k^{(i)} + p \cdot \nabla_r k^{(i)} \right] + \mathbb{P}^\perp \left[\frac{\partial}{\partial t} h^{(i-1)} + p \cdot \nabla_r h^{(i-1)} \right]$$
$$- \sum_{j=1}^{i-1} B_1(f^{(j)}, f^{(i-j)}) - \sum_{j,k=0, 0<j+k<i}^{i-1} B_2(f^{(j)}, f^{(k)}, f^{(i-j-k)}), \tag{8.1.58}$$

$$\mathbb{P} \left[\frac{\partial}{\partial t} k^{(i)} + p \cdot \nabla_r k^{(i)} \right] = -\mathbb{P} \left[\frac{\partial}{\partial t} h^{(i)} + p \cdot \nabla_r h^{(i)} \right]. \tag{8.1.59}$$

The above system (8.1.54)–(8.1.57) leads to

$$h' = \mathcal{L}_{22}^{-1} \left[\mathbb{P}^\perp \left[\frac{\partial}{\partial t} \mathfrak{E} + p \cdot \nabla_r \mathfrak{E} \right] \right],$$
$$h'' = \mathcal{L}_{22}^{-1} \left[\epsilon \mathbb{P}^\perp \left[\frac{\partial}{\partial t} k^{(1)} + p \cdot \nabla_r k^{(1)} \right] \right] \tag{8.1.60}$$

and

$$h^{(i)} = \mathcal{L}_{22}^{-1} \left[\epsilon \mathbb{P}^\perp \left[\frac{\partial}{\partial t} k^{(i)} + p \cdot \nabla_r k^{(i)} \right] + \mathbb{P}^\perp \left[\frac{\partial}{\partial t} h^{(i-1)} + p \cdot \nabla_r h^{(i-1)} \right] \right.$$
$$\left. - \sum_{j=1}^{i-1} B_1(f^{(j)}, f^{(i-j)}) - \sum_{j,k=0; 0<j+k<i}^{i-1} B_2(f^{(j)}, f^{(k)}, f^{(i-j-k)}) \right], \tag{8.1.61}$$

for $i = 2, 3, \cdots$

Using the same argument that leads to (8.1.50), we deduce from Equation (8.1.60) that

$$\mathbb{P}\left[\frac{\partial}{\partial t}\mathfrak{E} + p \cdot \nabla_r \mathfrak{E}\right] = -\epsilon \mathbb{P}\left[\frac{\partial}{\partial t}h' + p \cdot \nabla_r h'\right]$$

$$= -\epsilon \mathbb{P}\left[\frac{\partial}{\partial t}\mathcal{L}_{22}^{-1}\mathbb{P}^\perp\left[\frac{\partial}{\partial t}\mathfrak{E} + p \cdot \nabla_r \mathfrak{E}\right]\right. \tag{8.1.62}$$

$$\left. + p \cdot \nabla_r \mathcal{L}_{22}^{-1}\mathbb{P}^\perp\left[\frac{\partial}{\partial t}\mathfrak{E}\right]\right].$$

Equation (8.1.62) leads to the Navier–Stokes approximation. First, observe that

$$\mathbb{P}^\perp \Pi \mathfrak{E} = \frac{\mathfrak{E}^2}{\mathcal{G}}\sum_{i,j=1}^{3}\left\{(p^i - v_{ni})(p^j - v_{nj}) - \frac{1}{3}|p - v_n|^2\delta_{i,j}\right\}\frac{1}{\tau}\frac{\partial v_{nj}}{\partial x_i}$$

$$+ \frac{\mathfrak{E}^2}{\mathcal{G}}\left\{|p - v_n|^2 - \frac{10\tau\Omega_2(\gamma)}{3\Omega_1(\gamma)}\right\}\sum_{i=1}^{3}(p^i - v_{ni})\frac{1}{2\tau^2}\frac{\partial \tau}{\partial r_i}. \tag{8.1.63}$$

Classical techniques for the classical Boltzmann collision operator can be applied (cf. [158, pp. 456–457]), to get

$$-\mathbb{P}\Pi\mathcal{L}_{22}^{-1}\mathbb{P}^\perp\Pi\mathfrak{E} = \frac{\mathfrak{E}^2}{\mathcal{G}}\left(\sum_{k=1}^{3}\frac{\Theta_k}{\Xi_k}\left(\sum_{i=1}^{3}\frac{\partial}{\partial r_i}\left(\sigma(\gamma,\tau)\left(\frac{\partial v_{nk}}{\partial r_i} + \frac{\partial v_{ni}}{\partial r_k}\right)\right)\right)\right.$$

$$- \frac{2}{3}\frac{\partial}{\partial r_k}\left(\sigma(\gamma,\tau)\sum_{i=1}^{3}\frac{\partial v_{ni}}{\partial r_i}\right)\right) + 2\frac{\Theta_4}{\Xi_4}\left(\sum_{i=1}^{3}\frac{\partial}{\partial r_i}\left(\zeta(\gamma,\tau)\frac{\partial \tau}{\partial r_i}\right)\right. \tag{8.1.64}$$

$$\left.- \frac{2}{3}\zeta(\gamma,\tau)\left(\sum_{i=1}^{3}\frac{\partial v_{ni}}{\partial r_i}\right)^2 + \sigma(\gamma,\tau)\sum_{i,k=1}^{3}\frac{\partial v_{nk}}{\partial r_i}\left(\frac{\partial v_{nk}}{\partial r_i} + \frac{\partial v_{ni}}{\partial r_k}\right)\right)\right),$$

where $\sigma(\gamma,\tau) = -\frac{1}{\tau}\int_{\mathbb{R}^3}\xi_1\xi_2\mathfrak{C}_{12}(\xi)d\xi$, $\zeta(\gamma,\tau) = -\frac{1}{4\tau^2}\int_{\mathbb{R}^3}|\xi|^2\xi_1\mathfrak{C}_1(\xi)d\xi$, with ξ_1, ξ_2 being the components of the vectors $\xi = (\xi_1,\xi_2,\xi_3)$ and $\mathfrak{C}_1$, $\mathfrak{C}_{12}$ being as defined in (8.1.35).

From the above computations, the Navier–Stokes approximation then follows

$$\partial_t n_n + \nabla_r \cdot (n_n v_n) = 0,$$

$$n_n \partial_t \nabla v_{nj} + n_n v_n \cdot \nabla v_{nj} + \frac{2}{3}(e_n n_n) = \epsilon \left[\sum_{i=1}^{3} \frac{\partial}{\partial r_i} \left(\sigma \left(\frac{\partial v_{nj}}{\partial r_i} + \frac{\partial v_{ni}}{\partial r_j} \right) \right) \right. $$
$$\left. - \frac{2}{3} \frac{\partial}{\partial r_j} \left(\sigma \sum_{i=1}^{3} \frac{\partial v_{ni}}{\partial r_i} \right) \right], \quad j = 1, 2, 3,$$

$$\partial_t e_n + v_n \cdot \nabla_r e_n + \frac{2}{3} e_n \nabla_r \cdot v_n = \frac{\epsilon}{\mathfrak{T}} \left[\sum_{i=1}^{3} \frac{\partial}{\partial r_i} \left(\sigma^{(1)} \frac{\partial e_n}{r_i} + \sigma^{(2)} \frac{\partial n_n}{r_i} \right) \right.$$
$$+ \sigma \sum_{i,j=1}^{3} \frac{\partial v_{nj}}{\partial r_i} \left(\frac{\partial v_{nj}}{\partial r_i} + \frac{\partial v_{ni}}{\partial r_j} \right)$$
$$\left. - \frac{2}{3} \left(\sum_{i=1}^{3} \frac{\partial v_{ni}}{r_i} \right)^2 \right],$$

$$(8.1.65)$$

where $\mathfrak{T} = 2^{\frac{5}{2}} \pi \tau^{\frac{3}{2}} \gamma \Omega_1$, $\sigma^{(1)} = \zeta \frac{\partial}{\partial e_n} \tau$, and $\sigma^{(2)} = \zeta \frac{\partial}{\partial n_n} \tau$.

8.1.2 The Coupling Condensate-Thermal Cloud System

8.1.2.1 Properties of C_{12}

The collision operator C_{12} takes $\Upsilon_i(p) - v_{ci}$, $i = 1, 2, 3$, as collisional invariants

$$\int_{\mathbb{R}^3} (\Upsilon_i(p) - v_{ci}) C_{12}[f] \mathrm{d}p = \qquad (8.1.66)$$

$$\int_{\mathbb{R}^3} \left(\Upsilon_4(p) + 2U - 2\mu_c - v_c^2 \right) C_{12}[f] \mathrm{d}p = 0, \quad i = 1, 2, 3.$$

Notice that $\mathfrak{E}$ is an equilibrium of C_{22}, but it is not an equilibrium of C_{12}.

In other words, C_{12} and C_{22} do not share the same equilibrium

$$\Gamma_{12}[\mathfrak{E}] := \int_{\mathbb{R}^3} C_{12}[\mathfrak{E}] \mathrm{d}p \qquad (8.1.67)$$

$$= -n_c [1 - e^{-\beta(\mu - \mu_c - (v_n - v_c)^2/2)}] \iiint_{\mathbb{R}^3 \times \mathbb{R}^3 \times \mathbb{R}^3} \delta(v_c + p_1 - p_2 - p_3)$$
$$\times \delta(\omega_c + \omega_{p_1} - \omega_{p_2} - \omega_{p_3})(1 + \mathfrak{E}(t, r, p_1)) \mathfrak{E}(t, r, p_2) \mathfrak{E}(t, r, p_3) \mathrm{d}p_1 \mathrm{d}p_2 \mathrm{d}p_3.$$

Expanding $\mathfrak{E}$ into a Taylor series of $\mathcal{G}$, the above integral can be simplified as

$$\Gamma_{12}[\mathfrak{E}] \;=\; -n_c[1 - e^{-\beta(\mu-\mu_c-(u-v_c)^2/2)}] \tag{8.1.68}$$

$$\times \sum_{k_2,k_3\in\mathbb{N}\cup\{0\},k_1\in\mathbb{N}} \gamma^3 e^{-\frac{|v_c-u|^2(k_1+k_2+k_3)}{2\tau}} e^{\frac{(-2\omega_c+2U+v_c^2)k_1}{2\tau}}$$

$$\times \int_{x\cdot y=\frac{v_c^2}{2}+U-\omega_c} e^{-(k_1+k_2)[|x|^2+x\cdot(v_c-u)]-(k_1+k_3)[|y|^2+y\cdot(v_c-u)]/(2\tau)}\,\mathrm{d}x\mathrm{d}y,$$

where, from (8.1.41) and (8.1.42), γ and τ are functions of n_n and $\mathbb{E}_n$.

8.1.2.2 The Euler Limit

Now, we derive the Euler limit of the rescaled system (8.1.18). Since ρ is just a parameter, we set $\rho = 1$. Let ϵ be any parameter. By exactly the same process used to obtain (8.1.50), we also get

$$\mathbb{P}\mathcal{D}\mathfrak{E} \;=\; 0, \tag{8.1.69}$$

which leads to the desired Euler system

$$\frac{\partial}{\partial t}n_c + \nabla_r\cdot(n_c v_c) = -\,\Gamma_{12}[\mathfrak{E}],$$

$$\frac{\partial}{\partial t}v_c + \frac{\nabla_r v_c^2}{2} = -\nabla_r\mu_c,$$

$$\frac{\partial}{\partial t}n_n + \nabla_r\cdot(n_n v_n) = \Gamma_{12}[\mathfrak{E}], \tag{8.1.70}$$

$$n_n\left(\frac{\partial}{\partial t}+v_n\cdot\nabla\right)v_{nj} = -\nabla_r\tilde{\mathbb{E}}_n - n_n\nabla_r U - (v_{nj}-v_{cj})\Gamma_{12}[\mathfrak{E}],$$

$$\frac{\partial}{\partial t}\tilde{\mathbb{E}}_n + \nabla_r\cdot(\tilde{\mathbb{E}}_n v_n) = -\frac{2}{3}\tilde{\mathbb{E}}_n\nabla_r\cdot v_n + \frac{2}{3}\left(\frac{(v_n-v_c)^2}{2}+\mu_c-U\right)\Gamma_{12}[\mathfrak{E}].$$

8.1.2.3 The Two-Fluid Navier–Stokes Quantum Hydrodynamic Approximation

This section is devoted to the derivation of the Navier–Stokes approximation of the system (8.1.18) using the Chapman–Enskog expansion. The approximation is valid under the physical assumption that the collisions between excited atoms are fast enough to establish a local equilibrium within the non-condensate component and $\tilde{\tau}_{22} << \tilde{\tau}_{12}$. We suppose $\frac{\tilde{\tau}_{22}}{\tilde{\tau}_{12}} = \epsilon^2$ and ϵ will be used as the small parameter in the usual Chapman–Enskog expansion process. Adopting the procedure used to derive the Navier–Stokes approximation for the Boltzmann–Nordheim equation, we write

$$f = \sum_{i=0}^{n} \epsilon^i f^{(i)} + \epsilon^l \varsigma, \qquad (8.1.71)$$

in which n and l are positive integers.

It then follows that $f^{(0)}$ is also a Bose–Einstein distribution $f^{(0)} = \mathfrak{E}$. We also decompose $f^{(i)}$ into two parts $f^{(i)} = h^{(i)} + k^{(i)}$, where $h^{(i)} \in \mathcal{R}_{22}$, $k^{(i)} \in \mathcal{N}_{22}$. The quantity $h^{(1)}$ is also decomposed into the sum of h' and h'', where h' and h'' satisfy the following system of equations

$$\mathcal{L}h' = \mathbb{P}^{\perp}\mathcal{D}\mathfrak{E}, \qquad (8.1.72)$$

$$\mathbb{P}\mathcal{D}\mathfrak{E} = -\epsilon\mathbb{P}\mathcal{D}h', \qquad (8.1.73)$$

$$\mathcal{L}h'' = \epsilon\mathbb{P}^{\perp}\mathcal{D}k^{(1)}, \qquad (8.1.74)$$

$$\mathbb{P}\mathcal{D}k^{(1)} = -\mathbb{P}\mathcal{D}h'', \qquad (8.1.75)$$

and

$$\mathcal{L}h^{(i)} = \epsilon\mathbb{P}^{\perp}\mathcal{D}k^{(i)} + \mathbb{P}^{\perp}\mathcal{D}h^{(i-1)}$$

$$- \sum_{j=1}^{i-1} Q_1(f^{(j)}, f^{(i-j)}) - \sum_{j,k=0, 0<j+k<i}^{i-1} Q_2(f^{(j)}, f^{(k)}, f^{(i-j-k)}),$$

$$(8.1.76)$$

with $\mathbb{P}\mathcal{D}k^{(i)} = -\mathbb{P}\mathcal{D}h^{(i)}$.

By the same procedure used for the Boltzmann–Nordheim equation, it also follows that

$$\mathbb{P}\mathcal{D}\mathfrak{E} = -\epsilon\mathbb{P}\mathcal{D}h' = -\epsilon\mathbb{P}\mathcal{D}\mathcal{L}^{-1}\mathbb{P}^{\perp}\mathcal{D}\mathfrak{E}. \qquad (8.1.77)$$

Equation (8.1.77) indeed leads to the Navier–Stokes approximation. First, observe that

$$\mathbb{P}^{\perp}\Pi\mathfrak{E} = \frac{\mathfrak{E}^2}{\mathcal{M}} \sum_{i,j=1}^{3} \left\{ (p^i - v_{ni})(p^j - v_{nj}) - \frac{1}{3}|p - v_n|^2 \delta_{i,j} \right\} \frac{1}{\tau} \frac{\partial v_{nj}}{\partial x_i}$$

$$+ \frac{\mathfrak{E}^2}{\mathcal{M}} \left\{ |p - v_n|^2 - \frac{10\tau\Omega_2(\gamma)}{3\Omega_1(\gamma)} \right\} \sum_{i=1}^{3} (p^i - v_{ni}) \frac{1}{2\tau^2} \frac{\partial \tau}{\partial r_i}.$$

$$(8.1.78)$$

Similar to the Boltzmann–Nordheim case, we also have

$$
- \mathbb{P}\Pi\mathcal{L}^{-1}\mathbb{P}^{\perp}\Pi\mathfrak{E} = \frac{\mathcal{F}^2}{\mathcal{M}} \left(\sum_{k=1}^{3} \frac{\psi_k}{\omega_k} \left(\sum_{i=1}^{3} \frac{\partial}{\partial r_i} \left(\varpi(\gamma,\tau) \left(\frac{\partial v_{nk}}{\partial r_i} + \frac{\partial v_{ni}}{\partial r_k} \right) \right) \right) \right.
$$

$$
- \frac{2}{3} \frac{\partial}{\partial r_k} \left(\varpi(\gamma,\tau) \sum_{i=1}^{3} \frac{\partial v_{ni}}{\partial r_i} \right) \right) + 2\frac{\psi_4}{\omega_4} \left(\sum_{i=1}^{3} \frac{\partial}{\partial r_i} \left(\varrho(\gamma,\tau) \frac{\partial \tau}{\partial r_i} \right) \right) \qquad (8.1.79)
$$

$$
- \frac{2}{3} \varrho(\gamma,\tau) \left(\sum_{i=1}^{3} \frac{\partial v_{ni}}{\partial r_i} \right)^2 + \varpi(\gamma,\tau) \sum_{i,k=1}^{3} \frac{\partial v_{nk}}{\partial r_i} \left(\frac{\partial v_{nk}}{\partial r_i} + \frac{\partial v_{ni}}{\partial r_k} \right) \right),
$$

where

$$
\varpi(\gamma,\tau) = -\frac{1}{\tau} \int_{\mathbb{R}^3} \xi_1 \xi_2 \mathfrak{C}_{12}(\xi) d\xi, \qquad (8.1.80)
$$

$$
\varrho(\gamma,\tau) = -\frac{1}{4\tau^2} \int_{\mathbb{R}^3} |\xi|^2 \xi_1 \mathfrak{C}_1(\xi) d\xi. \qquad (8.1.81)
$$

Here ξ_1, ξ_2 are the components of the vectors $\xi = (\xi_1, \xi_2, \xi_3)$ and $\mathfrak{C}_1$, $\mathfrak{C}_{12}$ are as defined in (8.1.35).

Now, since g is also a small parameter, we can assume that $g = \epsilon^{\delta_0}$, $(0 < \delta_0 < 1)$. As

$$
\mathcal{D}\mathfrak{E} = \Pi\mathfrak{E} + O(\epsilon^{\delta_0}),
$$

the first-order approximation in terms of ϵ of the quantity $\epsilon\mathbb{P}\mathcal{D}\mathcal{L}^{-1}\mathbb{P}^{\perp}\mathcal{D}\mathfrak{E}$ is then $\epsilon\mathbb{P}\Pi\mathcal{L}^{-1}\mathbb{P}^{\perp}\Pi\mathfrak{E}$.

The Navier–Stokes system (8.1.62) becomes

$$
\mathbb{P}\mathcal{D}\mathfrak{E} = -\epsilon\mathbb{P}\Pi\mathcal{L}^{-1}\mathbb{P}^{\perp}\Pi\mathfrak{E}, \qquad (8.1.82)
$$

which, thanks to the identity (8.1.64), leads to

$$\partial_t n_n \;+\; \nabla_r \cdot (n_n v_n) \;=\; \epsilon \Gamma_{12}[\mathfrak{E}],$$

$$n_n \left(\partial_t + v_n \cdot \nabla\right) v_{nj} \;+\; \partial_{r_j}(n_n e_n) \;=\; -n_n \nabla_r \epsilon^{\delta_0} U \;-\; (v_{nj} - v_{cj}) \epsilon \Gamma_{12}[\mathfrak{E}]$$

$$+ \;\epsilon \left[\sum_{i=1}^{3} \frac{\partial}{\partial r_i} \left(\bar{\varpi}(n_n, e_n) \left(\frac{\partial v_{nj}}{r_i} + \frac{\partial v_{ni}}{r_j} \right) \right) \right.$$

$$\left. - \frac{2}{3} \frac{\partial}{\partial r_j} \left(\bar{\varpi}(n_n, e_n) \sum_{i=1}^{3} \frac{\partial v_{ni}}{r_i} \right) \right],$$

$$\partial_t e_n \;+\; \nabla_r \cdot (e_n v_n) \;+\; \frac{2}{3} e_n \nabla_r \cdot v_n =$$

$$= \frac{1}{n_n} \left[\frac{2}{3} \left(\frac{(v_n - v_c)^2}{2} + \mu_c - \epsilon^{\delta_0} U + e_n \right) \epsilon \Gamma_{12}[\mathfrak{E}] \right]$$

$$+ \frac{\epsilon}{\mathcal{G}(n_n, e_n)} \left[\sum_{i=1}^{3} \frac{\partial}{\partial r_i} \left(\varrho_1(n_n, e_n) \frac{\partial e_n}{\partial r_i} + \varrho_2(n_n, e_n) \frac{\partial n_n}{\partial r_i} \right) \right.$$

$$+ \bar{\varpi}(n_n, e_n) \sum_{i,k=1}^{3} \frac{\partial v_{nk}}{\partial x_i} \left(\frac{\partial v_{nk}}{\partial x_i} + \frac{\partial v_{ni}}{\partial x_k} \right)$$

$$\left. - \frac{2}{3} \varpi(n_n, e_n) \left(\sum_{i=1}^{3} \frac{\partial v_{ni}}{\partial x_i} \right)^2 \right],$$

$$(8.1.83)$$

where

$$\bar{\varpi}(n_n, e_n) \;=\; \varpi(\gamma, \tau),$$

$$\mathcal{G}(n_n, e_n) \;=\; 2^{5/2} \pi \tau^{3/2} \gamma \Omega_1(\gamma),$$

$$\varrho_1(n_n, e_n) \;=\; \varrho(\gamma, \tau) \frac{\partial \tau}{\partial e_n},$$

$$(8.1.84)$$

$$\varrho_2(n_n, e_n) \;=\; \varrho(\gamma, \tau) \frac{\partial \tau}{\partial n_n}.$$

Combining (8.1.83) and (8.1.24), which is

$$\partial_t n_c \;+\; \nabla_r \cdot (n_c v_c) \;=\; - \;\Gamma_{12}[f],$$

$$\partial_t v_c + \frac{\nabla_r v_c^2}{2} \;=\; - \nabla_r \mu_c,$$

we get the "closed system" of the Navier–Stokes approximation.

The Navier–Stokes system of the excitations is very different from the Navier–Stokes system obtained from the classical Boltzmann equation (cf. [31, 158]) in several respects:

- First, in the classical Navier–Stokes system, the viscosity coefficient $\bar{\varpi}$ and the heat conduction coefficient ϱ_1 depend only on e_n. In the above quantum Boltzmann system, they depend on both e_n and n_n.
- Second, different from the classical Navier–Stokes system, the second derivatives of n_n also appear in the system.

- Third, the Navier–Stokes system of the excitations is coupled with the system of the BEC super fluid via the quantity $\epsilon\Gamma_{12}[\mathfrak{E}]$, computed in (8.1.68).

As a consequence, the Navier–Stokes system for the excitations has a completely different nature, in comparison with the classical one. Hence, more complicated behaviors would be expected.

8.2 Microscopic Hydrodynamic Modes – The Ultra Low Temperature Regime

As discussed earlier in Section 6.3, the Zaremba et al./Pomeau et al. model has been very successfully applied to several Bose–Einstein Condensate (BEC) dynamical problems. The model gives a clear description of what physical processes are taking place in a partially condensed Bose gas, and also gives a precise prediction of the superfluid properties of the system. However, being based on a picture of an approximate particlelike Hartree–Fock excitation spectrum, the model can only give a precise description of the systems in high temperature ranges $T_{BE} > T \geq 0.5T_{BE}$ [90]. Predictions for transport properties at ultra low temperature, for instance, the hydrodynamic modes at $T = 0.01T_{BE}$, become inaccurate. The Reichl et al. model (cf. [78, 79, 80, 142, 143]) is an attempt to fix this inaccuracy. The new model can be used to correctly describe the collective phonon-like excitations, which become important at ultra low temperatures [3]. In this modified model, the important change is that a new collision operator C_{31}, taking into account 1↔3 type collisions between the bogolons, is added, besides the standard 1↔2 and 2↔2 types of collisions.

This section is devoted to the computations of hydrodynamic modes of the Reichl et al. model. The relevant physical experiments will also be discussed.

To obtain the dispersion relation for the hydrodynamic modes of the BEC, we consider one frequency component of the linearized kinetic equations by writing

$$f(t, q, p) \sim e^{i\omega t} \, \tilde{f}(\omega, q, p)$$
$$\text{and} \quad \phi(t, q) \sim e^{i\omega t} \, \tilde{\varphi}(\omega, q). \tag{8.2.1}$$

Equation (5.0.20) can then be rewritten as

$$\omega\tilde{f}(\omega, q, p) = \frac{\hbar}{m}p{\cdot}q\frac{\epsilon(p) + \Lambda_0}{E_p} \, \tilde{f}(\omega, q, p) - i\frac{\hbar}{m} \, q^2\tilde{\varphi}(\omega, q)\mathcal{N}_p^{\text{eq}} + i\mathbf{G}[\tilde{f}](\omega, q, p), \tag{8.2.2}$$

where we have used the fact that

$$v_c(\omega, q) = -i\frac{\hbar}{m}q\tilde{\varphi}(\omega, q).$$

Equation (5.0.22) also gives

$$\omega^2 \tilde{\varphi}(\omega, q) = i\frac{g}{m} \frac{1}{(2\pi)^3} \int_{\mathbb{R}^3} q \cdot k \; \tilde{f}(\omega, q, p) \mathrm{d}p + q^2 \frac{g}{m} \tilde{\varphi}(\omega, q) n^{eq}. \quad (8.2.3)$$

Equations (5.0.20)–(5.0.21) and (5.0.22)–(5.0.23) are the bogolon kinetic equations describing the hydrodynamic behavior of a dilute BEC. Combining (8.2.2) and (8.2.3) gives

$$\omega \; \tilde{f}(\omega, p, q) - \frac{q^2}{\omega^2 - v_B^2 q^2} \frac{g\hbar}{m^2} \mathcal{N}_p^{eq} \frac{1}{(2\pi)^3} \int_{\mathbb{R}^3} q \cdot p_1 \; \tilde{f}(\omega, p_1, q) \mathrm{d}p_1$$

$$= q \cdot p \; \frac{\hbar}{m} \frac{(\epsilon_k + \Delta)}{E_k} \tilde{f}(\omega, p, q,) + i\mathbf{G}[\tilde{f}](\omega, q, p),$$

where

$$v_B = \sqrt{\frac{gn^{eq}}{m}}.$$

Let us set

$$\tilde{f}(\omega, q, p) = \mathcal{N}_p^{eq} + \mathcal{N}_p^{eq}(1 + \mathcal{N}_p^{eq})h(\omega, q, p), \quad (8.2.4)$$

then the operator $\mathbf{G}[\mathbf{f}]$ can be replaced by the linearized operator $\mathcal{G}[h]$. In the hydrodynamic regime where spatial variations have very long wavelengths, the wave vector $|q|$ is very small. We formulate below the form of the collision operator $\mathcal{G}[h]$

$$\mathcal{G}[h](p_1) = \qquad\qquad\qquad\qquad\qquad\qquad\qquad\qquad\qquad (8.2.5)$$

$$- \mathcal{N}^{eq}(p_1)(1 + \mathcal{N}^{eq}(p_1)) \left(M(p_1)h(p_1) + \int_{\mathbb{R}^3} \frac{\mathcal{N}^{eq}(p_2)}{1 + \mathcal{N}^{eq}(p_1)} K(p_1, p_2)h(p_2)\mathrm{d}p_2 \right)$$

with

$$M(p_1) = \int_{\mathbb{R}^3} \frac{\mathcal{N}^{eq}(p_2)}{\mathcal{N}^{eq}(p_1) + 1} \left\{ 2A_0 T_A(p_1, p_2) + A_0 \frac{1 + \mathcal{N}^{eq}(p_3)}{\mathcal{N}^{eq}(p_2)} T_B(p_1, p_2) \right.$$

$$\left. + B_0 Q_A(p_1, p_2) + B_0 Q_B(p_1, p_2) + \frac{1}{3}\frac{1 + \mathcal{N}^{eq}(p_2)}{\mathcal{N}^{eq}(p_2)} Q_C(p_1, p_2) \right\} \mathrm{d}p_3,$$

and

$$K(p_1, p_2) = \left\{ 2A_0 T_A(p_1, p_2) - 2A_0 \frac{1 + \mathcal{N}^{eq}(p_2)}{\mathcal{N}^{eq}(p_2)} T_B(p_1, p_2) \right.$$

$$- 2A_0 \frac{1 + \mathcal{N}^{eq}(p_1)}{\mathcal{N}^{eq}(p_1)} T_B(p_2, p_1) + B_0 Q_A(p_1, p_2)$$

$$- 2B_0 \frac{1 + \mathcal{N}^{eq}(p_2)}{\mathcal{N}^{eq}(p_2)} R_A(p_1, p_2) + 2B_0 Q_B(p_1, p_2)$$

$$\left. - B_0 \frac{1 + \mathcal{N}^{eq}(p_2)}{\mathcal{N}^{eq}(p_2)} Q_C(p_1, p_2) - B_0 \frac{1 + \mathcal{N}^{eq}(p_1)}{\mathcal{N}^{eq}(p_1)} Q_C(p_2, p_1) \right\}.$$

The functions appearing in the above expressions are defined as

$$A_0 = \frac{4\pi N_c g^2}{(2\pi)^3 \hbar V}, \qquad B_0 = \frac{4\pi g^2}{(2\pi)^6 \hbar}, \qquad (8.2.6)$$

and

$$T_A(p_1, p_2) = \int_{\mathbb{R}^3} \delta(1+2-3)(W_{1,2,3}^{12})^2(\mathcal{N}^{\mathrm{eq}}(p_3)+1)\mathrm{d}p_3,$$

$$T_B(p_1, p_2) = \int_{\mathbb{R}^3} \delta(1-2-3)(W_{3,2,1}^{12})^2(\mathcal{N}^{\mathrm{eq}}(p_3)+1)\mathrm{d}p_3,$$

$$T_B(p_2, p_1) = \int_{\mathbb{R}^3} \delta(2-1-3)(W_{3,1,2}^{12})^2(\mathcal{N}^{\mathrm{eq}}(p_3)+1)\mathrm{d}p_3,$$

$$Q_A(p_1, p_2) = \int_{\mathbb{R}^3} \delta(1+2-3-4)(W_{1,2,3,4}^{22})^2(\mathcal{N}^{\mathrm{eq}}(p_3)+1)$$
$$\times\,(\mathcal{N}^{\mathrm{eq}}(p_4)+1)\mathrm{d}p_3\mathrm{d}p_4,$$

$$R_B(p_1, p_2) = \int_{\mathbb{R}^3} \delta(1+2-3-4)(W_{1,3,2,4}^{22})^2\mathcal{N}^{\mathrm{eq}}(p_3)(\mathcal{N}^{\mathrm{eq}}(p_4)+1)\mathrm{d}p_3\mathrm{d}p_4,$$

$$Q_B(p_1, p_2) = \int_{\mathbb{R}^3} \delta(1+2+3-4)(W_{4,3,2,1}^{31})^2\mathcal{N}^{\mathrm{eq}}(p_3)(\mathcal{N}^{\mathrm{eq}}(p_4)+1)\mathrm{d}p_3\mathrm{d}p_4,$$

$$Q_C(p_1, p_2) = \int_{\mathbb{R}^3} \delta(1-2-3-4)(W_{1,2,3,4}^{31})^2(\mathcal{N}^{\mathrm{eq}}(p_3)+1)$$
$$\times\,(\mathcal{N}^{\mathrm{eq}}(p_4)+1)\mathrm{d}p_3\mathrm{d}p_4,$$

$$(8.2.7)$$

where

$$\delta(1+2-3-4)\equiv\delta(p_1+p_2-p_3-p_4)\delta(E(p_1)+E(p_2)-E(p_3)-E(p_4))$$

(with similar definitions for $\delta(1+2-3)$ and $\delta(1+2+3-4)$, etc).

Plugging the linearized operator $\mathcal{G}$ back into (8.2.4), we obtain

$$\omega\,h(\omega,q,p) - \frac{q^2}{\omega^2-v_B^2 q^2}\frac{g\hbar}{m^2}\frac{1}{\mathcal{F}_p^{\mathrm{eq}}}\frac{1}{(2\pi)^3}\int_{\mathbb{R}^3} q{\cdot}p_1\,\mathcal{N}_{p_1}^{\mathrm{eq}}\mathcal{F}_{p_1}^{\mathrm{eq}}h(\omega,q,p_1)\mathrm{d}p_1$$

$$= q{\cdot}p\,\frac{\hbar}{m}\frac{(\epsilon_p+\Delta)}{E_k}h(\omega,q,p)$$

$$+ i\int_0^\infty\int_{\mathbb{S}^2}\sqrt{\frac{p_1^2\mathcal{N}_{p_1}^{\mathrm{eq}}\mathcal{F}_{p_1}^{\mathrm{eq}}}{p^2\mathcal{N}_p^{\mathrm{eq}}\mathcal{F}_p^{\mathrm{eq}}}}\,\mathcal{C}(p,p_1)h(\omega,p_1,q)\mathrm{d}|p_1|\mathrm{d}\Omega_1,$$

$$(8.2.8)$$

where $\mathcal{C}(p,p_1)$ is the bogolon collision operator, which is symmetric and has a complete set of orthonormal eigenfunctions.

The eigenvalues $\lambda_{\beta,\ell}$ and eigenstates $\Sigma_{\beta,\ell,m}(p_1)$, $\beta,\ell,m \in \mathbb{Z}$, of the operator $\mathcal{C}(p_1,p_2)$ follow the equations

$$\int_{\mathbb{R}^3} \mathcal{C}(p_1,p_2)\Sigma_{\beta,\ell,m}^{(0)}(p_2)\mathrm{d}p_2 = \lambda_{\beta,\ell}\Sigma_{\beta,\ell,m}^{(0)}(p_1) \qquad (8.2.9)$$

and

$$\int_{\mathbb{R}^3} \Sigma_{\beta,\ell,m}^{(0)}(p_1)\, \mathcal{C}(p_1,p_2)\mathrm{d}p_1 = \lambda_{\beta,\ell}\Sigma_{\beta,\ell,m}^{(0)}(p_2).\tag{8.2.10}$$

Due to the angular symmetry of the collision operator, $\lambda_{\beta,\ell}$ are independent of m.

In general, all the eigenstates of $\mathcal{C}(p_1,p_2)$ can be written, using spherical harmonic coordinates, in the form

$$\Sigma_{\beta,\ell,m}^0(p) = \Sigma_{\beta,\ell}(|p|)\, Y_\ell^m(\hat{p}),\tag{8.2.11}$$

with $\hat{p} = \frac{p}{|p|}$.

The eigenstates can be orthonormalized such that

$$\int_0^\infty \int_{\mathbb{S}^2} \Sigma_{\beta_1,\ell_1,m_1}^{(0)*}(p_1)\Sigma_{\beta_2,\ell_2,m_2}^{(0)}(p_1)\mathrm{d}|p_1|\mathrm{d}\Omega_1 = \delta_{\beta_1,\beta_2}\delta_{\ell_1,\ell_2}\delta_{m_1,m_2}\tag{8.2.12}$$

and

$$\int_0^\infty \Sigma_{\beta_1,\ell}^*(k_1)\Sigma_{\beta_2,\ell}(|p_1|)\mathrm{d}|p_1| = \delta_{\beta_1,\beta_2}.\tag{8.2.13}$$

The spectral decomposition of the collision operator can now be written

$$\mathcal{C}(p_1,p_2) = \sum_{\beta=0}^\infty\sum_{\ell=0}^\infty\sum_{m=-\ell}^\ell \lambda_{\beta,\ell}\,\Sigma_{\beta,\ell,m}^{(0)}(p_1)\Sigma_{\beta,\ell,m}^{(0)*}(p_2)$$

$$= \sum_{\ell=0}^\infty\sum_{m=-\ell}^\ell \mathcal{C}_\ell(k_1,k_2)\, Y_\ell^m(\hat{p}_1)Y_\ell^{m*}(\hat{p}_2),\tag{8.2.14}$$

where

$$\mathcal{C}_\ell(|p_1|,|p_2|) = \sum_{\beta=0}^\infty \lambda_{\beta,\ell}\,\Sigma_{\beta,\ell}(|p_1|)\Sigma_{\beta,\ell}^*(|p_2|).$$

The bogolon momentum and energy are conserved during collisions, although the bogolon number is not. Therefore, the collision operator $\mathcal{G}_{p_1}\{h\}$, acting on four conserved quantities, $h = E_p$, $h = p_x$, $h = p_y$, and $h = p_z$, gives zero. This can be used to form four eigenstates of $\mathcal{C}(p_1,p_2)$ as follows

$$\Sigma_{0,0,0}^{(0)}(p_1) = \Sigma_{0,0}(|p_1|)Y_0^0(\hat{p}_1),$$
$$\Sigma_{0,1,0}^{(0)}(p_1) = \Sigma_{0,1}(|p_1|)Y_1^0(\hat{p}_1),$$
$$\Sigma_{0,1,1}^{(0)}(p_1) = \Sigma_{0,1}(|p_1|)Y_1^1(\hat{p}_1),\tag{8.2.15}$$
$$\text{and } \Sigma_{0,1,-1}^{(0)}(p_1) = \Sigma_{0,1}(|p_1|)Y_1^{-1}(\hat{p}_1),$$

where $\hat{p}$ denotes $p/|p| \in \mathbb{S}^2$ for all $p \in \mathbb{R}^3$,

$$\Sigma_{0,0}(|p|) = D_{0,0}E_p\sqrt{p^2\mathcal{N}_p^{\mathrm{eq}}\mathcal{F}_p^{\mathrm{eq}}}$$

and

$$\Sigma_{0,1}(|p|) = D_{0,1}|p|\sqrt{p^2 \mathcal{N}_p^{eq} \mathcal{F}_p^{eq}}.$$

The quantities $D_{\beta,\ell}$, are normalized constants written as

$$D_{0,0} = \left(\int_0^\infty p^2 E_p^2 \mathcal{N}_p^{eq} \mathcal{F}_p^{eq} \mathrm{d}|p| \right)^{-1/2} \qquad (8.2.16)$$

$$\text{and} \quad D_{0,1} = \left(\int_0^\infty p^4 \mathcal{N}_p^{eq} \mathcal{F}_p^{eq} \mathrm{d}|p| \right)^{-1/2}.$$

The corresponding eigenvalues $\lambda_{\beta,\ell}$ are independent of m and degenerate. The two constants $\lambda_{0,0} = \lambda_{0,1} = 0$ and the constant $\lambda_{0,1}$ is three-fold degenerate.

Notice that (8.2.8) is an eigenvalue equation with an unusual structure, which can be rewritten in the form

$$h(\omega, q, p) = \frac{1}{\sqrt{p^2 \mathcal{N}_p^{eq} \mathcal{F}_p^{eq}}} \Sigma(\omega, q, p). \qquad (8.2.17)$$

The eigenvalue equation then follows

$$\omega \Sigma(\omega, q, p) - \frac{q^2}{\omega^2 - v_B^2 q^2} \frac{g \hbar |p|}{m^2} \sqrt{\frac{\mathcal{N}_p^{eq}}{\mathcal{F}_p^{eq}}} \frac{1}{(2\pi)^3} \int_{\mathbb{R}^3} \frac{1}{|p_1|} q \cdot p_1 \sqrt{\mathcal{N}_{p_1}^{eq} \mathcal{F}_{p_1}^{eq}} \Sigma(\omega, q, p_1)$$

$$= q \cdot p \frac{\hbar}{m} \frac{(\epsilon_p + \Delta)}{E_p} \Sigma(\omega, q, p) + i \int_0^\infty \int_{\mathbb{S}^2} \mathcal{C}(p, p_1) \Sigma(\omega, q, p) \mathrm{d}p_1 \mathrm{d}\Omega_1.$$

Since the hydrodynamic behavior occurs for long wavelength (small $|q|$) processes, it is sufficient to consider (8.2.18) for small $|q|$. Without loss of generality, we suppose that

$$q = |q| \hat{e}_z,$$

with $\hat{e}_z$ being the unit vector along the z-direction.

Using perturbation theory to solve (8.2.18) to second order in $|q|$, we expand

$$\omega = \omega^{(0)} + |q| \omega^{(1)} + |q|^2 \omega^{(2)} + \dots$$
$$\Sigma(|q|, |p|) = \Sigma^{(0)}(|p|) + |q| \Sigma^{(1)}(|p|) + |q|^2 \Sigma^{(2)}(|p|) + \dots \qquad (8.2.18)$$

Substituting (8.2.20) into (8.2.8), we obtain

$$(\omega^{(0)} + |q|\omega^{(1)} + ...) \left(\Sigma^{(0)}(|p|) + q\Sigma^{(1)}(|p|) + ... \right)$$

$$= |q|p_z \frac{\hbar}{m} \frac{(\epsilon_p + \Delta)}{E_p} \left(\Sigma^{(0)}(|p|) + q\Sigma^{(1)}(|p|) + ... \right)$$

$$+ \frac{q^3}{(\omega^{(0)} + |q|\omega^{(1)} + ...)^2 - v_B^2 |q|^2} \frac{g\hbar|p|}{m^2} \sqrt{\frac{\mathcal{N}_p^{\mathrm{eq}}}{\mathcal{F}_p^{\mathrm{eq}}}}$$

$$\times \frac{1}{(2\pi)^3} \int_0^\infty \int_{\mathbb{S}^2} p_{1,z} \sqrt{p_1^2 \mathcal{N}_{p_1}^{\mathrm{eq}} \mathcal{F}_{p_1}^{\mathrm{eq}}} \left(\Sigma^{(0)}(|p|) + |q|\Sigma^{(1)}(|p|) + ... \right) dp_1 d\Omega_1$$

$$+ i \int_0^\infty \int_{\mathbb{S}^2} \mathcal{C}(p, p_1) \left(\Sigma^{(0)}(|p|) + q\Sigma^{(1)}(|p|) + ... \right) dp_1 d\Omega_1.$$

$$(8.2.19)$$

While trying to obtain the dispersion relation of the hydrodynamic modes as a perturbation expansion in terms of $|q|$, we notice that at zeroth order the modes are four-fold degenerate. At zeroth order in $|q|$, equation (8.2.19) can be reduced to

$$\omega^{(0)} \, \Sigma^{(0)}(p) = +i \int_0^\infty \int_{\mathbb{S}^2} \mathcal{C}(p, p_1) \Sigma^{(0)}(p) dp_1 d\Omega_1, \qquad (8.2.20)$$

where

$$\Sigma^{(0)}(p) \equiv \Sigma(0, p).$$

As a consequence, to zeroth order in $|q|$, $\Sigma^{(0)}(p)$ is an eigenvector of the collision operator with eigenvalue $\omega^{(0)}$. As discussed above, the collision operator has four eigenvalues equal to zero with eigenfunctions that depend on the bogolon energy and momentum, being conserved during the collisions between bogolons.

Different from the classical particle case, the bogolon collision operator would have a fifth zero eigenvalue corresponding to the conservation of the particle number during collisions. The other nonzero eigenvalues of the collision operator are negative. In a classical gas, the five zero eigenvalues are the source of the five hydrodynamic modes. In our case, there are only four zero eigenvalues, but the BEC has six hydrodynamic modes. The remaining two hydrodynamic modes follow from the nonlinear dependence on ω in the eigenvalue equation (8.2.18). This, in turn, comes from the coupling of the bogolon kinetic equation to the macroscopic phase. In a classical monatomic gas, one pair of propagating hydrodynamic modes are sound modes. On the other hand, a BEC has two pairs of propagating modes. The additional pair of hydrodynamic sound modes in a BEC is the result of the coupling of the bogolon kinetic equation to the macroscopic phase. Below, we present the first and second order perturbation theories of the system. Based on these ideas, the hydrodynamic modes could be computed numerically.

8.2.1 First-Order Perturbation Theory

In order to compute the first-order and second-order contributions in $|q|$, it is necessary to find the correct combination of zeroth order eigenstates, since the eigenvalues of the collision operator are four-fold degenerate at zeroth order. We must lift the degeneracy of the zeroth order eigenstates. To first order in $|q|$, (8.2.18) can be expressed as

$$\omega^{(1)} \, \Sigma^{(0)}(p)-$$

$$\frac{1}{(\omega^{(1)})^2 - v_B^2} \frac{gp}{m^2} \sqrt{\frac{\mathcal{N}_p^{\mathrm{eq}}}{\mathcal{F}_p^{\mathrm{eq}}}} \frac{1}{(2\pi)^3} \int_{\mathbb{R}^3} \frac{1}{|p_1|} \, p_{1,z} \, \sqrt{\mathcal{N}_{p_1}^{\mathrm{eq}} \mathcal{F}_{p_1}^{\mathrm{eq}}} \Sigma^{(0)}(p_1) \mathrm{d}p_1$$

$$= k_z \, \frac{\hbar}{m} \frac{(\epsilon_p + \Delta)}{E_p} \Sigma^{(0)}(p) + i \int_0^\infty \int_{\mathbb{S}^2} \mathcal{C}(p, p_1) \Sigma^{(1)}(p) \mathrm{d}|p_1| \mathrm{d}\Omega_1.$$

$$(8.2.21)$$

Let us write

$$\Sigma_R^{(0)}(p) = \sum_{\ell=0}^{1} \sum_{m=-\ell}^{\ell} \Gamma_{0,\ell,m}^R \Sigma_{0,\ell,m}^{(0)}(p) \quad \text{and} \tag{8.2.22}$$

$$\Sigma_L^{(0)}(p) = \sum_{\ell=0}^{1} \sum_{m=-\ell}^{\ell} \Gamma_{0,\ell,m}^L \Sigma_{0,\ell,m}^{(0)}(p), \tag{8.2.23}$$

being left and right states of $\mathcal{C}(p, p_1)$ with eigenvalue zero.

Therefore, when multiplying (8.2.21) on the left by $\Sigma_L^{(0)}(p)$, the contribution from the collision operator drops out, and we find the matrix equation

$$\bar{\Gamma}_L^\dagger \cdot \bar{M} \cdot \bar{\Gamma}_R = 0.$$

In this equation $\bar{\Gamma}_R^\dagger$ is the row matrix

$$\bar{\Gamma}_L^\dagger = \{\Gamma_{0,0,0}^{L*}, \Gamma_{0,1,0}^{L*}, \Gamma_{01,1}^{L*}, \Gamma_{0,1,-1}^{L*}\}$$

and $\bar{\Gamma}_R$ is a column matrix whose transpose is

$$\bar{\Gamma}_R^T = \{\Gamma_{0,0,0}^R, \Gamma_{0,1,0}^R, \Gamma_{01,1}^R, \Gamma_{0,1,-1}^R\}.$$

The matrix $\bar{M}$ takes the form

$$\bar{M} = \begin{pmatrix} -\omega^{(1)} & \alpha + \frac{\gamma}{(\omega^{(1)})^2 - v_B^2} & 0 & 0 \\ \alpha & -\omega^{(1)} & 0 & 0 \\ 0 & 0 & -\omega^{(1)} & 0 \\ 0 & 0 & 0 & -\omega^{(1)} \end{pmatrix} \tag{8.2.24}$$

where

$$\alpha = \int_{\mathbb{R}^3} \Sigma_{0,0,0}^{(0)*}(p) p_z \frac{\hbar}{m} \frac{(\epsilon_p + \Delta)}{E_p} \Sigma_{0,0,1}^{(0)}(p) dp \qquad (8.2.25)$$

and

$$\gamma = \frac{g\hbar}{m^2 (2\pi)^3} \iint_{\mathbb{R}^3 \times \mathbb{R}^3} |p| \Sigma_{0,0,0}^{(0)*}(p) \sqrt{\frac{\mathcal{N}_p^{\mathrm{eq}}}{\mathcal{F}_p^{\mathrm{eq}}}} \frac{p_{1,z}}{|p_1|} \sqrt{\mathcal{N}_{|p_1|}^{\mathrm{eq}} \mathcal{F}_{p_1}^{\mathrm{eq}}} \Sigma_{0,0,1}^{(0)}(p_1) dp dp_1.$$

$$(8.2.26)$$

The values of $\omega^{(1)}$ can be solved explicitly using

$$\mathrm{Det} \begin{pmatrix} -\omega^{(1)} & \alpha + \frac{\gamma}{(\omega^{(1)})^2 - v_B^2} & 0 & 0 \\ \alpha & -\omega^{(1)} & 0 & 0 \\ 0 & 0 & -\omega^{(1)} & 0 \\ 0 & 0 & 0 & -\omega^{(1)} \end{pmatrix} = 0.$$

This gives

$$\omega_2^{(1)} = -\omega_1^{(1)} = \frac{1}{\sqrt{2}} \sqrt{v_B^2 + \alpha^2 - \sqrt{(v_B^2 - \alpha^2)^2 + 4\alpha\gamma}},$$

$$\omega_4^{(1)} = -\omega_3^{(1)} = \frac{1}{\sqrt{2}} \sqrt{v_B^2 + \alpha^2 + \sqrt{(v_B^2 - \alpha^2)^2 + 4\alpha\gamma}},$$

$$\omega_6^{(1)} = \omega_5^{(1)} = 0. \qquad (8.2.27)$$

At first order there are six hydrodynamic frequencies. The frequencies $\omega_2^{(1)} = -\omega_1^{(1)}$ and $\omega_4^{(1)} = -\omega_3^{(1)}$ correspond to fast sound modes. The frequencies $\omega_6^{(1)} = -\omega_5^{(1)} = 0$ correspond to non-propagating transverse viscous modes in the BEC.

The zeroth order left and right eigenstates can be obtained from

$$\bar{\Gamma}_L^\dagger \cdot \bar{M} = 0$$

and

$$\bar{M} \cdot \bar{\Gamma}_R = 0,$$

respectively. For the transverse modes, the left and right eigenstates are complex conjugates of one another and

$$\Sigma_R^{(0)}(p) = \Sigma_{0,1,\pm 1}^{(0)}(p).$$

For the longitudinal modes, the left and right eigenstates can be computed as

$$\Sigma_R^{(0)}(p) = \Gamma_{0,0,0} \Sigma_{0,0,0}^{(0)}(p) + \Gamma_{0,1,0} \Sigma_{0,1,0}^{(0)}(p).$$

The expression for $\Gamma_{0,0,0}$ and $\Gamma_{0,0,0}$ for a given mode can be found by numerically solving

$$\bar{M} \cdot \bar{\Gamma}_R = 0.$$

8.2.2 Second-Order Perturbation Theory

To second order in $|q|$, (8.2.18) can be written as

$$
\omega^{(2)}\ \Sigma^{(0)}(p) + \omega^{(1)}\ \Sigma^{(1)}(p)
$$

$$
-\frac{1}{(\omega^{(1)})^2 - v_B^2}\frac{g\hbar|p|}{m^2}\sqrt{\frac{\mathcal{N}_p^{\mathrm{eq}}}{\mathcal{F}_p^{\mathrm{eq}}}}\frac{1}{(2\pi)^3}\int_{\mathbb{R}^3}\frac{p_{1,z}}{|p_1|}\ \sqrt{\mathcal{N}_{p_1}^{\mathrm{eq}}\mathcal{F}_{p_1}^{\mathrm{eq}}}\Sigma^{(1)}(p_1)\mathrm{d}p_1
$$

$$
+\frac{2\omega^{(1)}\omega^{(2)}}{(\omega^{(1)})^2 - v_B^2)^2}\frac{g\hbar k}{m^2}\sqrt{\frac{\mathcal{N}_p^{\mathrm{eq}}}{\mathcal{F}_p^{\mathrm{eq}}}}\frac{1}{(2\pi)^3}\int_{\mathbb{R}^3}\frac{p_{1,z}}{|p_1|}\ \sqrt{\mathcal{N}_{p_1}^{\mathrm{eq}}\mathcal{F}_{p_1}^{\mathrm{eq}}}\Sigma^{(0)}(p_1)\mathrm{d}p_1
$$

$$
= p_z\frac{\hbar}{m}\frac{(\epsilon_p + \Delta)}{E_p}\Sigma^{(1)}(p) + i\int_0^\infty\int_{\mathbb{S}^2}\mathcal{C}(p,p_1)\Sigma^{(2)}(p_1)\mathrm{d}|p_1|\mathrm{d}\Omega_1. \tag{8.2.28}
$$

The state $\Sigma^{(1)}(0)$ can be derived from (8.2.21). Multiplying on the left by $\Sigma_L^{(0)}(p)$ and integrating, we can eliminate the collision operator and obtain values for $\omega^{(2)}$ for each of the hydrodynamic modes. These quantities are the decay rates of the modes.

In order to write explicit expressions for the decay rates, it is useful to introduce an abstract notation. We express the operator $\mathcal{C}_\ell(p_1, p_2)$ in the "bra-ket" notation as

$$
\hat{\mathcal{C}}_\ell = \sum_{\beta=0}^\infty\lambda_{\beta,\ell}|\Sigma_{\beta,\ell}\rangle\langle\Sigma_{\beta,\ell}|, \tag{8.2.29}
$$

so that

$$
\langle|p_1||\hat{\mathcal{C}}_\ell||p_2|\rangle = \mathcal{C}_\ell(|p_1|, |p_2|)
$$

and

$$
\langle|p||\Sigma_{\beta,\ell}\rangle = \Sigma_{\beta,\ell}(|p|),
$$

where

$$
\langle k_1|k_2\rangle = \delta(k_1 - k_2)
$$

and

$$
\int_0^\infty ||p|\rangle\langle|p||\mathrm{d}|p| = \hat{1},
$$

with $\hat{1}$ being the unit operator.

The decay rate for the viscous modes is now

$$
\omega_6^{(2)} = \frac{1}{5}\sum_{\beta=0}^\infty\frac{1}{\lambda_{\beta,2}}|\langle\Sigma_{\beta,\ell}||p|B_p|\Sigma_{\beta,2}\rangle|^2, \tag{8.2.30}
$$

where

$$
B_p = \frac{\hbar}{m}\frac{\epsilon_p + \Lambda_0}{E_p}. \tag{8.2.31}
$$

The viscosity η of the BEC is related to $\omega_6^{(2)}$ via the equation

$$\eta = \rho_n \omega_6^{(2)},$$

where ρ_n is the density of the normal (non-condensate) part of the BEC.

The decay rates for the sound modes is then

$$\omega^{(2)} = \frac{i}{1+\mathcal{S}} \left\{ \frac{1}{6}C_{0;1} + \frac{1}{6}C_{1;0} + \frac{2}{15}C_{1;2} + C_0' \right\}, \qquad (8.2.32)$$

where

$$C_{\ell',\ell} = \sum_{\beta=0}^{\infty} \frac{1}{\lambda_{\beta,\ell}} |\langle \Sigma_{0,\ell'} || p | B_p | \Sigma_{\beta,\ell} \rangle|^2, \qquad (8.2.33)$$

$$C_0' = \frac{g\hbar}{12\pi^2 m^2 D_{0,1}} \frac{1}{((\omega^{(1)})^2 - v_B^2)}$$

$$\times \sum_{\beta=0}^{\infty} \frac{1}{\lambda_{\beta,0}} \langle \Sigma_{0,\ell'} || p | B_p | \Sigma_{\beta,0} \rangle \int |p| \psi_{\beta,0}^*(p) \sqrt{\frac{\mathcal{N}_p^{\mathrm{eq}}}{\mathcal{F}_p^{\mathrm{eq}}}} \, \mathrm{d}|p| \qquad (8.2.34)$$

and

$$\mathcal{S} = \left[\frac{(\omega^{(1)})^2}{((\omega^{(1)})^2 - v_B^2)^2} \right] \frac{g\hbar}{m^2\alpha} \frac{1}{\sqrt{3}} \frac{1}{2\pi^2} \frac{1}{D_{0,1}} \int_{\mathbb{R}^3} |p| \sqrt{\frac{\mathcal{N}_p^{\mathrm{eq}}}{\mathcal{F}_p^{\mathrm{eq}}}} \psi_{0,0}(p) \mathrm{d}p. \qquad (8.2.35)$$

The lifetime of the sound modes in the BEC is given by $(\omega^{(2)} q^2)^{-1}$, which depends on the speed $\omega^{(1)}$ of the sound mode. The first three terms on the right-hand side of (8.2.32) are current-current correlation functions similar to those that determine the decay of sound modes in classical gases. The factor of $\mathcal{S}$ in the denominator is a consequence of the macroscopic phase resulting from the broken gauge symmetry in the BEC below $T = T_{BE}$.

8.2.3 Discussion

In the Steinhauer experiment [147], a sound mode was excited in an ^{87}Rb BEC, and observed to decay. The wavelength of the sound wave was approximately 18×10^{-6} m ($q = 0.35 \mu\mathrm{m}^{-1}$). The particle density of the BEC was approximately $n^{\mathrm{eq}} = 9.71 \times 10^{19}$ m^{-3}, giving a critical temperature of about $T_{BE} = 3.90 \times 10^{-7}$ K. The Bogoliubov speed in this case is around $v_B \approx 1.887$mm/s, which is approximately the sound speed observed in the experiment [147]. Around $\langle N \rangle = 5 \times 10^5$ atoms in the trap were used giving the critical temperature $T_{BE} = \frac{h}{m} \left(\frac{\langle N \rangle \mathbf{f}_1 \mathbf{f}_2 \mathbf{f}_3}{1.202} \right)^{1/3} \approx 3.9 \times 10^{-7}$. A harmonic trap with frequencies $\mathbf{f}_1 = \mathbf{f}_2 = 224$ Hz and $\mathbf{f}_3 = 26$ Hz was used to create a 1D sound mode. The sound mode had a lifetime $\tau_d \sim 9$ ms and a wavevector $q = 0.35 \mu$m^{-1}. The temperature of the BEC in the experiment was $T = 21 \pm 20$ nK. Using (8.2.32), the decay rate of sound modes in BECs can be computed numerically. It was observed that the value of the sound mode lifetime, predicted by (5.0.20)–(5.0.21), is consistent with that reported in [147].

In addition, we can compute the speed of both the fast mode and the slow mode, which turn out to approach finite values in the limit $T{\to}0$ K. This behavior of the sound speeds is consistent with the findings of Lee and Yang [103] using a very different approach. It is a consequence of the fact that the bogolon spectrum becomes phonon-like at very low temperature. Moreover, the sound speeds undergo an avoided crossing as the temperature is lowered. The temperature at which the avoided crossing occurs increases with increasing density of the gas. For $n^{eq}a^3 = 10^{-5}$ it occurs at $T/T_{BE}{\approx}0.05$. For $n^{eq}a^3 = 10^{-4}$ it occurs at $T/T_{BE}{\approx}0.11$.

8.3 Conclusion

In this chapter, several hydrodynamic aspects of the Zaremba et al./Pomeau et al. and Reichl et al. models have been discussed, including hydrodynamic modes and Euler and Navier–Stokes approximations. The derived hydrodynamic approximations of (5.0.3)–(5.0.2) are expected to have completely different behaviors in comparison with the classical Euler and Navier–Stokes equations. The rigorous mathematical study of these two systems, including short-time existence, blow-up results, etc., would be an interesting subject. Recent theoretical and experimental findings, in agreement with previous results obtained by Steinhauer and Lee–Yang, support the Reichl et al. model to be a possible complement of the Zaremba et al./Pomeau et al. model at ultra-low temperatures. Further comparisons of the two models would be an important research topic.

Chapter 9
Equilibrium Properties of a Dilute Bose Gas with Small Coupling at First Order

This chapter presents a derivation of the equilibrium properties of a dilute Bose gas in the limit of small coupling at first order with respect to the coupling constant, measured by the scattering length. The calculation is not completely trivial. It show that the BE transition is first order in the thermodynamic sense. This is because of the increase of the energy of small momenta due to the Bogoliubov renormalization of the particles' energy, which depletes the density of thermal particles and so yields a negative feedback for the growth of the condensate, compared to what happens without interactions.

9.1 Thermodynamics of the Dilute Bose Gas with an Energy Perturbed to First Order

Let us first recall the calculation exposed by Huang [85], under the assumption that the interaction is a perturbation of the energy of the non-interacting particles. As a consequence, we assume that the first correction to the energy coming from the interaction is provided by terms that are diagonal in states with a fixed number of particles at each momentum: with respect to those states, the unperturbed energy operator is diagonal. This assumption yields qualitatively correct results but not all of its conclusions follow from the analysis, as shown in this chapter. It is proved in [6] that this operator with the first-order perturbation is as follows

$$H[n_p] = \sum_p \frac{p^2}{2m} n_p + \frac{4\pi g \hbar^2}{m\Omega} \left(N^2 - \frac{n_0^2}{2} \right), \qquad (9.1.1)$$

in which n_p is the number of particles of momentum p, g is the positive scattering length characterizing the potential under the assumption that the temperature low enough to give only s-wave scattering in the long wave limit, m is the mass of the identical particles, Ω is the total volume, N is the total

© Springer Nature Switzerland AG 2019
Y. Pomeau, M.-B. Tran, *Statistical Physics of Non Equilibrium Quantum Phenomena*,
Lecture Notes in Physics 967, https://doi.org/10.1007/978-3-030-34394-1_9

number of particles and n_0 is the number of particles in the ground state. With this energy, the thermodynamic sum can be performed explicitly.

Based on (9.1.1), the canonical partition function reads

$$Q_N = \sum_{n_p} \exp\left(-\frac{1}{k_B T}\left(\sum_p \frac{p^2}{2m} n_p + \frac{4\pi g\hbar^2}{m\Omega}\left(N^2 - \frac{n_0^2}{2}\right)\right)\right),$$

where k_B is the Boltzmann constant and T is the absolute temperature. This function can be computed by taking the sum $\sum_{n_p}$ over states where $p \neq 0$, for

$$\sideset{}{'}\sum_p n_p = \sum_{p\neq 0} n_p = N - n_0$$

fixed and equal to N'. The sum is as in a non-interacting BE gas:

$$\begin{aligned}
Q'_N &= \sum_{n_p, p\neq 0} \exp\left(-\frac{1}{k_B T}\sideset{}{'}\sum_p \frac{p^2}{2m} n_p\right) \\
&= \exp\left(\frac{\mu(N - n_0)}{k_B T} - \frac{\Omega}{(2\pi\hbar)^3}\int_{\mathbb{R}^3} \ln\left(1 - e^{-\frac{\omega(p)-\mu}{k_B T}}\right) dp\right),
\end{aligned}$$

where $\omega(p) = \frac{p^2}{2m}$. The chemical potential μ is the Lagrange multiplier with respect to N', and satisfies

$$\frac{\partial Q_{N'}}{\partial \mu} = 0.$$

Therefore

$$\rho'(\mu) = \frac{1}{(2\pi\hbar)^3}\int_{\mathbb{R}^3} \frac{1}{e^{\frac{\omega(p)-\mu}{k_B T}} - 1} dp \equiv \left(\frac{mk_B T}{2\pi\hbar^2}\right)^{\frac{3}{2}} \sum_{k=1}^{\infty} \frac{e^{\mu k/k_B T}}{k^{3/2}}, \qquad (9.1.2)$$

in which

$$\rho'(\mu) = \frac{N'}{\Omega}$$

represents the number density of the particles outside of the condensate. The next step is to trace over the momentumless particles. In other words, we need to sum over all values of n_0 while fixing N

$$\begin{aligned}
Q_N = \sum_{n_0} \exp\Big[&-\frac{1}{k_B T}\left(\frac{4\pi g\hbar^2}{m\Omega}\left(N^2 - \frac{n_0^2}{2}\right) + \mu(N - n_0)\right. \\
&\left. + \frac{\Omega k_B T}{(2\pi\hbar)^3}\int_{\mathbb{R}^3} \ln\left(1 - e^{-\frac{\omega(p)-\mu}{k_B T}}\right) dp\right)\Big].
\end{aligned}$$

$$(9.1.3)$$

The sum is centered around a saddle at $\bar{n}_0$ and can be found by solving

$$\frac{\partial \ln Q_N}{\partial \bar{n}_0} = 0,$$

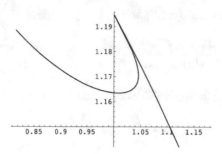

Fig. 9.1 The equilibrium pressure P/P_c is represented as a function of the specific volume ρ_c/ρ for $i_1 = 0.1$.

under the constraint that

$$\frac{4\pi g\hbar^2}{m\Omega}\bar{n}_0 + \mu = 0.$$

Let us remark that μ depends on n_0 by (9.1.2). The partition function (9.1.3) can be approximated by the value of Q_N at its maximum in the thermodynamic limit. Therefore, the total energy is

$$-k_BT\ln(Q_N) = \qquad\qquad\qquad\qquad\qquad (9.1.4)$$

$$\mu(N - \bar{n}_0) + \frac{4\pi g\hbar^2}{m\Omega}\left(N^2 - \frac{n_0^2}{2}\right) + \frac{\Omega k_BT}{(2\pi\hbar)^3}\int_{\mathbb{R}^3}\ln\left(1 - e^{-\frac{\omega(p)}{k_BT} - \frac{\mu}{k_BT}}\right)dp.$$

Under variations around n_0 for N fixed, the total energy is stationary. This implies that there is no exchange of mass between the condensate and the thermal particles at equilibrium. The number of particles outside the condensate (and inside, since N is fixed) is imposed by the relation between the chemical potentials (Lagrange multiplier), and the particle density of the condensate is then

$$\mu + \frac{4\pi g\hbar^2}{m}\rho_s = 0, \qquad\qquad\qquad (9.1.5)$$

in which

$$\rho_s = \frac{\bar{n}_0}{\Omega} = \rho - \rho'(\mu)$$

and

$$\rho = \frac{N}{\Omega}$$

represents the total number density.

Combining (9.1.5) and (9.1.2) solves the problem. We get

$$\tilde{\rho} - \tilde{\rho}_s = \frac{4}{\sqrt{\pi}\zeta(3/2)} \int_0^\infty \frac{x^2 dx}{e^{x^2 + \alpha\tilde{\rho}_s} - 1} \equiv \frac{1}{\zeta(3/2)} \sum_{k=1}^\infty \frac{e^{-\alpha\tilde{\rho}_s}}{k^{3/2}}, \qquad (9.1.6)$$

in which $\zeta(s)$ is the Riemann ζ function and

$$\tilde{\rho} = \frac{\rho}{\rho_c}, \qquad \tilde{\rho}_s = \frac{\rho_s}{\rho_c},$$

with ρ_c being the number density at the transition in the non-interacting BE case

$$\rho_c = \left(\frac{mk_BT}{2\pi\hbar^2}\right)^{\frac{3}{2}} \zeta\left(\frac{3}{2}\right).$$

The quantity

$$\alpha = 2(\zeta(3/2))^{2/3} g\rho_c^{1/3}$$

is the only dimensionless parameter of the problem, and is small. Since this quantity is proportional to the scattering length g, g can be seen as the small parameter.

An explicit representation of $\frac{\rho}{\rho_c}$ can be given

$$\sum_{k=1}^\infty \frac{e^{-zk}}{k^s} = \Gamma(1-s)z^{s-1} + \sum_{r=0}^\infty \frac{(-z)^r}{\Gamma(r+1)}\zeta(z-r), \qquad (9.1.7)$$

which then implies

$$\tilde{\rho} - \tilde{\rho}_s = 1 - \frac{2\sqrt{\pi}}{\zeta(3/2)}(\alpha\tilde{\rho}_s)^{1/2} - \frac{\zeta(1/2)}{\zeta(3/2)}\alpha\tilde{\rho}_s + \cdots. \qquad (9.1.8)$$

Setting $\delta\tilde{\rho} = \tilde{\rho} - 1$, we get

$$(\delta\tilde{\rho} - \tilde{\rho}_s)^2 = \frac{4\pi}{\zeta^2(3/2)}\alpha\tilde{\rho}_s.$$

Finally, let us derive the state equation that describes the relation between ρ and the pressure P. This can be obtained from the general expression for the pressure:

$$\begin{aligned}
P &= \frac{4\pi f\hbar^2}{m}\left(\rho^2 - \frac{\rho_s^2}{2}\right) - \frac{k_BT}{(2\pi\hbar)^3}\int_{\mathbb{R}^3} \ln\left(1 - e^{-\frac{\omega(p)-\mu}{k_BT}}\right) dp \\
&= \frac{4\pi f\hbar^2}{m}\left(\rho^2 - \frac{\rho_s^2}{2}\right) + \frac{P_c}{\zeta(5/2)}\sum_{k=1}^\infty \frac{e^{-\alpha\tilde{\rho}_s k}}{k^{5/2}},
\end{aligned} \qquad (9.1.9)$$

in which P_c is the pressure at the BE transition in a perfect gas

$$P_c = k_B T \left(\frac{mk_B T}{2\pi\hbar^2} \right)^{\frac{3}{2}} \zeta\left(\frac{5}{2}\right).$$

The dimensionless ratio P/P_c reads

$$P/P_c = \alpha\frac{\zeta(3/2)}{\zeta(5/2)}\left(\rho^2 - \frac{\rho_s^2}{2}\right) + \frac{1}{\zeta(5/2)}\sum_{k=1}^{\infty}\frac{e^{-\alpha\tilde{\rho}_s k}}{k^{5/2}}, \qquad (9.1.10)$$

which gives at first order with respect to α

$$\tilde{P} = 1 - \frac{\zeta(3/2)}{\zeta(5/2)}\alpha\tilde{\rho}_s + \frac{\zeta(3/2)}{\zeta(5/2)}\alpha\left(\rho^2 - \frac{\rho_s^2}{2}\right).$$

9.2 Thermodynamics of the Dilute Bose Gas at the Bogoliubov Order

The results obtained in the previous section of this chapter are indeed correct, in the following sense: the orders of magnitude of macroscopic quantities like P and ρ_s are correct to the first-order corrections, at least not too close to the transition point. However, the thermodynamical properties at first order in g are not captured by the calculations of the previous section, due to the fact that the effect of the Bogoliubov renormalization on the energy spectrum at low momenta is missing. At finite temperatures, the Bogoliubov renormalization is indeed quite different from the original zero-temperature renormalization. The assumption that the interaction appears as a first-order perturbation is one essential ingredient of this derivation. However part of the spectrum responsible for the expansion of $\rho'(\mu)$ near $\mu = 0$ is not quite exact. Beyond ρ_c, the first relevant term is of order $(-\mu)^{\frac{1}{2}}$. This term can be derived from the contribution to the integral over the momenta under the constraint that $\frac{p^2}{m}$ is of order μ, or equivalently p is of order $(-\mu)^{\frac{1}{2}}$. The interaction and kinetic energy are indeed of the same order, based on the fact that those momenta are small enough. From this argument, one can compare the magnitudes of the kinetic and interaction energy for this range of momenta. For any particle, Bogoliubov's theory gives at once the order of magnitude of the interaction energy. According to Bogoliubov's theory [28], the energy of the quasiparticles of momentum p is given by

$$\omega(p) = \sqrt{\frac{p^4}{4m^2} + \frac{4\pi\hbar^2 g}{m^2}\rho_s p^2},$$

which can be combined with (9.1.5), and under the restriction of momenta satisfying

$$\frac{p^2}{2m} \approx -\mu,$$

the kinetic energy $\frac{p^2}{2m}$ and the energy of the interaction with the condensate $\frac{2\pi\hbar^2 g}{m^2}\rho_s$ are of the same order of magnitude. As a consequence, all the calculations of the previous section should be reconsidered, due to the fact that the interaction cannot be seen as a perturbation in this range of momenta, whereas the interaction was assumed to be a uniformly small perturbation of the kinetic energy. In order to apply the principles outlined by Bogoliubov at zero temperature, one needs to split the interaction part of the energy into terms that involve the condensate and terms that do not involve the condensate. Let $\mathbf{a}_i^\dagger(\mathbf{a}_i)$ be the creation (annihilation) operator in the state of momentum p_i. The interaction part of the energy operator can be written as follows

$$V_{int} = \frac{2\pi g\hbar^2}{m\Omega} \sum_{i_1,i_2,i_3,i_4} \mathbf{a}_{i_1}^\dagger \mathbf{a}_{i_2}^\dagger \mathbf{a}_{i_3} \mathbf{a}_{i_4} \delta(p_{i_1} + p_{i_2} - p_{i_3} - p_{i_4}).$$

Following Bogoliubov's theory, under the presence of the condensation in the state of zero momentum, the operators of index zero become c-numbers

$$\mathbf{a}_0 = \Psi\Omega^{\frac{1}{2}}, \quad \mathbf{a}_0^\dagger = \bar{\Psi}_0\Omega^{\frac{1}{2}},$$

with Ψ_0 being the ground state wavefunction and $\bar{\Psi}_0$ the complex conjugate. This is because the commutation relations introduce quantum fluctuations of order one in the occupation number, whereas the mean value of this occupation number in the ground state is very large.

We will now derive the expression (9.1.1) for the interaction energy from V_{int}. The interaction energy is the average value of V_{int}, at first order, calculated with the unperturbed equilibrium ensemble under the condition that the occupation numbers at different momenta p are uncorrelated. The terms with nonzero momentum are finite quantities in the thermodynamic limit. As a consequence, the terms where the four momenta are equal $i_1 = i_2 = i_3 = i_4 \neq 0$ have a finite contribution to V_{int} and they are negligible since, in the thermodynamic limit, one expects a contribution of order Ω (or N'). When there is no condensate, the average value of V_{int} is equal to

$$\frac{2\pi g\hbar^2}{m\Omega} 2 \sum_{i_1} \mathbf{a}_{i_1}^\dagger \mathbf{a}_{i_1} \sum_{i_2} \mathbf{a}_{i_2}^\dagger \mathbf{a}_{i_2}.$$

Notice that the factor 2 appears in front of the sum since there are two ways of choosing either $i_3 = i_1$ or $i_4 = i_1$ to satisfy the Kronecker condition. In the case when the state of momentum zero is macroscopically occupied, there is no such factor 2, since it would amount to counting the

$$\frac{2\pi g\hbar^2}{m\Omega} \mathbf{a}_0^\dagger \mathbf{a}_0^\dagger \mathbf{a}_0 \mathbf{a}_0$$

twice. Therefore, one subtracts this last contribution once to avoid this double counting. This yields the sought average value of V_{int}

$$\langle V_{int} \rangle = \frac{2\pi g\hbar^2}{m\Omega}(2N^2 - n_0^2).$$

One gets four types of terms, listed below, by singling out the contribution of the ground state to the sum over the momenta giving V_{int}:

(I) The terms where all four wavenumbers $i_1, i_2, i_3, i_4 \neq 0$. Denote the corresponding contribution to V_{int} by $V_{int,1}$. This contribution is a small perturbation, and can be estimated under the assumption that it is equal to its average value on the unperturbed state. This is indeed a straightforward consequence of the general formula for estimating the first-order perturbation, as presented earlier for the pure BE system. This means that one needs to neglect the quantum correlations between states of different wavenumbers. Therefore, the only surviving terms in the sum are terms whose indices are paired in such a way that the same index appears once in a creation and in an annihilation operator

$$V_{int,1} = \frac{2\pi g\hbar^2}{m\Omega} 2 \sum_{i_1,i_2,i_3,i_4 \neq 0} \mathbf{a}_{i_1}^\dagger \mathbf{a}_{i_2}^\dagger \mathbf{a}_{i_3} \mathbf{a}_{i_4} \delta(p_{i_1} + p_{i_2} - p_{i_3} - p_{i_4})$$

$$\approx \frac{4\pi g\hbar^2}{m\Omega} \left(\sum_{i_1 \neq 0} \mathbf{a}_{i_1}^\dagger \mathbf{a}_{i_1} \right) \left(\sum_{i_2 \neq 0} \mathbf{a}_{i_2}^\dagger \mathbf{a}_{i_2} \right) = \frac{4\pi g\hbar^2}{m\Omega}(N - n_0)^2.$$

This result has not changed at the dominant order, under the effect of the Bogoliubov renormalization. Indeed, the Bogoliubov renormalization only changes the contribution to the energy of particles with a small momentum, and this is just a small proportion of the total number of particles. As a consequence, we can safely use the above formula for $V_{int,1}$, when estimating the perturbation by $V_{int,1}$ to the thermodynamical properties, as we are considering the dominant corrections.

(II) The terms where one wavenumber is zero

$$\frac{4\pi g\hbar^2}{m\Omega} \sum_{i_2,i_3,i_4 \neq 0} (\mathbf{a}_0^\dagger \mathbf{a}_{i_2}^\dagger \mathbf{a}_{i_3} \mathbf{a}_{i_4} + \mathbf{a}_0 \mathbf{a}_{i_3}^\dagger \mathbf{a}_{i_4}^\dagger \mathbf{a}_{i_2}) \delta(p_{i_2} - p_{i_3} - p_{i_4}).$$

(III) The terms where two wavenumbers are zero

$$\frac{4\pi g\hbar^2}{m\Omega} \sum_{i \neq 0} (\mathbf{a}_0^2 \mathbf{a}_i^\dagger \mathbf{a}_{-i}^\dagger + (\mathbf{a}_0^\dagger)^2 \mathbf{a}_i \mathbf{a}_{-i}) + 4|\mathbf{a}_0|^2 \mathbf{a}_i^\dagger \mathbf{a}_i.$$

(IV) The terms with three zero wavenumbers do not exist due to the Kronecker delta.

(V) The terms with four zero wavenumbers

$$\frac{4\pi g\hbar^2}{m\Omega}|\mathbf{a}_0|^4.$$

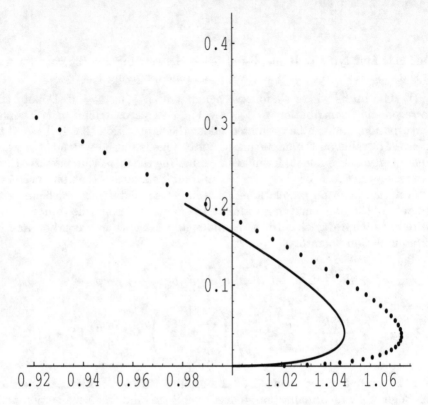

Fig. 9.2 The condensate density ρ_s/ρ_c is represented as a function of ρ_c/ρ. The full curve is the computation up to first order following equation (9.1.8). The dotted curve is the numerical solution of (9.2.6) and (9.2.8) for $i_1 = 0.1$.

In the rest of this section, we will study the thermodynamics of a system at the "Bogoliubov" order, where the interactions are among the five cases (I)–(V). Except near the transition, interaction (II) is always negligible compared with the other interaction terms. The reason for this is that this interaction has the lowest order in a_0, with respect to other terms that have a_0, and therefore it becomes the most important containing a_0 when the superfluid density tends to zero. On the other hand, this term may be neglected outside the neighborhood of the transition. Although we shall deal with terms of at most first order in i_1, the effect of (II) would require one to go to the second order in i_1 in a regular perturbation scheme since any combination cubic in creation-annihilation operators brings no first-order contribution.

Therefore, the energy operator can be written as

$$H = \sum_p \omega(p) \mathbf{a}_p^\dagger \mathbf{a}_p + \frac{2\pi\hbar^2 g}{m\Omega}[2(N - n_0)^2 + n_0^2]$$

$$+ \frac{2\pi\hbar^2 g}{m\Omega} \sum_{p\neq 0} \left(\Psi_0^2 \mathbf{a}_p^\dagger \mathbf{a}_{-p}^\dagger + \bar{\Psi}_0^2 \mathbf{a}_p \mathbf{a}_{-p} + 4|\Psi_0|^2 \mathbf{a}_p^\dagger \mathbf{a}_p \right). \tag{9.2.1}$$

Notice that since $|\Psi_0|^2 \equiv \rho_s$, one recovers formula (9.1.1) when neglecting the off-diagonal terms. Despite the fact that lots of work has been done to study the energy operators whose forms are similar to (9.2.1), see [11, 33], they are restricted to cases where the energy is exactly quadratic in the operators $\mathbf{a}_p^\dagger$ and $\mathbf{a}_p$, with $p \neq 0$. The contribution of

$$\frac{4\pi\hbar^2 g}{m\Omega}(N - n_0)^2$$

makes our case different from the others and is necessary to get all the interaction effects consistently at first order in the smallness parameter. This is the reason why important differences between our formula can be deduced in comparison with the results of [33]. For instance, the explicit relation between the number densities (of the condensate and of the excited particles) and the chemical potential are different. Following now the same method outlined in the previous section, we will compute the partition function

$$Q_N = \mathrm{Tr}\left(e^{-\frac{H}{k_B T}} \right),$$

with respect to a given total number of particles N in a box of volume Ω. Due to the fact that the energy operator is a quadratic functional of the operators $\mathbf{a}_p^\dagger$ and $\mathbf{a}_p$ and a function of N and n_0, the calculation is not difficult but not straightforward. We split the computation into several main steps. First, we compute the contribution to the partition function that depends explicitly on $\mathbf{a}_p^\dagger$ and $\mathbf{a}_p$, with a fixed amount of condensate particles. We then calculate the corresponding grand canonical partition function

$$Q^B(\mu) = \sum_{N'} e^{-\frac{\mu N'}{k_B T}} Q_{N'}^B,$$

to get Q_N, and express $Q_{N'}^B$ in the form $e^{+\frac{\mu N'}{k_B T}}$, multiplying by this grand canonical partition function in which we neglect the fluctuations of N'. $Q^B(\mu)$ can be expressed as the exponential of a quadratic form in the creation-annihilation operators:

$$H = \mathrm{Tr}\exp\left[-\frac{1}{k_B T} \left(\sum_{p\neq 0}(\omega(p) - \mu)\mathbf{a}_p^\dagger \mathbf{a}_p \right. \right.$$

$$\left. \left. + \frac{2\pi\hbar^2 g}{m} \sum_{p\neq 0} \left(\Psi_0^2 \mathbf{a}_p^\dagger \mathbf{a}_{-p}^\dagger + \bar{\Psi}_0^2 \mathbf{a}_p \mathbf{a}_{-p} + 2|\Psi_0|^2 \mathbf{a}_p^\dagger \mathbf{a}_p \right) \right) \right]. \tag{9.2.2}$$

The chemical potential term

$$\mu N' = \mu \sum_{p \neq 0} \mathbf{a}_p^\dagger \mathbf{a}_p$$

is included together with the standard kinetic energy term, altogether with the combination $\omega(p) - \mu$ in the diagonal term in the occupation numbers $n_p = \mathbf{a}_p^\dagger \mathbf{a}_p$. Since the Bogoliubov transformation requires a factor 2 only in the diagonal part of the interaction term, the sum

$$\frac{4\pi\hbar^2 g}{m} \sum_{p \neq 0} \sum_{p \neq 0} |\Psi_0|^2 \mathbf{a}_p^\dagger \mathbf{a}_p$$

is a constant, giving that N' is fixed. Therefore this sum is included in the full free energy at the end in the form

$$\frac{4\pi\hbar^2 g}{m\Omega} n_0(N - n_0).$$

With a fixed number of particles at each momentum, the full energy operator is not diagonal in the representation on the basis of states; however, in states with a fixed number of quasiparticles, it is diagonal. As a consequence, we cannot apply the Bogoliubov theory to a first-order perturbation of the interaction operator diagonal in states with a fixed number of particles, which was indeed fundamental to the computation in the previous section. Due to the noncommutation between the particles' and quasiparticles' operators, while computing the contribution of the thermally excited states to the partition function, one has to take into account the Lagrange constraint of a fixed number of particles (not of quasiparticles). The work [104] is based on the assumption that the total number of particles and of quasiparticles are the same (equation (19) in [104]). This assumption does not follow from the operator algebra. From our argument, it is clear that the Lagrange constraint for the number of particles formally means adding $-\mu$ to the energy per particle.

Following the technique of Bogoliubov, one can diagonalize the resulting operator (that is, the energy minus $\mu N'$), but by replacing everywhere $\omega(p)$ by $\omega(p) - \mu$ in the final expression. After a Bogoliubov transformation, we can compute the trace in the grand canonical ensemble and in the basis where the Hamiltonian is diagonal. The last trace is formally over an arbitrary number of quasiparticles, resulting in

$$Q^B(\mu) = \exp\left[-\frac{\mu(N - n_0)}{k_B T} - \frac{\Omega}{(2\pi\hbar)^3} \int_{\mathbb{R}^3} \ln\left(1 - e^{-\omega_B(p,\mu,\rho_s)}\right) \right],$$

where

$$\omega_B(p, \mu, \rho_s) = \sqrt{\left(\frac{p^2}{2m} - \mu\right)^2 + 2\left(\frac{p^2}{2m} - \mu\right)\frac{4\pi g\hbar^2 \rho_s}{m}}. \tag{9.2.3}$$

In this expression for $Q^B_{N'}(\mu)$, the contribution coming from the change in the ground state energy arising from the Bogoliubov transformation is neglected since it only leads to higher order corrections, which are responsible for the term of order $\alpha^{3/2}$ (in our notation) in the expansion of the ground state energy. These terms correspond to the second-order term in equation (17) of [104] for instance, and would similarly yield corrections, though smaller than the one used above. By deriving the free energy with respect to μ, the density of thermal particles follows

$$\rho'(\mu) \;=\; -\frac{1}{(2\pi\hbar)^3} \int_{\mathbb{R}^3} \frac{1}{e^{\frac{\omega_B(p,\mu,\rho_s)}{k_B T}}-1} \frac{\partial \omega_B(p,\mu,\rho_s)}{\partial \mu}\,\mathrm{d}p. \tag{9.2.4}$$

Since the chemical potential enters in a non-trivial way in quantities related to the quasiparticles, the thermodynamical expression used in our book may look different from other expressions in the literature. In the case where the quasiparticles are phonons in a solid, they would be associated only to the conservation of energy, while their thermodynamics would depend only on the temperature without involving chemical potentials. However, in a condensed Bose gas, the quasiparticles are not independent of the true particles. As a result, if the occupation number of the quasiparticles changes, the occupation number of the particles themselves also changes. The thermodynamics of quasiparticles should depend on both the temperature, which is related to the energy conservation, and the chemical potential, which is related to the conservation of the number of particles, in a nontrivial way. At this Bogoliubov approximation, the total partition function is

$$Q^B_N \;=\; \sum_{n_0} \exp\left\{ -\frac{1}{k_B T}\Big[\mu(N-n_0) \right.$$
$$\left. +\; \frac{2\pi\hbar^2 g}{m\Omega} \int_{\mathbb{R}^3} \ln\left(1 - e^{-\frac{\omega_B(p,\mu,\rho_s)}{k_B T}}\right)\mathrm{d}p \Big]\right\}. \tag{9.2.5}$$

As in the previous section, $\bar{n}_0$ can be obtained by finding the saddle point of the sum. We recall that $\omega_B(p,\mu,\rho_s)$ depends explicitly on n_0 since $n_0 = \rho_s(\Omega)$. We have

$$\mu + \frac{4\pi\hbar^2 g}{m}(\rho - \rho_s) \;=\; \frac{1}{(2\pi\hbar)^3} \int_{\mathbb{R}^3} \frac{1}{e^{\frac{\omega_B(p,\mu,\rho_s)}{k_B T}}-1} \frac{\partial \omega_B(p,\mu,\rho_s)}{\partial \rho_s}\,\mathrm{d}p. \tag{9.2.6}$$

The two equations (9.2.4) and (9.2.6) solve the problem in principle. Computing $\frac{\partial \omega_B}{\partial \mu}$ and $\frac{\partial \omega_B}{\partial \rho_s}$ from (9.2.3) and plugging the results into (9.2.4), we can rewrite equation (9.2.6) as follows

$$\tilde{\mu} \;\equiv\; \frac{\mu}{k_B T} \;=\; -\alpha^2 \tilde{\rho}_s \frac{4}{\sqrt{\pi}\zeta(3/2)} \int_0^\infty \frac{1}{\epsilon(x)-1} \frac{x^2}{\epsilon(x)}\,\mathrm{d}x, \tag{9.2.7}$$

with

$$\epsilon(x) \;=\; \sqrt{(x^2-\tilde{\mu})^2 + 2\alpha(x^2-\tilde{\mu})\tilde{\rho}_s}.$$

As a consequence, (9.2.4) becomes

$$\tilde{\rho} = \tilde{\rho}_s + \frac{\tilde{\mu}}{\alpha} + \frac{4}{\sqrt{\pi}\zeta(3/2)} \int_0^\infty \frac{1}{\epsilon(x) - 1} \frac{x^2(x^2 - \tilde{\mu})}{\epsilon(x)} \mathrm{d}x. \qquad (9.2.8)$$

These coupled equations have been solved for $\alpha = 0.1$. The relation between $\tilde{\rho}_s$ and $\tilde{\mu}$ can be derived from (9.2.7).

We directly obtain $\tilde{\rho}$ as a function of $\tilde{\mu}$. We have plotted $\tilde{\rho}_s$ versus $1/\tilde{\rho}$ as a parametric plot of $\tilde{\mu}$ in Figure 9.2.

ρ_s and μ vanish near the BE transition. Using the singular behavior of the integral (9.2.7) near $x = 0$, we obtain the relation:

$$\int_0^\infty \frac{x^2 \mathrm{d}x}{(x^2 - \tilde{\mu})(x^2 - \tilde{\mu} + 2\alpha\tilde{\rho}_s)} = \frac{\pi(\sqrt{2\alpha\tilde{\rho}_s - \mu} - \sqrt{-\mu})}{4\alpha\tilde{\rho}_s}.$$

Near the transition (ρ_s and μ close to zero) we have

$$(-\tilde{\mu})^{\frac{3}{2}} = \frac{\sqrt{\pi}}{\zeta(3/2)} \alpha^2 \rho_s. \qquad (9.2.9)$$

As a consequence, we get, up to Bogoliubov order

$$\tilde{\mu} \approx -\tilde{\rho}_s^{\frac{2}{3}} >> \tilde{\rho}_s$$

near the transition. This means that the dominant order is not the one found by the computation of the previous section in this range.

By the same assumptions, we approximate (9.2.8) by

$$\delta\tilde{\rho} = \tilde{\rho} - 1 = \tilde{\rho}_s + \frac{\tilde{\mu}}{i_1} - \frac{\sqrt{\pi}}{\zeta(3/2)} \sqrt{2\alpha\tilde{\rho}_s - \tilde{\mu}},$$

which implies $\delta\tilde{\rho}$ and $\tilde{\rho}_s$ scale as α.

Finally, the state equation can be obtained and plotted as

$$\frac{P}{P_c} = \alpha \frac{\zeta(3/2)}{2\zeta(5/2)} \left((\tilde{\rho} - \tilde{\rho}_s)^2 + \tilde{\rho}^2 \right) - \frac{4}{\sqrt{\pi}\zeta(5/2)} \int_0^\infty \ln\left(1 - e^{-\epsilon(x)}\right) x^2 \mathrm{d}x. \qquad (9.2.10)$$

9.3 Conclusions

Under the Bogoliubov transformation, at the dominant order where g or α is small, the condition that the total partition function is stationary under exchange of particles between the condensate and the thermal particles near $\rho_s = 0$ can be written as

$$\mu \approx -\alpha^{\frac{4}{3}} \rho_s^{\frac{2}{3}}.$$

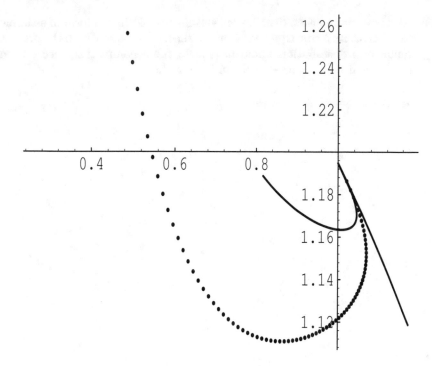

Fig. 9.3 The equilibrium pressure P/P_c is represented as a function of the specific volume ρ_c/ρ. The full curve corresponds to equation (9.1.10). The dotted curve corresponds to equation (9.2.10) for $\alpha = 0.1$.

These considerations result in a consistent derivation of the first correction to the thermodynamic parameters, arising from the short-range interaction, of the BE gas, in the dilute gas limit.

In conclusion:

I Near $\rho_s = 0$, the chemical potential μ is given at the dominant order by

$$\mu \approx -\alpha^{\frac{4}{3}} \rho_s^{\frac{2}{3}}.$$

As ρ_s vanishes with fixed α, this quantity is much larger than $\alpha \rho_s$. If ρ_s is small, but much larger than $\alpha \rho_s$, the above relation for the chemical potential is valid. Terms other than the ones we have considered in this section to get the energy operator of equation (9.2.2) become relevant if ρ_s becomes less than $\alpha \rho_s$.

II Changing the momentum distribution for small momenta is among the important effects of the Bogoliubov renormalization. The form of the momentum distribution of quasiparticles at small momenta is as follows

$$-\frac{1}{(2\pi\hbar)^3} \frac{1}{e^{\frac{\omega_B(p,\mu,\rho_s)}{k_B T}} - 1} \frac{\partial \omega_B(p,\mu,\rho_s)}{\partial \rho_s}.$$

The above momentum distribution cannot be cast in the form of a simple Bose factor. As a consequence, a kinetic theory consistent with the present equilibrium distribution would have a far more complex structure than the classical Boltzmann–Nordheim kinetic theory.

Chapter 10
Mathematical Analysis of the Coupling Condensate – Thermal Cloud Systems

Quantum kinetic theory has recently become an important topic of research in the mathematics community. While the mathematical analysis and rigorous derivations of the Boltzmann–Nordheim equation and the related 4-wave kinetic equation has quite a large body of literature and breakthroughs have been made (see [19, 20, 21, 22, 34, 35, 37, 39, 40, 43, 58, 59, 60, 61, 64, 65, 66, 68, 75, 105, 107, 108, 109, 110, 111, 112, 113, 116, 117, 122, 137, 152, 154] and references therein), this chapter is only devoted to the discussion of recent research on the coupling condensate-thermal cloud systems in both regimes: low temperatures and ultra-low temperatures, with a focus on the works of the authors, in which a full mathematical proof of a new result is given in detail. Since both systems in the two regimes are very complicated, recent mathematical works are only devoted to partial approximations of them.

10.1 Approximations of the Coupling Condensate – Thermal Cloud Systems

10.1.1 First Approximation: Systems Without the $2 \leftrightarrow 2$ and $1 \leftrightarrow 3$ Collisions

In the first work in this direction [14], the system is assumed to be homogeneous in space and the $2 \leftrightarrow 2$ and $1 \leftrightarrow 3$ collisions are dropped. The simplified model reads

$$\frac{\partial f}{\partial t}(t,p) = C_{12}[f](t,p), \quad f(0,p) = f_0(p), \quad (t,p) \in \mathbb{R}_+ \times \mathbb{R}^3$$

$$\frac{\partial n_c(t)}{\partial t} = -\int_{\mathbb{R}^3} C_{12}[f](t,p)\mathrm{d}p, \quad n_c(0) = n_{c,0}, \quad (10.1.1)$$

in which

© Springer Nature Switzerland AG 2019
Y. Pomeau, M.-B. Tran, *Statistical Physics of Non Equilibrium Quantum Phenomena*,
Lecture Notes in Physics 967, https://doi.org/10.1007/978-3-030-34394-1_10

$$C_{12}[f] := n_c(t) \int_{\mathbb{R}^3} \int_{\mathbb{R}^3} \big[R(p, p_1, p_2) - R(p_1, p, p_2) - R(p_2, p_1, p) \big] \mathrm{d}p_1 \mathrm{d}p_2 \,,$$

$$R(p, p_1, p_2) :=$$
$$|W_{1,2,3}^{12}|^2 \big[\delta \left(E(p) - E(p_1) - E(p_2) \right) \delta(p - p_1 - p_2) \big]$$
$$\times \big[f(p_1)f(p_2)(1 + f(p)) - (1 + f(p_1))(1 + f(p_2))f(p) \big] \,.$$
$$\tag{10.1.2}$$

The dispersion relation $E(p)$ is the Bogoliubov relation

$$E(p) = |p| \sqrt{\frac{p^2}{2m^2} + \frac{gn_c}{m}}. \tag{10.1.3}$$

The kernel $|W_{1,2,3}^{12}|^2$ is approximated by

$$|W_{1,2,3}^{12}|^2 \approx \left(\min\left\{ \frac{|p|}{n_c}, 1 \right\} \right) \left(\min\left\{ \frac{|p_1|}{n_c}, 1 \right\} \right) \left(\min\left\{ \frac{|p_2|}{n_c}, 1 \right\} \right). \tag{10.1.4}$$

In [14], the authors provided the first global well-posedness theory of strong and radial symmetric solutions of this approximation.

A modification of the above approximation was investigated in a later work [5], in which $E(p)$ is approximated by the phonon dispersion relation

$$\mathcal{E}(p) = \sqrt{\frac{gn_c(t)}{m}} |p|. \tag{10.1.5}$$

The transition probability $|W_{1,2,3}^{12}|^2$ is approximated by

$$|\mathcal{W}_{1,2,3}^{12}|^2 = \kappa |p||p_1||p_2|, \tag{10.1.6}$$

where

$$\kappa = \frac{9c}{64\pi^2 mn_c^2} = \frac{9m^{1/2}}{64\pi^2 (gn_c)^{3/2}}. \tag{10.1.7}$$

This approximation makes the analysis more complicated. Below, we will state the main theorem of [5]. To do that, we recall the definitions of some of the notations that will be used later.

The k-moment ($k \geq 0$) of a radial symmetric function $f(p) = f(|p|)$ is defined to be

$$\mathfrak{M}_k \langle f \rangle = \int_{\mathbb{R}_+} |f(p)||p|^k \mathrm{d}|p|. \tag{10.1.8}$$

We define the functional spaces

$$L^k(\mathbb{R}^3) := \left\{ f \text{ measurable} \mid \|f\|_{L^k} = \int_{\mathbb{R}^3} |f(p)|^k \mathrm{d}p < \infty \right\}, \quad 1 \leq k < \infty, \tag{10.1.9}$$

$$L^\infty(\mathbb{R}^3) := \left\{ f \text{ measurable} \mid \|f\|_{L^\infty} = \operatorname*{ess\,sup}_{p \in \mathbb{R}^3} |f(p)| < \infty \right\}, \tag{10.1.10}$$

and

$$L^1(\mathbb{R}_+, |p|^k) := \left\{ f \text{ measurable} \mid \int_{\mathbb{R}_+} |f(p)||p|^k \mathrm{d}|p| < \infty, \ k \geq 0 \right\}. \quad (10.1.11)$$

In addition, for any normed space $(X, \|\cdot\|_X)$, we define

$$C([0,T), X) := \left\{ f : [0,T) \to X \mid f \text{ is continuous from } [0,T) \text{ to } X \right\} \quad (10.1.12)$$

and

$$C^1((0,T), X) := \left\{ f : (0,T) \to X \mid f \text{ is differentiable from } (0,T) \text{ to } X \right\}, \quad (10.1.13)$$

for any $T \in (0, \infty]$. The above definitions can also be extended to the spaces $C([0,T], X)$, $C^1([0,T), X)$ for any $T \in (0, \infty)$.

We also define the set

$$\mathcal{S} := \Big\{ f \in L^1(\mathbb{R}^3) \mid \textbf{i.} \ f \text{ nonnegative \& radially symmetric},$$

$$\textbf{ii.} \ \mathfrak{M}_3\langle f \rangle = \int_{\mathbb{R}_+} f(|p|)|p|^3 \mathrm{d}|p| = \mathfrak{h}_3,$$

$$\textbf{iii.} \ \mathfrak{M}_8\langle f \rangle = \int_{\mathbb{R}_+} f(|p|)|p|^8 \mathrm{d}|p| \leq \mathfrak{h}_8, \quad (10.1.14)$$

$$\textbf{iv.} \ \big\| f(\cdot) \,|\cdot|^2 \big\|_\infty \leq \mathfrak{h}_\infty < \infty \Big\}.$$

Theorem 10.1.1 (cf. [5]). *Let* $f_0(p) = f_0(|p|) \in \mathcal{S}$, $n_{c,0} > 0$. *Let* θ *be a strictly positive constant. Then, there exists a constant* $\mathcal{B}(f_0) > 0$, *depending on* f_0, *such that for any initial BEC having mass*

$$n_{c,0} \geq \mathcal{B}(f_0) - \mathfrak{M}_2\langle f_0 \rangle + \theta, \qquad \theta > 0, \quad (10.1.15)$$

the system (10.1.1)–(10.1.5)–(10.1.6) *has a unique classical solution* (f, n_c) *such that*

$$0 \leq f(t,p) = f(t, |p|) \in C\big([0,T]; \mathcal{S}\big) \cap C^1\big((0,T]; L^1(\mathbb{R}^3)\big),$$
$$\theta \leq n_c(t) = n_c[f](t) \in C\big([0,T]\big) \cap C^1\big((0,T]\big), \quad (10.1.16)$$

for any $T > 0$. *Momentum and energy are conserved for* $f(t, \cdot)$, *and the total mass of the system is also conserved*

$$\mathfrak{M}_2\langle f(t) \rangle + n_c[f](t) = \mathfrak{M}_2\langle f_0 \rangle + n_0.$$

Inequality (10.1.15) provides a constraint for the stability of the condensate. The condition essentially implies that if the density of the condensate is larger than a fixed constant, the system has a strong, global solution. We would like to thank Ricardo Alonso and Irene M. Gamba for pointing out this important fact. This result has been extended in [71] for near resonance wave turbulence equations of stratified flows in the ocean.

Another approximation has also been proposed in [124]. Based on the existence of a solution (f, n_c) satisfying $0 < \delta \leq n_c(t) \leq \mathfrak{M}_2\langle f_0\rangle + n_0$, proved in Theorem 10.1.1, we can replace the quantity n_c in (10.1.3)–(10.1.6) by some strictly positive constant N_c

$$\mathcal{Z}(p) \;=\; |p|\sqrt{\frac{p^2}{2m^2} + \frac{gN_c}{m}} \tag{10.1.17}$$

and

$$|\mathfrak{W}^{12}_{1,2,3}|^2 = \frac{9m^{1/2}}{64\pi^2(gN_c)^{3/2}}|p||p_1||p_2|. \tag{10.1.18}$$

The goal of this work is to understand the effect of the Bogoliubov dispersion relation. Indeed, due to the Delta functions in (10.1.1), the conservation of momentum and energy reads

$$\mathcal{Z}(p) = \mathcal{Z}(p_1) + \mathcal{Z}(p_2), \quad p = p_1 + p_2, \tag{10.1.19}$$

which shows that p, p_1, p_2 belong to the so-called resonance manifold. This creates a technical issue in the mathematical analysis of the system.

Let us now present the main result of [124]. For $m \geq 1$, we introduce the function space $\mathbb{L}^1_m(\mathbb{R}^3)$, defined by its finite norm

$$\|f\|_{\mathbb{L}^1_m} := \int_{\mathbb{R}^3} \left(1 + \mathcal{Z}(p)^{m/2}\right)|f(p)|\mathrm{d}p. \tag{10.1.20}$$

Theorem 10.1.2 (cf. [124]). Let $f_0(p) = f_0(|p|) \geq 0$ be a positive radial initial data in $\mathbb{L}^1_N(\mathbb{R}^3) \cap C(\mathbb{R}^3)$, for some $N \geq 2$. Suppose that the Cauchy problem (10.1.1)–(10.1.18)–(10.1.19) has a unique classical positive radial solution $f(t, p) = f(t, |p|) \geq 0$ in $C([0, \infty), \mathbb{L}^1_N(\mathbb{R}^3) \cap C(\mathbb{R}^3)) \cap C^1([0, \infty), \mathbb{L}^1_N(\mathbb{R}^3) \cap C(\mathbb{R}^3))$, $n_c \in C([0, \infty)) \cap C^1((0, \infty))$ and n_c satisfies $\delta \leq n_c(t)$, for some strictly positive constant δ. Assume that $f_0(p) \geq \rho_0$ on $B_{R_0} = \{|p| \leq R_0\}$ for some positive constants ρ_0, R_0. Then, for any time $T > 0$, there exist positive constants ρ_1, ρ_2 independent of p such that

$$f(t, p) \geq \rho_1 \exp(-\rho_2|p|^2), \qquad \forall\, t \geq T, \quad \forall\, p \in \mathbb{R}^3. \tag{10.1.21}$$

Results on the relaxation in time of solutions of linearized models of (10.1.1) around the Bose–Einstein distribution have been provided in [15, 62, 63].

There are several advantages of this type of approximation. In the first place, (10.1.1) is much simpler than the other two full models of Zaremba et al./Pomeau et al. and Reichl et al. Therefore, it provides an accessible model that can be studied using tools from mathematical analysis. Although it is simple, in this model, the coupling between the condensate and the thermal cloud is still kept. As a consequence, by studying (10.1.1), we expect to have some idea of how the condensate and the thermal cloud interact. An important feature of this approximation is the preservation of the conservation of mass

$$\frac{\partial}{\partial t}\left(\int_{\mathbb{R}^3} f(t,p)\mathrm{d}p + n_c(t)\right) = 0 \quad \forall t > 0.$$

The physical assumption behind this model is that the temperature of the system is sufficiently low and the $2 \leftrightarrow 2$ and $1 \leftrightarrow 3$ collisions are negligible in comparison with the $1 \leftrightarrow 2$ collisions. In these models, the approximations on the dispersion relation are the Bogoliubov (10.1.3) or the phonon dispersion approximation (10.1.5). The higher temperature case, in which the dispersion relation is of the classical type $|p|^2$, has been deeply studied in [46].

The first attempt to study a simplified spatial inhomogeneous system was made in [16]. The existence and uniqueness of solutions, as well as their asymptotic properties, were shown. In this work, the solutions $f(t,r,p)$ and $\Phi(t,r)$ of (5.0.2)–(5.0.3) are assumed to be 2π-periodic in r and $f(t,r,p)$ is assumed to be cylindrically symmetric in p. A later work [149], in which the periodic assumption in r is removed, but the collision operator is replaced by a linear one, leads to an analysis based on the normal form transformation of the Gross–Pitaevskii equation.

Conclusion: Taking into account the complicated forms of the Zaremba et al./Pomeau et al. model and the Reichl et al. model, each approximated system provides some insight into one component of the complex mathematical structures of the full models. The first works [14, 15, 16] have mathematically explored the dynamics of the coupling structure of the two equations. The work [5] has tried to approximate $W_{1,2,3}^{12}$ by an unbounded function, leading to a moment analysis based on an abstract fixed point theorem. The mathematical structures of the resonance manifolds and the normal form transformation of the Gross–Pitaevskii part of the system have been studied in [124, 149].

10.1.2 Second Approximation: Spatially Homogeneous Systems with $2 \leftrightarrow 2$ and $1 \leftrightarrow 3$ Collisions

In Chapters 5 and 6, it is shown that after linearizing the very low temperature systems around the equilibrium, treating the perturbation of n_c as a dynamical variable has no effect on the evolution of the linearized system. It is then possible to approximate the condensate density function by a constant. The approximated coupling system, which is a version of (6.3.4)–(6.3.5), reads

$$\frac{\partial}{\partial t}f = \mathbf{C}_{12}[f] + \mathbf{C}_{22}[f] + \theta_{3,1}\mathbf{C}_{31}[f], \quad f(0,p) = f_0(p), \forall (t,p) \in \mathbb{R}_+ \times \mathbb{R}^3,$$

$$\frac{\partial}{\partial t}\tilde{n}_c = -\int_{\mathbb{R}^3} \mathbf{C}_{12}[f]\mathrm{d}p, \quad \tilde{n}_c(0) = \tilde{n}_{c,0}, \forall t \in \mathbb{R}_+,$$

$$(10.1.22)$$

where

$$
\begin{aligned}
\mathbf{C}_{12}[f_1] =& \lambda_1 N_c \iint_{\mathbb{R}^3 \times \mathbb{R}^3} (W^{12}_{1,2,3})^2 \delta(p_1 - p_2 - p_3)\delta(E_{p_1} - E_{p_2} - E_{p_3}) \\
& \times [(1 + f_1)f_2 f_3 - f_1(1 + f_2)(1 + f_3)]\mathrm{d}p_2 \mathrm{d}p_3 \\
& - 2\lambda_1 n_c \iint_{\mathbb{R}^3 \times \mathbb{R}^3} (W^{12}_{1,2,3})^2 \delta(p_2 - p_1 - p_3)\delta(E_{p_2} - E_{p_1} - E_{p_3}) \\
& \times [(1 + f_2)f_1 f_3 - f_2(1 + f_1)(1 + f_3)]\mathrm{d}p_2 \mathrm{d}p_3,
\end{aligned}
$$

$$(10.1.23)$$

and

$$
\begin{aligned}
\mathbf{C}_{31}[f_1] =& \\
& \frac{2g^2 N_c}{3(2\pi)^2 \hbar^4} \iiint_{\mathbb{R}^3 \times \mathbb{R}^3 \times \mathbb{R}^3} \delta(p_1 - p_2 - p_3 - p_4)\delta(E_1 - E_2 - E_3 - E_4) \\
& \times (W^{31}_{1,2,3,4})^2 [f_1(f_2 + 1)(f_3 + 1)(f_4 + 1) - (f_1 + 1)f_2 f_3 f_4]\mathrm{d}p_2 \mathrm{d}p_3 \mathrm{d}p_4 \\
& - \frac{2g^2 N_c}{(2\pi)^2 \hbar^4} \iiint_{\mathbb{R}^3 \times \mathbb{R}^3 \times \mathbb{R}^3} \delta(p_2 - p_1 - p_3 - p_4)\delta(E_2 - E_1 - E_3 - E_4) \\
& \times (W^{31}_{1,2,3,4})^2 [f_2(f_1 + 1)(f_3 + 1)(f_4 + 1) - (f_2 + 1)f_1 f_3 f_4]\mathrm{d}p_2 \mathrm{d}p_3 \mathrm{d}p_4,
\end{aligned}
$$

$$(10.1.24)$$

with N_c being the density of the condensate at equilibrium. Notice that $\tilde{n}_c$ is the perturbation of N_c around the equilibrium. This approximation also has the conservation of mass

$$
\frac{\partial}{\partial t} \left(\int_{\mathbb{R}^3} f(t, p)\mathrm{d}p + \tilde{n}_c(t) \right) = 0 \quad \forall t > 0.
$$

Since the equation for f is decoupled from the equation for $\tilde{n}_c$, we often consider only the equation for f

$$
\frac{\partial}{\partial t} f = \mathbf{C}_{12}[f] + \mathbf{C}_{22}[f] + \theta_{3,1}\mathbf{C}_{31}[f], \quad f(0, p) = f_0(p), \forall (t, p) \in \mathbb{R}_+ \times \mathbb{R}^3.
$$

$$(10.1.25)$$

The model (10.1.25) is based on the assumption that the system is very close to equilibrium, following the discussions in Chapters 5, 6, and 8.

There are several advantages of this type of approximation. It is clear that (10.1.25) is much simpler than the other two full models; and they can be studied using tools from mathematical analysis. Based on the same perturbative argument, the hydrodynamic modes are computed in Chapter 8. To be more precise, in order to calculate the eigenvalues and eigenfunctions of the bogolon collision operator $\mathcal{C}(p_1, p_2)$, n_c is approximated by N_c in the parameters given in (8.2.6). It is also shown in Chapter 8 that the obtained results are in agreement with the Steinhauer experiment and the Lee–Yang theory.

In [150], the existence and uniqueness of global classical positive radial solutions of the following model of (10.1.25) have been studied for $\theta_{3,1} = 0$

$$
\frac{\partial}{\partial t} f = \mathfrak{Q}_{12}[f] + \mathfrak{Q}_{22}[f], \quad f(0, p) = f_0(|p|), \forall p \in \mathbb{R}^3, \tag{10.1.26}
$$

where

$$\mathfrak{Q}_{12}[f](t,r,p_1) := \tag{10.1.27}$$

$$N_c \iint_{\mathbb{R}^3 \times \mathbb{R}^3} K^{12}(p_1,p_2,p_3)\delta(p_1 - p_2 - p_3)\delta(\mathcal{Z}_{p_1} - \mathcal{Z}_{p_2} - \mathcal{Z}_{p_3})$$

$$\times [(1+f_1)f_2 f_3 - f_1(1+f_2)(1+f_3)]dp_2 dp_3$$

$$- 2N_c \iint_{\mathbb{R}^3 \times \mathbb{R}^3} K^{12}(p_1,p_2,p_3)\delta(p_2 - p_1 - p_3)\delta(\mathcal{Z}_{p_2} - \mathcal{Z}_{p_1} - \mathcal{Z}_{p_3})$$

$$\times [(1+f_2)f_1 f_3 - f_2(1+f_1)(1+f_3)]dp_2 dp_3,$$

and

$$\mathfrak{Q}_{22}[f] = \tag{10.1.28}$$

$$\iiint_{\mathbb{R}_+ \times \mathbb{R}_+ \times \mathbb{R}_+} K^{22}(p_1,p_2,p_3,p_4)\frac{\min\{|p_1|,|p_2|,|p_3|,|p_4|\}|p_1||p_2||p_3||p_4|}{|p_1|^2}$$

$$\times \delta(\mathcal{Z}_{p_1} + \mathcal{Z}_{p_2} - \mathcal{Z}_{p_3} - \mathcal{Z}_{p_4})$$

$$\times [f_3 f_4 (1+f_1+f_2) - f_1 f_2 (1+f_3+f_4)]d|p_2|d|p_3|d|p_4|.$$

Since the temperature of the system is very low, the energy is assumed to be the same as (10.1.17)

$$\mathcal{Z}_p = \mathcal{Z}(p) = \sqrt{\kappa_1 |p|^2 + \kappa_2 |p|^4}, \tag{10.1.29}$$

$$\kappa_1 = \frac{gN_c}{m} > 0, \quad \kappa_2 = \frac{1}{4m^2} > 0.$$

The transition probability kernels are given by

$$K^{12}(p_1,p_2,p_3) = (W^{12}_{1,2,3})^2 \tag{10.1.30}$$

and

$$K^{22}(p_1,p_2,p_3,p_4) = (W^{22}_{1,2,3,4})^2 \chi_{\{|p_1|<\epsilon\}}\chi_{\{|p_2|<\epsilon\}}\chi_{\{|p_3|<\epsilon\}}\chi_{\{|p_4|<\epsilon\}}, \tag{10.1.31}$$

where $\chi_{\{|p|<\epsilon\}}$ is the characteristic function of $\{|p| < \epsilon\}$. Without loss of generality, we could assume $\epsilon = 1$,

$$K^{22}(p_1,p_2,p_3,p_4) = (W^{22}_{1,2,3,4})^2 \chi_{\{|p_1|<1\}}\chi_{\{|p_2|<1\}}\chi_{\{|p_3|<1\}}\chi_{\{|p_4|<1\}}. \tag{10.1.32}$$

We also define the following cut-off version of $(W^{31}_{1,2,3,4})^2$

$$K^{31}(p_1,p_2,p_3,p_4) = (W^{31}_{1,2,3,4})^2 \chi_{\{|p_1|<1\}}\chi_{\{|p_2|<1\}}\chi_{\{|p_3|<1\}}\chi_{\{|p_4|<1\}}. \tag{10.1.33}$$

This cut-off assumption is also valid in the near-equilibrium regime. The following theorem was obtained in [150].

Theorem 10.1.3 (cf. [150]). *Suppose that $f_0(p) = f_0(|p|) \geq 0$. For any positive numbers $R, T > 0$, there exist two constants $n, n^* > 1$ and $c_{n_*}(R, T)$ such that $c_{n_*}(R, T)$ goes to infinity as T or R goes to infinity and if the initial condition satisfies*

$$\int_{\mathbb{R}^3} \mathcal{Z}_p^{n^*} f_0(p) \mathrm{d}p < \mathfrak{c}_{n_*}(R, 0), \quad and \quad \int_{\mathbb{R}^3} f_0(p) \mathrm{d}p < R,$$

then there exists a unique classical positive radial solution

$$f(t, p) = f(t, |p|) \in C([0, T], \mathbb{L}^1_{2n}(\mathbb{R}^3)) \cap C^1((0, T), \mathbb{L}^1_{2n}(\mathbb{R}^3))$$

of (10.1.26).

In [50], the authors considered Discrete Velocity Models of (10.1.25), in which the kinetic equation is discretized on a finite lattice. The discretized system has a structure very similar to dynamical systems of chemical reaction networks. The quest to understand the qualitative behavior of deterministically modeled chemical reaction systems, such as the existence of positive equilibria, stability properties of equilibria, and the non-extinction, or persistence of species, has become a central research area [7, 12, 69, 70, 83]. Early works of Feinberg, Horn, and Jackson [69, 82, 84] have shown that there exists a unique complex-balanced equilibrium within the interior of each positive compatibility class or invariant manifold if a reaction network with deterministic mass-action kinetics admits a complex-balanced equilibrium. The name "toric dynamical system" has been proposed in [48], to underline the tight connection with the algebraic study of toric varieties. The Global Attractor Conjecture (cf. [6, 9]), which says that the complex balanced equilibrium of a toric dynamical system is a globally attracting point within each linear invariant subspace, is considered the most important problem in the theory of toric dynamical systems. A proof of the conjecture for low-dimensional systems was proposed in [49], and the complete proof was given by Craciun in [47]. Using the techniques developed in [47], convergence to equilibrium results have been obtained for Discrete Velocity Models of (10.1.25) as well as other models in wave turbulence theory [50, 157]. Several results on the well-posedness, boundary layers and related issues on Discrete Velocity Models for $\theta_{31} = 0$ have been provided in [23, 24, 25]. Below, we only recall the result obtained in [50].

Define the lattice

$$\mathcal{L}_{\varsigma, R} = \{p = (\varsigma k_1, \varsigma k_2, \varsigma k_3) \quad | \quad k_1, k_2, k_3 \in \mathbb{Z}, |k_1|, |k_2|, |k_3| < R\}.$$

We obtain a simplified and discretized system on $\mathcal{L}_{\varsigma, R}$ of (10.1.25) for $\theta_{3.1} = 1$ (ultra-low temperature regime)

$$\dot{f}_{p_1} = \sum_{\substack{p_2, p_3 \in \mathcal{L}_{\varsigma, R}, \\ p_1 - p_2 - p_3 = 0, \\ E(p_1) - E(p_2) - E(p_3) = 0}} K^{12} \left\{ (f_{p_1} + 1) f_{p_2} f_{p_3} - f_{p_1} (f_{p_2} + 1)(f_{p_3} + 1) \right\}$$

$$- 2 \sum_{\substack{p_2, p_3 \in \mathcal{L}_{\varsigma, R}, \\ p_1 + p_2 - p_3 = 0, \\ E(p_1) + E(p_2) - E(p_3) = 0}} K^{12} \left\{ (f_{p_3} + 1) f_{p_1} f_{p_2} - f_{p_3} (f_{p_1} + 1)(f_{p_2} + 1) \right\}$$

$$+ \sum_{\substack{p_2, p_3, p_4 \in \mathcal{L}_{\varsigma, R}, \\ p_1 + p_2 - p_3 - p_4 = 0, \\ E(p_1) + E(p_2) - E(p_3) - E(p_4) = 0}} K^{22} \left\{ (f_{p_1} + 1)(f_{p_2} + 1) f_{p_3} f_{p_4} - f_{p_1} f_{p_2} (f_{p_3} + 1)(f_{p_4} + 1) \right\}$$

$$+ \sum_{\substack{p_2, p_3, p_4 \in \mathcal{L}_{\varsigma, R}, \\ p_1 - p_2 - p_3 - p_4 = 0, \\ E(p_1) - E(p_2) - E(p_3) - E(p_4) = 0}} K^{31} \left\{ (f_{p_1} + 1) f_{p_2} f_{p_3} f_{p_4} - f_{p_1} (f_{p_2} + 1)(f_{p_3} + 1)(f_{p_4} + 1) \right\}$$

$$- 3 \sum_{\substack{p_2, p_3 \in \mathcal{L}_{\varsigma, R}, \\ p_1 + p_2 + p_3 - p_4 = 0, \\ E(p_1) + E(p_2) + E(p_3) - E(p_4) = 0}} K^{31} \left\{ (f_{p_4} + 1) f_{p_1} f_{p_2} f_{p_4} - f_{p_4} (f_{p_1} + 1)(f_{p_2} + 1)(f_{p_3} + 1) \right\},$$

$$(10.1.34)$$

$$\forall p_1 \in \mathcal{L}_{\varsigma, R}.$$

Theorem 10.1.4 (cf. [50]). *Suppose that the initial condition is radial and positive $f_0(p) = f_0(|p|) \geq 0$ and $E(p)$ is approximated by the phonon dispersion relation $E(p) = c_P |p|$, $(c_P > 0)$. The solution f_p, $p \in \mathcal{L}_{\varsigma, R}$, of the system (10.1.34) relaxes to the unique equilibrium $\frac{c''}{e^{c'|p|} - 1}$ exponentially in time*

$$\sup_{p \in \mathcal{L}_{\varsigma, R}} \left| f_p - \frac{c''}{e^{c'|p|} - 1} \right| \leq C_{D,1} e^{-C_{D,2} t},$$

where $C_{D,1}, C_{D,2}, c', c'' > 0$ are universal constants.

Conclusion: In the above types of approximations, the collisions $1 \leftrightarrow 3$ and $2 \leftrightarrow 2$ are kept and the conservation of mass is also preserved. The assumption that the system is very close to equilibrium allows one to drop the equation of $\tilde{n}_c$ since it is decoupled from the kinetic equation.

10.2 A Well-Posedness Theory

According to the famous French mathematician Jacques Hadamard, mathematical models of physical phenomena should have solutions which are unique and their behaviors should change continuously with the initial conditions. This section is devoted such a theory of the following model of (10.1.25) in the case $\theta_{3,1} = 0$

$$\frac{\partial}{\partial t} f = Q[f] = \mathcal{Q}_{12}[f] + \mathcal{Q}_{22}[f], \quad f(0, p) = f_0(|p|), \forall p \in \mathbb{R}^3, \quad (10.2.1)$$

where

$$\mathcal{Q}_{12}[f](t,r,p_1) := \tag{10.2.2}$$

$$N_c \iint_{\mathbb{R}^3 \times \mathbb{R}^3} K^{12}(p_1,p_2,p_3)\delta(p_1 - p_2 - p_3)\delta(\mathcal{Z}_{p_1} - \mathcal{Z}_{p_2} - \mathcal{Z}_{p_3})$$

$$\times [(1+f_1)f_2f_3 - f_1(1+f_2)(1+f_3)]\mathrm{d}p_2\mathrm{d}p_3$$

$$- 2N_c \iint_{\mathbb{R}^3 \times \mathbb{R}^3} K^{12}(p_1,p_2,p_3)\delta(p_2 - p_1 - p_3)\delta(\mathcal{Z}_{p_2} - \mathcal{Z}_{p_1} - \mathcal{Z}_{p_3})$$

$$\times [(1+f_2)f_1f_3 - f_2(1+f_1)(1+f_3)]\mathrm{d}p_2\mathrm{d}p_3,$$

and

$$\mathcal{Q}_{22}[f] = \iiint_{\mathbb{R}^3 \times \mathbb{R}^3 \times \mathbb{R}^3} K^{22}(p_1,p_2,p_3,p_4)\delta(p_1 + p_2 - p_3 - p_4) \tag{10.2.3}$$

$$\times \delta(\mathcal{Z}_{p_1} + \mathcal{Z}_{p_2} - \mathcal{Z}_{p_3} - \mathcal{Z}_{p_4})$$

$$\times [f_3f_4(1+f_1+f_2) - f_1f_2(1+f_3+f_4)]\mathrm{d}p_2\mathrm{d}p_3\mathrm{d}p_4.$$

The kernels are given in (10.1.30) and (10.1.32). The dispersion relation is assumed to be the same as (10.1.17) and (10.1.29).

As discussed in the previous section. The equation is a part of the following model of (10.1.22)

$$\frac{\partial}{\partial t}f = \mathcal{Q}_{12}[f] + \mathcal{Q}_{22}[f], \quad f(0,p) = f_0(|p|), \forall (t,p) \in \mathbb{R}_+ \times \mathbb{R}^3,$$
$$\frac{\partial}{\partial t}\tilde{n}_c = -\int_{\mathbb{R}^3} \mathcal{Q}_{12}[f]\mathrm{d}p, \quad \tilde{n}_c(0) = \tilde{n}_{c,0}, \forall t \in \mathbb{R}_+. \tag{10.2.4}$$

The system (10.2.4) conserves the total particle number. As discussed in Chapters 5 and 6, the contribution of $\tilde{n}_c$ to the long time dynamics of the kinetic equation is negligible. Consequently, we only need to study (10.2.1). In Theorem 10.2.1, we show that for any positive radial initial condition, even far away from equilibrium, equation (10.2.1) always has a unique global positive and radial solution.

Let us start by defining the functional spaces

$$L_m^1(\mathbb{R}^3) = \left\{ f \ \middle| \ \|f\|_{L_m^1} := \int_{\mathbb{R}^3} |p|^m |f(p)|\mathrm{d}p < \infty \right\}, \tag{10.2.5}$$

$$\mathcal{L}_m^1(\mathbb{R}^3) = \left\{ f \ \middle| \ \|f\|_{\mathcal{L}_m^1} := \int_{\mathbb{R}^3} |f(p)|\mathcal{Z}_p^{m/2}\mathrm{d}p < \infty \right\}. \tag{10.2.6}$$

We also adopt the functional spaces (10.1.9), (10.1.10), (10.1.12), (10.1.13), (10.1.20) in this section.

The main result of this section is the following theorem, which is an extension of Theorem 10.1.3 obtained in [150].

Theorem 10.2.1. *Suppose that $f_0(p) = f_0(|p|) \geq 0$. Let n, n^* be two positive integers, $n > 1$, and n_* be an odd number, $n^* > n+4$. Let $[0, T]$ be an arbitrary time interval and $\mathcal{R}$ be an arbitrary positive number. There exists a positive constant $\mathfrak{c}_{n_*}(\mathcal{R}, T)$ depending on $\mathcal{R}$ and T, and increasing in both $\mathcal{R}$, T, such that if*

$$\int_{\mathbb{R}^3} \mathcal{Z}_p^{n^*} f_0(p) \mathrm{d}p < \mathfrak{c}_{n_*}(\mathcal{R}, 0) \text{ and } \int_{\mathbb{R}^3} f_0(p) \mathrm{d}p < \mathcal{R},$$

then there exists a unique classical positive radial solution

$$f(t,p) = f(t, |p|) \in C([0, T], \mathbb{L}_{2n}^1(\mathbb{R}^3)) \cap C^1((0, T), \mathbb{L}_{2n}^1(\mathbb{R}^3))$$

of (10.2.1).

The rest of this section is devoted to the full mathematical proof of Theorem 10.2.1. This proof is more complicated than the proof of Theorem 10.1.3 obtained in [150] for (10.1.26) since we need to analyze the resonance manifold of $\mathcal{Q}_{22}$. We thank Avy Soffer for fruitful discussions on this generalization.

In Subsection 10.2.1, the definitions of mass, moment and energy, and the resonance manifolds for both $\mathcal{Q}_{12}$, $\mathcal{Q}_{22}$ will be introduced. The conservation of momentum and energy will be introduced in Subsection 10.2.2. We will analyze the resonance manifolds for $\mathcal{Q}_{12}$ and $\mathcal{Q}_{22}$ in Subsections 10.2.3 and 10.2.4. An estimate of the mass will be given in Subsection 10.2.5. We will bound the two operators $\mathcal{Q}_{12}$, $\mathcal{Q}_{22}$ in Subsections 10.2.6 and 10.2.7. These bounds will be used to estimate moments of all orders greater than 1 of the solution on any time interval $[0, T]$ in Subsection 10.2.8. The two operators will be shown to be Hölder continuous in Subsection 10.2.9. The proof of the theorem will be given in the last subsection. The authors would like to thank Ricardo Alonso and Irene M. Gamba for the illuminating discussions on the topic.

10.2.1 Mass, Moment and Energy

In this section, we introduce the precise definitions of mass, moment and energy, as well as the resonance manifolds for both $\mathcal{Q}_{12}$, $\mathcal{Q}_{22}$ and various forms of collision operators, to be used later in the proof. We will make use of the following notation

$$m_k[f] = \int_{\mathbb{R}^3} \mathcal{Z}^k(p_1) f(p_1) \mathrm{d}p_1. \tag{10.2.7}$$

For convenience, we introduce

$$\mathcal{Q}_{12}[f] = Q_{12}^1[f] + Q_{12}^2[f], \tag{10.2.8}$$

in which

$$Q_{12}^1[f] :=$$

$$\iint_{\mathbb{R}^3 \times \mathbb{R}^3} \mathcal{K}^{12}(p_1, p_2, p_3) \Big[f(p_2)f(p_3) - f(p_1)(f(p_2) + f(p_3) + 1) \Big] \mathrm{d}p_2 \mathrm{d}p_3,$$

$$Q_{12}^2[f] :=$$

$$-2 \iint_{\mathbb{R}^3 \times \mathbb{R}^3} \mathcal{K}^{12}(p_2, p_1, p_3) \Big[f(p_1)f(p_3) - f(p_2)(f(p_1) + f(p_3) + 1) \Big] \mathrm{d}p_2 \mathrm{d}p_3,$$

where the collision kernel is as follows

$$\mathcal{K}^{12}(p_1, p_2, p_3) = N_c K^{12}(p_1, p_2, p_3)\Big(\delta(\mathcal{Z}(p_1) - \mathcal{Z}(p_2) - \mathcal{Z}(p_3))\delta(p_1 - p_2 - p_3)\Big).$$

The energy surfaces/resonance manifolds of $\mathcal{Q}_{12}$ are now

$$S_p^0 := \Big\{ x \in \mathbb{R}^3 \ : \ \mathcal{Z}(p - x) + \mathcal{Z}(x) = \mathcal{Z}(p) \Big\},$$

$$S_p^1 := \Big\{ x \in \mathbb{R}^3 \ : \ \mathcal{Z}(p + x) = \mathcal{Z}(p) + \mathcal{Z}(x) \Big\}, \qquad (10.2.9)$$

$$S_p^2 := \Big\{ x \in \mathbb{R}^3 \ : \ \mathcal{Z}(x) = \mathcal{Z}(p) + \mathcal{Z}(x - p) \Big\}$$

for all $p \in \mathbb{R}^3 \setminus \{0\}$.

We also define the functions for $x \in \mathbb{R}^3$

$$H_0^p(x) := \mathcal{Z}(p - x) + \mathcal{Z}(x) - \mathcal{Z}(p),$$
$$H_1^p(x) := \mathcal{Z}(p + x) - \mathcal{Z}(p) - \mathcal{Z}(x), \qquad (10.2.10)$$
$$H_2^p(x) := \mathcal{Z}(x) - \mathcal{Z}(p) - \mathcal{Z}(x - p).$$

We set

$$\bar{K}^{12}(p_1, p_2, p_3) = \lambda_1 n_c K^{12}(p_1, p_2, p_3).$$

The nature of the Dirac delta function allows us to express the collision operators in the form

$$Q_{12}^1[f] := \int_{S_{p_1}^0} \bar{K}_0^{12}(p_1, p_1 - p_3, p_3)$$

$$\times \Big[f(p_1 - p_3)f(p_3) - f(p_1)(f(p_1 - p_3) + f(p_3) + 1) \Big] \, \mathrm{d}\sigma(p_3),$$

$$Q_{12}^2[f] := 2 \int_{S_{p_1}^1} \bar{K}_1^{12}(p_1 + p_3, p_1, p_3)$$

$$\times \Big[f(p_1 + p_3)(f(p_1) + f(p_3) + 1) - f(p_1)f(p_3) \Big] \, \mathrm{d}\sigma(p_3),$$

where

$$\bar{K}_0^{12}(p_1, p_1 - p_3, p_3) = \frac{\bar{K}^{12}(p_1, p_1 - p_3, p_3)}{|\nabla H_p^0(p_3)|}$$

and

$$\bar{K}_1^{12}(p_1 + p_3, p_1, p_3) = \frac{\bar{K}^{12}(p_1 + p_3, p_1, p_3)}{|\nabla H_p^1(p_3)|}.$$

We also split the collision operator $\mathcal{Q}_{12}[f]$ as the sum of gain and loss terms:

$$\mathcal{Q}_{12}[f] = Q_{12}^{\mathrm{gain}}[f] - Q_{12}^{\mathrm{loss}}[f], \qquad (10.2.11)$$

in which

$$Q_{12}^{\mathrm{gain}}[f] := \int_{S_{p_1}^0} \bar{K}_0^{12}(p_1, p_1 - p_3, p_3) f(p_1 - p_3) f(p_3) \, d\sigma(p_3)$$

$$+ 2 \int_{S_{p_1}^1} \bar{K}_1^{12}(p_1 + p_3, p_1, p_3) f(p_1 + p_3)\Big(f(p_1) + f(p_3) + 1\Big) \, d\sigma(p_3),$$

$$Q_{12}^{\mathrm{loss}}[f] := f Q_{12}^-[f],$$

$$Q_{12}^-[f] := \int_{S_{p_1}^0} \bar{K}_0^{12}(p_1, p_1 - p_3, p_3)\Big(f(p_1 - p_3) + f(p_3) + 1\Big) \, d\sigma(p_3)$$

$$+ 2 \int_{S_{p_1}^1} \bar{K}_1^{12}(p_1 + p_3, p_1, p_3) f(p_3) \, d\sigma(p_3).$$

As for $\mathcal{Q}_{12}$, $\mathcal{Q}_{22}$ can be split into gain and loss operators, as follows

$$\mathcal{Q}_{22}[f] = Q_{22}^{\mathrm{gain}}[f] - Q_{22}^{\mathrm{loss}}[f], \qquad (10.2.12)$$

in which

$$Q_{22}^{\mathrm{gain}}[f] := \lambda_2 \iiint_{\mathbb{R}^{3\times 3}} \mathcal{K}^{22}(p_1, p_2, p_3, p_4)(1 + f(p_1))(1 + f(p_2))$$

$$\times f(p_3) f(p_4) dp_2 dp_3 dp_4,$$

$$Q_{22}^{\mathrm{loss}}[f] := f Q_{22}^-[f],$$

$$Q_{22}^-[f] := \lambda_2 \iiint_{\mathbb{R}^{3\times 3}} \mathcal{K}^{22}(p_1, p_2, p_3, p_4) f(p_2)$$

$$\times (1 + f(p_3))(1 + f(p_4)) dp_2 dp_3 dp_4,$$

and

$$\mathcal{K}^{22}(p_1, p_2, p_3, p_4) = \lambda_2 K^{22}(p_1, p_2, p_3, p_4)\delta(p_1 + p_2 - p_3 - p_4)$$

$$\times \delta(\mathcal{Z}_{p_1} + \mathcal{Z}_{p_2} - \mathcal{Z}_{p_3} - \mathcal{Z}_{p_4}).$$

Define the resonant manifold $\mathcal{S}_{p_1,p_2}$ of $\mathcal{Q}_{22}$ to be the zero set of

$$\mathcal{G}^{p_1,p_2}(x) := \mathcal{Z}(p_1 + p_2 - x) + \mathcal{Z}(x) - \mathcal{Z}(p_1) - \mathcal{Z}(p_2) = 0, \qquad (10.2.13)$$

which leads to the following representation

$$\iint_{\mathbb{R}^{2\times 3}} \mathcal{K}^{22}(p_1, p_2, p_3, p_4) F(p_3) dp_3 dp_4 = \qquad (10.2.14)$$

$$\int_{\mathcal{S}_{p_1,p_2}} \frac{\mathcal{K}^{22}(p_1,p_2,p_3,p_1+p_2-p_3)F(p_3)}{|\nabla\mathcal{G}^{p_1,p_2}(p_3)|} \mathrm{d}\mu(p_3)$$

where μ is the surface measure on $\mathcal{S}_{p,p_1}$.

Q is also split into the sum of a gain and a loss operator

$$Q[f] = Q^{\mathrm{gain}}[f] - Q^{\mathrm{loss}}[f], \tag{10.2.15}$$

in which

$$Q^{\mathrm{gain}}[f] = Q^{\mathrm{gain}}_{12}[f] + Q^{\mathrm{gain}}_{22}[f], \quad Q^{\mathrm{loss}}[f] = Q^{\mathrm{loss}}_{12}[f] + Q^{\mathrm{loss}}_{22}[f]$$

and $Q^{\mathrm{loss}}[f] = f Q^-[f]$, with $Q^-[f] = Q^-_{12}[f] + Q^-_{22}[f]$.

10.2.2 Conservation of Momentum and Energy

In this section, we present the basic properties of smooth solutions of (10.2.1), including the conservation of momentum and energy.

Lemma 10.2.1. *The equation*

$$\int_{\mathbb{R}^3} Q[f](p_1)\varphi(p_1)\mathrm{d}p_1$$

$$= \iiint_{\mathbb{R}^3\times\mathbb{R}^3\times\mathbb{R}^3} R_{12}[f](p_1,p_2,p_3)\big(\varphi(p_1) - \varphi(p_2) - \varphi(p_3)\big)\,\mathrm{d}p_1\mathrm{d}p_2\mathrm{d}p_3$$

$$+ \frac{1}{2}\iiiint_{\mathbb{R}^3\times\mathbb{R}^3\times\mathbb{R}^3\times\mathbb{R}^3} R_{22}[f](p_1,p_2,p_3,p_4)$$

$$\times \big(\varphi(p_1) + \varphi(p_2) - \varphi(p_3) - \varphi(p_4)\big)\,\mathrm{d}p_1\mathrm{d}p_2\mathrm{d}p_3\mathrm{d}p_4$$

holds for any smooth test function φ, in which

$$R_{12}[f](p_1,p_2,p_3)$$
$$= \lambda_1 N_c K^{12}(p_1,p_2,p_3)\delta(p_1 - p_2 - p_3)\delta(\mathcal{Z}_{p_1} - \mathcal{Z}_{p_2} - \mathcal{Z}_{p_3})$$
$$\times [(1 + f(p_1))f(p_2)f(p_3) - f(p_1)(1 + f(p_2))(1 + f(p_3))],$$
$$= R_{22}[f](p_1,p_2,p_3,p_4)$$
$$\lambda_2 K^{22}(p_1,p_2,p_3,p_4)\delta(p_1 + p_2 - p_3 - p_4)$$
$$\times \delta(\mathcal{Z}_{p_1} + \mathcal{Z}_{p_2} - \mathcal{Z}_{p_3} - \mathcal{Z}_{p_4})[(1 + f(p_1))(1 + f(p_2))f(p_3)f(p_4)$$
$$- f(p_1)f(p_2)(1 + f(p_3))(1 + f(p_4))].$$

Proof. In view of (10.2.1), we obtain

$$\int_{\mathbb{R}^3} \mathcal{Q}_{12}[f](p_1)\varphi(p_1)\mathrm{d}p_1 + \int_{\mathbb{R}^3} \mathcal{Q}_{22}[f](p_1)\varphi(p_1)\mathrm{d}p_1 = I_1 + I_2,$$

in which

$$I_1 := \iiint_{\mathbb{R}^3 \times \mathbb{R}^3 \times \mathbb{R}^3} \Big(R_{12}[f](p_1, p_2, p_3) - R_{12}[f](p_2, p_1, p_3) - R_{12}[f](p_3, p_2, p_1) \Big)$$
$$\times \varphi(p_1) \, \mathrm{d}p_1 \mathrm{d}p_2 \mathrm{d}p_3,$$

$$I_2 := \iiint_{\mathbb{R}^3 \times \mathbb{R}^3 \times \mathbb{R}^3 \times \mathbb{R}^3} R_{22}[f](p_1, p_2, p_3, p_4) \varphi(p_1) \, \mathrm{d}p_1 \mathrm{d}p_2 \mathrm{d}p_3 \mathrm{d}p_4.$$

The conclusion of the lemma can be proved simply by switching the variables $p_1 \leftrightarrow p_2$, $p_1 \leftrightarrow p_3$ in the integrals of I_1 and $(p_1, p_2) \leftrightarrow (p_2, p_1)$, $(p_1, p_2) \leftrightarrow (p_3, p_4)$ in the integrals of I_2, respectively, as in [124].   □

As a consequence, the following two corollaries can also be derived.

Corollary 1 (Conservation of momentum and energy). *Smooth solutions $f(t, p)$ of (10.2.1) satisfy the following conservation of energy and momentum*

$$\int_{\mathbb{R}^3} f(t, p) p \mathrm{d}p = \int_{\mathbb{R}^3} f_0(p) p \mathrm{d}p, \qquad (10.2.16)$$

$$\int_{\mathbb{R}^3} f(t, p) \mathcal{Z}(p) \mathrm{d}p = \int_{\mathbb{R}^3} f_0(p) \mathcal{Z}(p) \mathrm{d}p \qquad (10.2.17)$$

for all $t \geq 0$.

Proof. This is a consequence of Lemma 10.2.1 by taking $\varphi(p) = p$ or $\mathcal{Z}(p)$.   □

10.2.3 Resonance Manifold/Energy Surface Analysis for $\mathcal{Q}_{12}$

As discussed in Section 10.1, [124] proposes a mathematical framework for studying energy surface integrals. Below, we will establish estimates on the energy surface integrals on S_p^1 and S_p^2 following this framework. We would like to thank Toan T. Nguyen for his important contribution in this analysis.

Remark 10.2.2. *In the context of the phonon Boltzmann and the Hubbard–Boltzmann equations, the technical problems of manifolds are normally treated by oscillatory integral methods (cf. [114, 115, 117]).*

Lemma 10.2.2. *Let S_p^0 be the surface defined in (10.2.9). The following estimate holds*

$$\int_{S_p^0} \frac{K^{12}(p, w, p - w) |w|^{k_1} |p - w|^{k_2}}{|\nabla H_0^p(w)|} \mathrm{d}\sigma(w) \gtrsim |p|^{k_1 + k_2 + 1} \min\{1, |p|\}^{k_1 + k_2 + 7},$$

$$(10.2.18)$$

for some non-negative constants k_1, k_2.

Moreover, for any radial and positive function $F(\cdot) : \mathbb{R}^3 \to \mathbb{R}$

$$F(u) = F(|u|),$$

the following estimate holds true

$$\int_{S_p^0} \frac{F(|w|)}{|\nabla H_0^p(|w|)|} \, d\sigma(w) \lesssim \int_0^{|p|} |u| F(|u|) \, d|u|. \qquad (10.2.19)$$

Proof. By definition, S_p^0 is the surface of all vectors w solving the equation $\mathcal{Z}(p-w) + \mathcal{Z}(w) = \mathcal{Z}(p)$. For the special choices, $w = 0$ and $w = p$, the above equation is automatically satisfied, hence $\{0, p\} \subset S_p^0$. Consider $\mathcal{Z}(\varrho)$ as a function of $|\varrho|$: $\mathcal{Z}(\varrho) = \mathcal{Z}(|\varrho|)$. Then it follows that

$$\mathcal{Z}'(|\varrho|) = \frac{\kappa_1 + 2\kappa_2 |\varrho|^2}{\sqrt{\kappa_1 + \kappa_2 |\varrho|^2}} > 0,$$

which means that the function $\mathcal{Z}(|\varrho|)$ is strictly increasing in $|p|$. Since for all $w \in S_p^0 \backslash \{0, p\}$, $\mathcal{Z}(|p-w|) < \mathcal{Z}(|p|)$ and $\mathcal{Z}(|w|) < \mathcal{Z}(|p|)$, by the monotonicity of $\mathcal{Z}(|\varrho|)$, we find $|w| < |p|$ and $|p - w| < |p|$, for all $w \in S_p \backslash \{0, p\}$. As a result, the resonance manifold S_p^0 is a subset of $\overline{B(0, |p|)} \cap \overline{B(p, |p|)}$. Now, let us recall $H_0^p(w) := \mathcal{Z}(p - w) + \mathcal{Z}(w) - \mathcal{Z}(p)$, whose directional derivative in the direction of w can be computed as

$$\nabla_w H_0^p = \frac{w - p}{|p - w|} \mathcal{Z}'(|p - w|) + \frac{w}{|w|} \mathcal{Z}'(|w|). \qquad (10.2.20)$$

Let w be of the form $w = \gamma p + q e_0, \gamma, q \in \mathbb{R}_+, e_0 \cdot p = 0$. The derivative of H with respect to q is then

$$\partial_q H_0^p = \partial_q w \cdot \nabla_w H_0^p = e_0 \cdot \nabla_w H = q|e_0|^2 \left[\frac{\mathcal{Z}'(p-w)}{|p - w|} + \frac{\mathcal{Z}'(w)}{|w|} \right] > 0, \qquad (10.2.21)$$

which implies that $H(w)$ is strictly increasing in q.

Setting $q = 0$ and $\gamma \in (0, 1)$, we will show that

$$H_0^p(w) = H_0^p(\gamma p) < 0. \qquad (10.2.22)$$

Now, observe that

$$\sqrt{(\kappa_1 + \kappa_2 \gamma^2 |p|^2)(\kappa_1 + \kappa_2 (1 - \gamma)^2 |p|^2)} < \kappa_1 + \kappa_2 (\gamma^2 - \gamma + 2)|p|^2 \quad \text{for } p \neq 0.$$

Multiplying both sides of the above inequality by $2\gamma(1 - \gamma)|p|^2$, we find

$$2\sqrt{(\kappa_1 \gamma^2 + \kappa_2 \gamma^4 |p|^2)(\kappa_1 (1 - \gamma)^2 + \kappa_2 (1 - \gamma)^4 |p|^2)}|p|^2$$
$$< 2\kappa_1 \gamma(1 - \gamma)|p|^2 + 2\kappa_2 \gamma(1 - \gamma)(\gamma^2 - \gamma + 2)|p|^4. \qquad (10.2.23)$$

Adding $\kappa_1 \gamma^2 |p|^2 + \kappa_2 \gamma^4 |p|^4 + \kappa_1(1-\gamma)^2 |p|^2 + \kappa_2(1-\gamma)^4 |p|^4$ to both sides of the above inequality yields

$$
\kappa_1 \gamma^2 |p|^2 + \kappa_2 \gamma^4 |p|^4 + \kappa_1(1-\gamma)^2 |p|^2 + \kappa_2(1-\gamma)^4 |p|^4
$$
$$
+ 2\sqrt{(\kappa_1 \gamma^2 + \kappa_2 \gamma^4 |p|^2)(\kappa_1(1-\gamma)^2 + \kappa_2(1-\gamma)^4 |p|^2)}|p|^2
$$
$$
< \kappa_1 |p|^2 + \kappa_2 |p|^4.
$$

Rearranging the terms and taking the square root leads to

$$
\sqrt{\kappa_1 \gamma^2 |p|^2 + \kappa_2 \gamma^4 |p|^4} + \sqrt{\kappa_1(1-\gamma)^2 |p|^2 + \kappa_2(1-\gamma)^4 |p|^4} < \sqrt{\kappa_1 |p|^2 + \kappa_2 |p|^4},
$$

and (10.2.22) is proved. Therefore, for any unit vector e_0 orthogonal to p, the surface S_p and the set $\mathcal{P}_\gamma = \{\gamma p + q e_0, q \in \mathbb{R}_+\}$ intersect at only one point, for each $\gamma \in (0,1)$. Denote the intersection by $W_\gamma = \gamma p + q_\gamma e_0$. Since $\mathcal{Z}(p - \mathfrak{J}_\gamma) + \mathcal{Z}(\mathfrak{J}_\gamma) = \mathcal{Z}(p)$, we find $\mathcal{Z}(\mathfrak{J}_\gamma) < \mathcal{Z}(p)$ and then

$$
|\mathfrak{J}_\gamma| = \sqrt{\gamma^2 |p|^2 + |q_\gamma|^2} < |p|, \quad |\mathfrak{J}_\gamma - p| = \sqrt{(1-\gamma)^2 |p|^2 + |q_{1-\gamma}|^2} < |p|.
$$

The above inequalities yield $|q_\gamma| < |p|$, as well as $\gamma |p| < |\mathfrak{J}_\gamma| < |p|$, and $(1-\gamma)|p| < |p - \mathfrak{J}_\gamma| < |p|$. Taking the derivative with respect to γ of the identity $H_0^p(\mathfrak{J}_\gamma) = 0$, we find

$$
0 = \partial_\gamma \mathfrak{J}_\gamma \cdot \nabla_w H_0^p = \partial_\gamma \mathfrak{J}_\gamma \cdot \left(\frac{\mathfrak{J}_\gamma - p}{|p - \mathfrak{J}_\gamma|} \mathcal{Z}'(|p - \mathfrak{J}_\gamma|) + \frac{\mathfrak{J}_\gamma}{|\mathfrak{J}_\gamma|} \mathcal{Z}'(|\mathfrak{J}_\gamma|) \right)
$$
$$
= \frac{1}{2} \partial_\gamma |q_\gamma|^2 \left[\frac{\mathcal{Z}'(|p - \mathfrak{J}_\gamma|)}{|p - \mathfrak{J}_\gamma|} + \frac{\mathcal{Z}'(|\mathfrak{J}_\gamma|)}{|\mathfrak{J}_\gamma|} \right]
$$
$$
+ \gamma |p|^2 \frac{\mathcal{Z}'(|\mathfrak{J}_\gamma|)}{|\mathfrak{J}_\gamma|} - (1-\gamma)|p|^2 \frac{\mathcal{Z}'(|p - \mathfrak{J}_\gamma|)}{|p - \mathfrak{J}_\gamma|}, \qquad (10.2.24)
$$

where we have used the fact that $\partial_\gamma \mathfrak{J}_\gamma = p$ and $|\mathfrak{J}_\gamma|^2 = \gamma^2 |p|^2 + |q_\gamma|^2$. Since $\mathcal{Z}'(|\mathfrak{J}_\gamma|) > 0$, the above identity implies

$$
\frac{1}{2} \partial_\gamma |q_\gamma|^2 \leq (1-\gamma)|p|^2 \qquad (10.2.25)
$$

for all p and all $\gamma \in (0,1)$.

We now estimate q_γ by considering two cases $|p| \geq 1$ and $|p| < 1$.

- Case 1: $|p|$ is small. It is straightforward that

$$
-w \cdot (w - p)\left(\kappa_1 + \kappa_2 |w|^2 + \kappa_2 |w - p|^2 + \kappa_2 |p|^2 \right) \qquad (10.2.26)
$$
$$
= \mathcal{Z}(w)\mathcal{Z}(p - w) - \kappa_2 |w|^2 |p - w|^2
$$

for all $w \in S_p$. Let us compute explicitly the right-hand side

$$\mathcal{Z}(w)\mathcal{Z}(p-w) - \kappa_2|w|^2|p-w|^2$$

$$= |w||p-w|\frac{\kappa_1\Big(\kappa_1 + \kappa_2|w|^2 + \kappa_2|w-p|^2\Big)}{\sqrt{(\kappa_1 + \kappa_2|w|^2)(\kappa_1 + \kappa_2|w-p|^2)} + \kappa_2|w||p-w|}.$$

We will now develop an asymptotic expansion of $\mathcal{Z}(w)\mathcal{Z}(p-w) - \kappa_2|w|^2|p-w|^2$ in terms of $|p|$ by expanding

$$\sqrt{\left(1 + \frac{\kappa_2}{\kappa_1}|w|^2\right)\left(1 + \frac{\kappa_2}{\kappa_1}|w-p|^2\right)} = 1 + \frac{\kappa_2}{2\kappa_1}(|w|^2 + |w-p|^2) + \mathcal{O}(|p|^4),$$

which yields

$$\mathcal{Z}(w)\mathcal{Z}(p-w) - \kappa_2|w|^2|p-w|^2 \tag{10.2.27}$$

$$= |w||p-w|\Big(\kappa_1 + \kappa_2|w|^2 + \kappa_2|w-p|^2 + \kappa_2|p|^2\Big)$$

$$- \frac{\kappa_2}{2}|w||w-p|\Big(|w|^2 + |w-p|^2 + 2|w||w-p| + 2|p|^2\Big)\Big(1 + \mathcal{O}(|p|^2)\Big).$$

Let ρ_γ be the angle between $\mathfrak{J}_\gamma$ and $\mathfrak{J}_\gamma - p$. It follows that $\mathfrak{J}_\gamma \cdot (\mathfrak{J}_\gamma - p) = |\mathfrak{J}_\gamma||\mathfrak{J}_\gamma - p|\cos\rho_\gamma$, which, together with (10.2.26)–(10.2.27), leads to

$$1 + \cos\rho_\gamma =$$

$$\frac{\kappa_2}{2}\frac{\Big(|\mathfrak{J}_\gamma|^2 + |\mathfrak{J}_\gamma - p|^2 + 2|\mathfrak{J}_\gamma||\mathfrak{J}_\gamma - p| + 2|p|^2\Big)\Big(1 + \mathcal{O}(|p|^2)\Big)}{\kappa_1 + \kappa_2|\mathfrak{J}_\gamma|^2 + \kappa_2|\mathfrak{J}_\gamma - p|^2 + \kappa_2|p|^2} = \mathcal{O}(|p|^2).$$

Therefore, $\sin\rho_\gamma = \mathcal{O}(|p|)$. We compute the area of the parallelogram formed by $\mathfrak{J}_\gamma$ and $\mathfrak{J}_\gamma - p$ as $2|p||q_\gamma| = |\mathfrak{J}_\gamma \times (\mathfrak{J}_\gamma - p)| = |\mathfrak{J}_\gamma||\mathfrak{J}_\gamma - p|\sin\rho_\gamma$, which implies

$$\gamma(1-\gamma)|p|^2 \lesssim |q_\gamma| \lesssim |p|^2 \tag{10.2.28}$$

for all $\gamma \in (0,1)$.

- Case 2: $|p| \geq 1$. Notice that at $\gamma = \frac{1}{2}$, we have $|\mathfrak{J}_{1/2}| = |\mathfrak{J}_{1/2} - p|$, which leads to $2\mathcal{E}(\mathfrak{J}_{1/2}) = \mathcal{E}(p)$. Noting that $|\mathfrak{J}_{1/2}|^2 = \frac{1}{4}|p|^2 + |q_{1/2}|^2$, we have

$$4\left[\kappa_1\left(\frac{1}{4}|p|^2 + |q_{1/2}|^2\right) + \kappa_2\left(\frac{1}{4}|p|^2 + |q_{1/2}|^2\right)^2\right] = \kappa_1|p|^2 + \kappa_2|p|^4.$$

Hence,

$$\kappa_2\left(\frac{1}{4}|p|^2 + |q_{1/2}|^2\right)^2 + \kappa_1|q_{1/2}|^2 = \frac{\kappa_2}{4}|p|^4,$$

which implies

$$c_0|p|^2 = c_0|p|^2\min\left\{1, |p|^2\right\} \leq |q_{1/2}|^2 \leq C_0|p|^2\min\left\{1, |p|^2\right\} = C_0|p|^2,$$

$$\tag{10.2.29}$$

for some constants c_0, C_0, independent of $|p|$.

Combining (10.2.25), (10.2.29) and the identity

$$|q_\gamma|^2 = |q_{1/2}|^2 - \int_\gamma^{\frac{1}{2}} \partial_{\gamma'} |q_{\gamma'}|^2 \, d\gamma'$$

yields

$$|q_\gamma|^2 \geq c_0 |p|^2 - 2\left|\gamma - \frac{1}{2}\right| |p|^2 \geq \frac{1}{2} c_0 |p|^2 \qquad (10.2.30)$$

for all γ satisfying $\left|\gamma - \frac{1}{2}\right| \leq \frac{c_0}{4}$.

The two inequalities (10.2.30) and (10.2.28) are all we need to obtain (10.2.18). To continue, we parameterize the surface S_p^0. Let $p^\perp$ be a vector in $\mathcal{P}_0 = \{p \cdot q = 0\}$ and e_θ be the unit vector in $\mathcal{P}_0$ such that the angle between $p^\perp$ and e_θ is θ. The surface S_p is now

$$S_p^0 = \left\{ W(\gamma, \theta) = \gamma p + |q_\gamma| e_\theta \; : \; \theta \in [0, 2\pi], \; \gamma \in [0, 1] \right\}.$$

Since the vector $\partial_\theta e_\theta$ is orthogonal to both vectors p and e_θ, we can compute the surface area as

$$d\sigma(w) = |\partial_\gamma \mathfrak{I}_\gamma \times \partial_\theta \mathfrak{I}_\gamma| d\gamma d\theta = \left|(p + \partial_\gamma |q_\gamma| e_\theta) \times |q_\gamma| \partial_\theta e_\theta \right| d\gamma d\theta$$

$$= \left|(|q_\gamma| p + \frac{1}{2} \partial_\gamma |q_\gamma|^2 e_\theta) \times \partial_\theta e_\theta \right| d\gamma d\theta \qquad (10.2.31)$$

$$= \sqrt{|p|^2 |q_\gamma|^2 + \frac{1}{4} |\partial_\gamma(|q_\gamma|^2)|^2} d\gamma d\theta.$$

It follows from (10.2.24) that

$$\partial_\gamma |q_\gamma|^2 = 2|p|^2 \frac{\gamma \frac{\mathcal{Z}'(|\mathfrak{I}_\gamma|)}{|\mathfrak{I}_\gamma|} + (\gamma - 1)\frac{\mathcal{Z}'(|p - \mathfrak{I}_\gamma|)}{|p - \mathfrak{I}_\gamma|}}{\frac{\mathcal{Z}'(|p - \mathfrak{I}_\gamma|)}{|p - \mathfrak{I}_\gamma|} + \frac{\mathcal{Z}'(\mathfrak{I}_\gamma)}{|\mathfrak{I}_\gamma|}}. \qquad (10.2.32)$$

We now compute $|\nabla H_0^p|$ under this new parametrization

$$|\nabla H_0^p|^2 = |p|^2 \left[\gamma \frac{\mathcal{Z}'(|\mathfrak{I}_\gamma|)}{|\mathfrak{I}_\gamma|} + (\gamma - 1)\frac{\mathcal{Z}'(|p - \mathfrak{I}_\gamma|)}{|p - \mathfrak{I}_\gamma|}\right]^2$$

$$+ |q_\gamma|^2 \left[\frac{\mathcal{Z}'(|p - \mathfrak{I}_\gamma|)}{|p - \mathfrak{I}_\gamma|} + \frac{\mathcal{Z}'(|\mathfrak{I}_\gamma|)}{|\mathfrak{I}_\gamma|}\right]^2,$$

which, together with (10.2.32), yields

$$|\nabla H_0^p|^2 = \frac{|\partial_\gamma |q_\gamma|^2|^2}{4|p|^2} \left[\frac{\mathcal{Z}'(|p - \mathfrak{I}_\gamma|)}{|p - \mathfrak{I}_\gamma|} + \frac{\mathcal{Z}'(|\mathfrak{I}_\gamma|)}{|\mathfrak{I}_\gamma|}\right]^2$$

$$+ |q_\gamma|^2 \left[\frac{\mathcal{Z}'(|p - \mathfrak{I}_\gamma|)}{|p - \mathfrak{I}_\gamma|} + \frac{\mathcal{Z}'(|\mathfrak{I}_\gamma|)}{|\mathfrak{I}_\gamma|}\right]^2. \qquad (10.2.33)$$

Observe that $\mathcal{Z}'(x) \geq cx$ for all $x \in \mathbb{R}_+$. We get the following bound

$$|\nabla H_0^p| = \frac{\sqrt{\frac{|\partial_\gamma |q_\gamma|^2|^2}{4} + |q_\gamma|^2|p|^2}}{|p|} \left[\frac{\mathcal{Z}'(|p - \mathfrak{I}_\gamma|)}{|p - \mathfrak{I}_\gamma|} + \frac{\mathcal{Z}'(|\mathfrak{I}_\gamma|)}{|\mathfrak{I}_\gamma|} \right]. \qquad (10.2.34)$$

The two inequalities (10.2.30) and (10.2.28) will be our keys to estimating the integral

$$Z := \int_{S_p^0} \bar{K}_0^{12}(p, w, p - w)|w|^{k_1}|p - w|^{k_2} \mathrm{d}\sigma(w).$$

Since $K^{12}(p, w, p - w) \gtrsim (|p| \wedge p_0)(|p - w| \wedge p_0)(|w| \wedge p_0)$, Z can be bounded from below by Z', with

$$Z' := \int_{S_p^0} (|p| \wedge p_0)(|w| \wedge p_0)(|p - w| \wedge p_0)|w|^{k_1}|p - w|^{k_2} \mathrm{d}\sigma(w).$$

By (10.2.31), Z' can be rewritten as

$$\int_0^{2\pi} \int_0^1 \frac{|p|(|p| \wedge p_0)(|w| \wedge p_0)(|p - w| \wedge p_0)|w|^{k_1}|p - w|^{k_2}}{\frac{\mathcal{Z}'(|p-w|)}{|p-w|} + \frac{\mathcal{Z}'(|w|)}{|w|}} \mathrm{d}\gamma\mathrm{d}\theta.$$

Due to (10.2.30), for $|p|$ large, and $\gamma \in \left[\frac{2-c_0}{4}, \frac{2+c_0}{4}\right]$, $|p - \mathfrak{I}_\gamma|^2 \geq |q_\gamma|^2 \geq \frac{1}{2}c_0|p|^2$ and $|\mathfrak{I}_\gamma|^2 \geq |q_\gamma|^2 \geq \frac{1}{2}c_0|p|^2$. We can thus bound

$$\frac{|p|}{\frac{\mathcal{Z}'(|p-w|)}{|p-w|} + \frac{\mathcal{Z}'(|w|)}{|w|}} \geq \frac{|p|}{2\frac{\mathcal{Z}'(c_0|p|)}{c_0|p|}} \geq \frac{c_0|p|^2}{2\frac{\kappa_1 + 2\kappa_2|c_0 p|^2}{\sqrt{\kappa_1 + \kappa_2|c_0 p|^2}}},$$

where we have used the fact that $\frac{\mathcal{Z}'(|\varrho|)}{|\varrho|}$ is decreasing in $|\varrho|$. Since $|p|$ is large, it is clear that $\frac{|p|}{\frac{\mathcal{Z}'(|p-w|)}{|p-w|} + \frac{\mathcal{Z}'(|w|)}{|w|}} \gtrsim |p|$. As a consequence, Z' can be estimated as follows

$$Z' \gtrsim \int_0^{2\pi} \int_{\frac{1-c_0}{2}}^{\frac{1+c_0}{2}} (|p| \wedge p_0) \left(\left| \sqrt{\frac{c_0}{2}}|p| \right| \wedge p_0 \right)^2 \left| \sqrt{\frac{c_0}{2}}|p| \right|^{k_1+k_2} |p|\mathrm{d}\gamma\mathrm{d}\theta$$

$$\gtrsim |p|^{k_1+k_2+1}.$$

Thanks to (10.2.28), for p small, on the interval $\gamma \in \left[\frac{1}{3}, \frac{1}{2}\right]$, $|p - \mathfrak{I}_\gamma|^2 \geq |q_\gamma|^2 \geq c_1|p|^4$, and $|\mathfrak{I}_\gamma|^2 \geq |q_\gamma|^2 \geq c_1|p|^4$. Thus, we can bound

$$\frac{|p|}{\frac{\mathcal{Z}'(|p-w|)}{|p-w|} + \frac{\mathcal{Z}'(|w|)}{|w|}} \geq \frac{|p|}{2\frac{\mathcal{Z}'(c_1|p|^2)}{c_1|p|^2}} \geq \frac{c_1|p|^3}{2\frac{\kappa_1 + 2\kappa_2 c_1^2|p|^4}{\sqrt{\kappa_1 + \kappa_2 c_1^2|p|^4}}},$$

where we have used the fact that $\frac{Z'(|\varrho|)}{|\varrho|}$ is decreasing in $|\varrho|$. Since $|p|$ is small, it is clear that $\frac{|p|}{\frac{Z'(|p-w|)}{|p-w|} + \frac{Z'(|w|)}{|w|}} \gtrsim C|p|^3$. Therefore, Z' can be estimated as follows

$$Z' \gtrsim \int_0^{2\pi} \int_{\frac{1}{3}}^{\frac{1}{2}} (|p| \wedge p_0) \left(|p|^2 \wedge p_0\right)^2 |p|^{2k_1+2k_2} |p|^3 \mathrm{d}\gamma \mathrm{d}\theta \gtrsim |p|^{2k_1+2k_2+8}.$$

The above shows that (10.2.18) holds true.

As for the surface integral of a radial function $\mathcal{G}(|w|)$, let us introduce the radial variable $u = |\mathfrak{J}_\alpha| = \sqrt{\alpha^2|p|^2 + |q_\alpha|^2}$. We compute $2u\mathrm{d}u = \partial_\alpha |\mathfrak{J}_\alpha|^2 \mathrm{d}\alpha$ and

$$\frac{1}{|\nabla H_0^p|}\mathrm{d}\sigma(w) = \frac{|p|}{2\left[\frac{Z'(|p-\mathfrak{J}_\gamma|)}{|p-\mathfrak{J}_\gamma|} + \frac{Z'(|\mathfrak{J}_\gamma|)}{|\mathfrak{J}_\gamma|}\right] \partial_\alpha |\mathfrak{J}_\alpha|^2} u\mathrm{d}u\mathrm{d}\theta.$$

Using (10.2.24), we obtain

$$\frac{|p|}{2\left[\frac{Z'(|p-\mathfrak{J}_\gamma|)}{|p-\mathfrak{J}_\gamma|} + \frac{Z'(|\mathfrak{J}_\gamma|)}{|\mathfrak{J}_\gamma|}\right] \partial_\alpha |\mathfrak{J}_\alpha|^2} = \frac{1}{4|p|\frac{Z'(|p-\mathfrak{J}_\gamma|)}{|p-\mathfrak{J}_\gamma|}} \leq \frac{1}{4Z'(|p|)},$$

in which we have used $\frac{Z'(|p-\mathfrak{J}_\gamma|)}{|p-\mathfrak{J}_\gamma|} \geq \frac{Z'(|p|)}{|p|}$. This, together with the bound $Z'(|p|) \gtrsim 1$, shows that

$$\mathrm{d}\sigma(w) \lesssim u\mathrm{d}u\mathrm{d}\theta. \tag{10.2.35}$$

This yields the upper bound on the surface integral. ∎

Lemma 10.2.3. *Let S_p^1 be the surface defined in (10.2.9). For any radial and positive function $F(\cdot) : \mathbb{R}^3 \to \mathbb{R}$*

$$F(u) = F(|u|),$$

the following estimate

$$\int_{S_p^1} \frac{F(|w|)}{|\nabla H_1^{P_1}(|w|)|} \mathrm{d}\sigma(w) \lesssim \frac{\int_0^{|p|} |u|F(|u|)\,\mathrm{d}|u|}{|p|}$$

holds true uniformly in $p \in \mathbb{R}^3$.

Proof. First, observe that

$$\mathcal{Z}(p+w)^2 - \left(\mathcal{Z}(p) + \mathcal{Z}(w)\right)^2$$
$$= 2\kappa_1 w \cdot p + 2\kappa_2 w \cdot p(|p|^2 + |w|^2 + |p+w|^2) + 2\kappa_2|p|^2|w|^2 - 2\mathcal{Z}(p)\mathcal{Z}(w). \tag{10.2.36}$$

Now, since $\kappa_1 \neq 0$, it follows that $\kappa_2|p|^2|w|^2 < \mathcal{Z}(p)\mathcal{Z}(w)$. This, together with (10.2.36), shows that if $w \in S_p^1 \setminus \{0\}$, then $w \cdot p > 0$. We calculate the derivative

$$\nabla_w H_1^p = \frac{p+w}{|p+w|} \mathcal{Z}'(|p+w|) - \frac{w}{|w|} \mathcal{Z}'(|w|),$$

with $\mathcal{Z}'(|\varrho|) = \frac{2\kappa_1 + 4\kappa_2 |\varrho|^2}{\sqrt{\kappa_1 + \kappa_2 |\varrho|^2}}$. The derivative at $w = \gamma p$ with $\gamma \in \mathbb{R}_+$ can be determined using the previous identity

$$\partial_\gamma H_1^p = \partial_\gamma w \cdot \nabla_w H_1^p |_{w=\gamma p} = |p| \mathcal{Z}'((1+\gamma)p) - |p| \mathcal{Z}'(\gamma p).$$

By the monotonicity of $\mathcal{Z}(p)$ in $|p|$, it follows that $\mathcal{Z}'((1+\gamma)p) > \mathcal{Z}'(\gamma p)$, and therefore $\partial_\gamma H_1^p > 0$ for all $\gamma > 0$. Since $H_1^p(0) = 0$, we have $H_1^p(\gamma p) > 0$ for all positive γ. Let us now consider all the points $\mathfrak{J}_\gamma = \gamma p + q$, in which $q \cdot p = 0$, for each fixed $\gamma > 0$. The directional derivative of H_1^p at $\mathfrak{J}_\gamma = \gamma p + q$ in the direction of $q \neq 0$ satisfies

$$q \cdot \nabla_w H_1^p = |q|^2 \left[\frac{\mathcal{Z}'(p + \mathfrak{J}_\gamma)}{|p + \mathfrak{J}_\gamma|} - \frac{\mathcal{Z}'(\mathfrak{J}_\gamma)}{|\mathfrak{J}_\gamma|} \right] < 0,$$

where we have used the fact that $\mathcal{Z}'(p)/|p|$ is strictly decreasing in $|p|$. In view of (10.2.36), the sign of $H_1^p(w)$, with $\mathfrak{J}_\gamma = \gamma p + q$, is the same as the following quantity

$$\gamma |p|^2 \left(\kappa_1 + \kappa_2 (|p|^2 + |\mathfrak{J}_\gamma|^2 + |p + \mathfrak{J}_\gamma|^2) \right) + \kappa_2 |p|^2 |\mathfrak{J}_\gamma|^2 - \mathcal{Z}(p) \mathcal{Z}(\mathfrak{J}_\gamma)$$

$$= \gamma |p|^2 \left(\kappa_1 + 2\kappa_2 (|p|^2 + \gamma |p|^2 + |\mathfrak{J}_\gamma|^2) \right)$$

$$- \frac{\kappa_1^2 |\mathfrak{J}_\gamma|^2 |p|^2 + \kappa_1 \kappa_2 |\mathfrak{J}_\gamma|^2 |p|^4 + \kappa_1 \kappa_2 |p|^2 |\mathfrak{J}_\gamma|^4}{\sqrt{\kappa_1 |p|^2 + \kappa_2 |p|^4} \sqrt{\kappa_1 |\mathfrak{J}_\gamma|^2 + \kappa_2 |\mathfrak{J}_\gamma|^4} + \kappa_2 |p|^2 |\mathfrak{J}_\gamma|^2}.$$

This means $H_1^p(\gamma p + q) < 0$ whenever

$$\gamma < \frac{\kappa_1 (\kappa_1 + \kappa_2 |p|^2) + \kappa_1 \kappa_2 |\mathfrak{J}_\gamma|^2}{\sqrt{\kappa_1 |p|^2 + \kappa_2 |p|^4} \sqrt{\frac{\kappa_1}{|\mathfrak{J}_\gamma|^2} + \kappa_2} + \kappa_2 |p|^2} \frac{1}{\left(\kappa_1 + 2\kappa_2 (|p|^2 + \gamma |p|^2 + |\mathfrak{J}_\gamma|^2) \right)}.$$

Since $H_1^p(\gamma p) > 0$ and $q \cdot \nabla_w H_1^p < 0$, for a given direction q, there exists a q_γ such that $q_\gamma \cdot q > 0$ and q_γ is parallel with q if and only if

$$\lim_{q \to \infty} H_1^p(\gamma p + q) < 0. \tag{10.2.37}$$

Taking $q \to \infty$ (and so $|\mathfrak{J}_\gamma| \to \infty$), (10.2.37) is equivalent to

$$\gamma < \gamma_p := \frac{1}{2} \frac{\kappa_1}{\kappa_2 |p|^2 + 2\sqrt{\kappa_2} \sqrt{\kappa_1 |p|^2 + \kappa_2 |p|^4}}. \tag{10.2.38}$$

In particular, $\gamma_p |p| (1 + |p|) \lesssim 1$, $\forall\, p \in \mathbb{R}^3$. Thus, for positive values of the parameter γ satisfying (10.2.38), there is a unique $|q_v|$ such that $\overline{G}(\gamma p + q) = 0$, for any $|q| = |q_\gamma|$. Moreover, by the continuity of $H_1^p(\mathfrak{J}_\gamma)$, the function $|q_\gamma|$ is continuously differentiable in γ. When $\gamma \geq \gamma_p$, $H_1^p(\gamma p + q) > 0$, for all q such that $q \cdot p = 0$.

Now, we are able to parameterize the surface S_p^1

$$S_p^1 = \left\{ w(\gamma, \theta) = \gamma p + |q_\gamma| e_\theta \ : \ \gamma \in [0, \gamma_p), \ \theta \in [0, 2\pi] \right\}, \qquad (10.2.39)$$

in which γ_p and $|q_\gamma|$ are defined as above and e_θ is the unit vector rotating around p and orthogonal to p. As in (10.2.31), we find

$$d\sigma(w) = \sqrt{|p|^2 |q_\gamma|^2 + \frac{1}{4} |\partial_\gamma (|q_\gamma|^2)|^2} d\gamma d\theta$$

and thus, the surface integral can be rewritten as

$$\int_{S_p^1} \frac{F(|w|)}{|\nabla H_1^p(w)|} d\sigma(w) =$$

$$\iint_{[0,2\pi] \times [0,\gamma_p]} \frac{F(|\gamma p + q_\gamma|)}{|\nabla H_1^p(|\gamma p + q_\gamma|)|} \sqrt{|p|^2 |q_\gamma|^2 + \frac{1}{4} |\partial_\gamma (|q_\gamma|^2)|^2} d\gamma d\theta.$$

We introduce the variable $u = |\mathfrak{I}_\gamma| = \sqrt{\gamma^2 |p|^2 + |q_\gamma|^2}$ and compute $2u du = \partial_\gamma |\mathfrak{I}_\gamma|^2 d\gamma$, which yields

$$\int_{S_p^1} \frac{F(|w|)}{|\nabla H_1^p(w)|} d\sigma(w) \lesssim \pi \int_0^\infty F(u) \frac{\sqrt{|p|^2 |q_\gamma|^2 + \frac{1}{4} |\partial_\gamma (|q_\gamma|^2)|^2}}{2\partial_\gamma |\mathfrak{I}_\gamma|^2 |\nabla H_1^p(|\gamma p + q_\gamma|)|} u du.$$
$$(10.2.40)$$

Since $H_1^p(\mathfrak{I}_\gamma) = 0$, we find

$$0 = \partial_\gamma \mathfrak{I}_\gamma \cdot \nabla_w H_1^p = \frac{1}{2} \partial_\gamma |\mathfrak{I}_\gamma|^2 \left[\frac{\mathcal{Z}'(p + w_\gamma)}{|p + \mathfrak{I}_\gamma|} - \frac{\mathcal{Z}'(\mathfrak{I}_\gamma)}{|w_\gamma|} \right] + |p|^2 \frac{\mathcal{Z}'(p + w_\gamma)}{|p + \mathfrak{I}_\gamma|},$$

which leads to

$$|p|^2 \frac{\mathcal{Z}'(p + \mathfrak{I}_\gamma)}{|p + \mathfrak{I}_\gamma|} = \frac{1}{2} \partial_\gamma |\mathfrak{I}_\gamma|^2 \left[\frac{\mathcal{Z}'(\mathfrak{I}_\gamma)}{|\mathfrak{I}_\gamma|} - \frac{\mathcal{Z}'(p + \mathfrak{I}_\gamma)}{|p + \mathfrak{I}_\gamma|} \right] \qquad (10.2.41)$$

and

$$\partial_\gamma |q_\gamma|^2 = 2 \frac{-\gamma |p|^2 \frac{\mathcal{Z}'(|\mathfrak{I}_\gamma|)}{|\mathfrak{I}_\gamma|} + (1 + \gamma) |p|^2 \frac{\mathcal{Z}'(|p + \mathfrak{I}_\gamma|)}{|p + \mathfrak{I}_\gamma|}}{\left[\frac{\mathcal{Z}'(\mathfrak{I}_\gamma)}{|\mathfrak{I}_\gamma|} - \frac{\mathcal{Z}'(p + \mathfrak{I}_\gamma)}{|p + \mathfrak{I}_\gamma|} \right]}. \qquad (10.2.42)$$

It follows from (10.2.42) that

$$|\nabla H_1^p|^2 = \frac{|\partial_\gamma |q_\gamma|^2|^2}{4 |p|^2} \left[\frac{\mathcal{Z}'(|p + \mathfrak{I}_\gamma|)}{|p + \mathfrak{I}_\gamma|} - \frac{\mathcal{Z}'(|\mathfrak{I}_\gamma|)}{|\mathfrak{I}_\gamma|} \right]^2$$

$$+ |q_\gamma|^2 \left[\frac{\mathcal{Z}'(|p + \mathfrak{I}_\gamma|)}{|p + \mathfrak{I}_\gamma|} - \frac{\mathcal{Z}'(|\mathfrak{I}_\gamma|)}{|\mathfrak{I}_\gamma|} \right]^2,$$

which implies

$$\int_{S_p^1} \frac{F(|w|)}{|\nabla H_1^p(w)|}\, d\sigma(w) \lesssim \int_0^\infty \frac{F(u)|p|}{\partial_\gamma|\mathfrak{z}_\gamma|^2 \left[\frac{\mathcal{Z}'(|\mathfrak{z}_\gamma|)}{|\mathfrak{z}_\gamma|} - \frac{\mathcal{Z}'(|p+\mathfrak{z}_\gamma|)}{|p+\mathfrak{z}_\gamma|}\right]}\, u\,du.$$

The above inequality and (10.2.41) yield

$$\int_{S_p^1} \frac{F(|w|)}{|\nabla H_1^p(w)|}\, d\sigma(w) \lesssim \int_0^\infty \frac{F(u)}{2|p|\frac{\mathcal{Z}'(|p+w|)}{|p+w|}}\, u\,du.$$

Using the fact that $\frac{\mathcal{Z}'(|p+w|)}{|p+w|} \gtrsim 1$, we obtain

$$\int_{S_p^1} \frac{F(|w|)}{|\nabla H_1^p(w)|}\, d\sigma(w) \lesssim \frac{1}{|p|} \int_0^\infty F(u)u\,du.$$

$\square$

Lemma 10.2.4. *Let S_p^2 be the surface defined in (10.2.9). For any radial and positive function $F(\cdot) : \mathbb{R}^3 \to \mathbb{R}$*

$$F(u) = F(|u|),$$

the following estimate

$$\int_{S_p^2} \frac{F(|w|)}{|\nabla H_2^p(|w|)|}\, d\sigma(w) \lesssim \frac{\int_0^{|p|} |u|F(|u|)\, d|u|}{|p|}$$

holds true uniformly in $p \in \mathbb{R}^3$.

Proof. Observe that

$$\begin{aligned}
S_p^2 &= \{p_* \mid \mathcal{Z}(p_*) = \mathcal{Z}(p) + \mathcal{Z}(p_* - p)\} \\
&= \{p_* + p \mid \mathcal{Z}(p_* + p) = \mathcal{Z}(p) + \mathcal{Z}(p_*)\} \\
&= p + S_p^1.
\end{aligned}$$

The above identity shows that the same argument of Lemma 10.2.3 could be applied and the conclusion of the lemma follows. $\square$

10.2.4 Resonance Manifold/Energy Surface Analysis for $\mathcal{Q}_{22}$

Below, we will establish estimates on the energy surface integrals of $\mathcal{Q}_{22}$.

Proposition 10.2.1. *Let $\mathcal{S}_{p_1,p_2}$ be the resonance manifold of $\mathcal{Q}_{22}$. For any radial and positive function $F(\cdot) : \mathbb{R}^3 \to \mathbb{R}$*

$$F(u) = F(|u|),$$

the following estimate

$$\int_{\mathcal{S}_{p_1,p_2}} \frac{F(|w|)}{|\nabla \mathcal{G}^{p_1,p_2}(|w|)|} \, \mu(w) \lesssim \frac{\int_0^{c_{22} \sqrt{\mathcal{Z}_{p_1} + \mathcal{Z}_{p_2}}} |u| F(|u|) \, \mathrm{d}|u|}{|p_1 + p_2|}$$

holds true uniformly in $p_1, p_2 \in \mathbb{R}^3$, for some universal constant $c_{22} > 0$.

Proof. Setting $p_1 + p_2 = \rho$, we now parameterize the resonant manifold $\mathcal{S}_{p_1,p_2}$. To do this, we compute the derivative of $\mathcal{G}^{p_1,p_2}$

$$\nabla_x \mathcal{G}^{p_1,p_2} = \frac{x - \rho}{|x - \rho|} \mathcal{Z}'(|\rho - x|) + \frac{x}{|x|} \mathcal{Z}'(|x|).$$

In particular, let q be any vector orthogonal to ρ, i.e. $\rho \cdot q = 0$. The directional derivative of $\mathcal{G}^{p_1,p_2}$ in the direction of q, with $x = \alpha\rho + q, \alpha \in \mathbb{R}$, satisfies

$$q \cdot \nabla_x \mathcal{G}^{p_1,p_2} = |q|^2 \left[\frac{\mathcal{Z}'(|\rho - x|)}{|\rho - x|} + \frac{\mathcal{Z}'(|x|)}{|x|} \right] > 0,$$

which means that $\mathcal{G}^{p_1,p_2}(x)$ is strictly increasing in any direction that is orthogonal to ρ. This proves that the intersection between the surface $\mathcal{S}_{p,p_1}$ and the plane $\mathcal{P}_\alpha = \left\{ \alpha\rho + q, \rho \cdot q = 0 \right\}$ is either empty or the circle centered at $\alpha\rho$ with finite radius r_α, for $\alpha \in \mathbb{R}$. As a consequence, we can parameterize $\mathcal{S}_{p,p_1}$ as follows. Let $\rho^\perp$ be the vector orthogonal to both ρ and a fixed vector e of $\mathbb{R}^3$ and let e_θ be the unit vector in $\mathcal{P}_0 = \{\rho \cdot q = 0\}$ such that the angle between $\rho^\perp$ and e_θ is θ. We parameterize $\mathcal{S}_{p,p_1}$ by

$$\left\{ x = \alpha\rho + r_\alpha e_\theta \; : \; \theta \in [0, 2\pi], \; \alpha \in A_{p,p_1} \right\}, \tag{10.2.43}$$

where A_{p,p_1} is the set of α for which a solution to $\mathcal{G}^{p_1,p_2}(x) = 0$ exists. We can think of $\mathcal{G}$ as a function of α and r: $\mathcal{G}^{p_1,p_2} = \mathcal{G}^{p_1,p_2}(r, \alpha)$. We just saw that $\partial_r \mathcal{G}^{p_1,p_2} > 0$ for $r > 0$. Therefore, by the implicit function theorem, the zero set of $\mathcal{G}^{p_1,p_2}$ can be parameterized as

$$\{(\alpha, r = r_\alpha), \alpha \in A_{p,p_1}\},$$

where $\alpha \mapsto r_\alpha$ is a smooth function on A_{p,p_1} vanishing on its boundary.

Next, we have by definition that $\mathcal{G}^{p_1,p_2}(x_\alpha) = 0$ for all α and therefore, keeping θ fixed,

$$
\begin{aligned}
0 = \partial_\alpha x_\alpha \cdot \nabla_x \mathcal{G}^{p_1,p_2} &= \partial_\alpha x_\alpha \cdot \left(\frac{x_\alpha - \rho}{|x_\alpha - \rho|} \mathcal{Z}'(|x_\alpha - \rho|) + \frac{x_\alpha}{|x_\alpha|} \mathcal{Z}'(|x_\alpha|) \right) \\
&= \frac{1}{2} \partial_\alpha |x_\alpha|^2 \left[\frac{\mathcal{Z}'(|\rho - x_\alpha|)}{|\rho - x_\alpha|} + \frac{\mathcal{Z}'(|x_\alpha|)}{|x_\alpha|} \right] - |\rho|^2 \frac{\mathcal{Z}'(|\rho - x_\alpha|)}{|\rho - x_\alpha|}.
\end{aligned}
$$
(10.2.44)

Therefore,

$$
\partial_\alpha |c_\alpha|^2 = 2 \frac{\frac{\mathcal{Z}'(|\rho - z_\alpha|)}{|\rho - x_\alpha|} |\rho|^2}{\frac{\mathcal{Z}'(|\rho - x_\alpha|)}{|\rho - x_\alpha|} + \frac{\mathcal{Z}'(|x_\alpha|)}{|x_\alpha|}}.
$$
(10.2.45)

This implies in particular that $\alpha \mapsto |x_\alpha|$ is increasing on A_{p,p_1}. Defining r to be zero on the complement of A_{p,p_1}, we get that $\alpha \mapsto |x_\alpha|$ is an increasing function on $\mathbb{R}$; therefore, the change of coordinates $\alpha \to |x_\alpha|$ is well defined.

Since $\partial_\theta e_\theta$ is orthogonal to both ρ and e_θ, we compute the surface area

$$
d\mu(x) = \sqrt{|\rho|^2 r_\alpha^2 + \frac{1}{4} |\partial_\alpha (r_\alpha^2)|^2} d\alpha d\theta.
$$
(10.2.46)

Using $|x|^2 = \alpha^2 |\rho|^2 + r_\alpha^2$, we learn from the last line of (10.2.44) that

$$
\partial_\alpha r_\alpha^2 = 2|\rho|^2 \frac{-\alpha \frac{\mathcal{Z}'(|x_\alpha|)}{|x_\alpha|} + (-\alpha + 1) \frac{\mathcal{Z}'(|\rho - x_\alpha|)}{|\rho - x_\alpha|}}{\frac{\mathcal{Z}'(|\rho - x_\alpha|)}{|\rho - x_\alpha|} + \frac{\mathcal{Z}'(x_\alpha)}{|x_\alpha|}}.
$$
(10.2.47)

Now, let us compute $|\nabla_x \mathcal{G}^{p_1,p_2}|$ under the new parameterization:

$$
\begin{aligned}
|\nabla_x \mathcal{G}^{p_1,p_2}|^2 = |\rho|^2 & \left[\alpha \frac{\mathcal{Z}'(|x_\alpha|)}{|x_\alpha|} + (\alpha - 1) \frac{\mathcal{Z}'(|\rho - x_\alpha|)}{|\rho - x_\alpha|} \right]^2 \\
& + r_\alpha^2 \left[\frac{\mathcal{Z}'(|\rho - x_\alpha|)}{|\rho - x_\alpha|} + \frac{\mathcal{Z}'(|x_\alpha|)}{|x_\alpha|} \right]^2.
\end{aligned}
$$

In addition to (10.2.47), this implies that

$$
\begin{aligned}
|\nabla_x \mathcal{G}^{p_1,p_2}|^2 = & \frac{|\partial_\alpha r_\alpha^2|^2}{4|\rho|^2} \left[\frac{\mathcal{Z}'(|\rho - x_\alpha|)}{|\rho - x_\alpha|} + \frac{\mathcal{Z}'(|x_\alpha|)}{|x_\alpha|} \right]^2 \\
& + r_\alpha^2 \left[\frac{\mathcal{Z}'(|\rho - x_\alpha|)}{|\rho - x_\alpha|} + \frac{\mathcal{Z}'(|x_\alpha|)}{|x_\alpha|} \right]^2.
\end{aligned}
$$
(10.2.48)

Therefore

$$
\frac{d\mu(x)}{|\nabla_x \mathcal{G}^{p_1,p_2}|} = \frac{|\rho|}{\frac{\mathcal{Z}'(|\rho - x_\alpha|)}{|\rho - x_\alpha|} + \frac{\mathcal{Z}'(|x_\alpha|)}{|x_\alpha|}} d\alpha \, d\theta.
$$
(10.2.49)

Introducing the variable $u = |x_\alpha| = \sqrt{\alpha^2 |\rho|^2 + r_\alpha^2}$, by (10.2.45) we get

$$\frac{\mathrm{d}\mu(x)}{|\nabla_x \mathcal{G}^{p_1,p_2}|} = \frac{|\rho - x_\alpha|}{\mathcal{Z}'(|\rho - x_\alpha|)|\rho|} u \, \mathrm{d}u \, \mathrm{d}\theta.$$

Since $\frac{|\rho - x_\alpha|}{\mathcal{Z}'(|\rho - x_\alpha|)} \lesssim 1$,

$$\frac{\mathrm{d}\mu(x)}{|\nabla_x \mathcal{G}^{p_1,p_2}|} \lesssim \frac{u}{|\rho|} \mathrm{d}u \mathrm{d}\theta, \tag{10.2.50}$$

which implies the conclusion of the proposition.

$$\square$$

10.2.5 Boundedness of the Total Mass for the Kinetic Equation

The following proposition establishes a bound in time for the total mass of the kinetic equation.

Proposition 10.2.2. *Suppose $f_0(p) = f_0(|p|)$ satisfies*

$$\int_{\mathbb{R}^3} f_0(p_1)\mathrm{d}p_1 < \infty, \int_{\mathbb{R}^3} f_0(p_1)\mathcal{Z}(p_1)\mathrm{d}p_1 < \infty.$$

There exist universal positive constants C_1, C_2 such that the mass of the positive radial solution $f(t,p) = f(t,|p|)$ of (10.2.1) is bounded from above as

$$\int_{\mathbb{R}^3} f(t,p)\mathrm{d}p \le C_1 e^{C_2 t}.$$

Proof. First, we observe that the constant function 1 can be used as the test function for (10.2.1)

$$\frac{d}{dt} \int_{\mathbb{R}^3} f(p_1)\mathrm{d}p_1 = \int_{\mathbb{R}^3} \mathcal{Q}_{12}[f](p_1)\mathrm{d}p_1 + \int_{\mathbb{R}^3} \mathcal{Q}_{22}[f](p_1)\mathrm{d}p_1, \tag{10.2.51}$$

in which we notice that

$$\int_{\mathbb{R}^3} \mathcal{Q}_{22}[f](p_1)\mathrm{d}p_1 = 0$$

and

$$\int_{\mathbb{R}^3} \mathcal{Q}_{12}[f](p_1)\mathrm{d}p_1 =$$

$$N_c \iiint_{\mathbb{R}^3 \times \mathbb{R}^3 \times \mathbb{R}^3} K^{12}(p_1, p_2, p_3)\delta(p_1 - p_2 - p_3)\delta(\mathcal{Z}_{p_1} - \mathcal{Z}_{p_2} - \mathcal{Z}_{p_3})$$

$$\times [f(p_1) + 2f(p_1)f(p_2) - f(p_2)f(p_3)]\mathrm{d}p_1 \mathrm{d}p_2 \mathrm{d}p_3.$$

The above computations show that the control of the total mass really comes from estimating the collision operator $\mathcal{Q}_{12}$, since the integral of $\mathcal{Q}_{22}$ is already 0. Setting

$$J_1 = N_c \iiint_{\mathbb{R}^3 \times \mathbb{R}^3 \times \mathbb{R}^3} K^{12}(p_1, p_2, p_3)\delta(p_1 - p_2 - p_3)$$
$$\times \delta(\mathcal{Z}_{p_1} - \mathcal{Z}_{p_2} - \mathcal{Z}_{p_3})f(p_1)\mathrm{d}p_1\mathrm{d}p_2\mathrm{d}p_3$$

and

$$J_2 = 2N_c \iiint_{\mathbb{R}^3 \times \mathbb{R}^3 \times \mathbb{R}^3} K^{12}(p_1, p_2, p_3)\delta(p_1 - p_2 - p_3)$$
$$\times \delta(\mathcal{Z}_{p_1} - \mathcal{Z}_{p_2} - \mathcal{Z}_{p_3})f(p_1)f(p_2)\mathrm{d}p_1\mathrm{d}p_2\mathrm{d}p_3,$$

we get

$$\frac{\partial}{\partial t} \int_{\mathbb{R}^3} f(p_1)\mathrm{d}p_1 = \int_{\mathbb{R}^3} Q[f](p_1)\mathrm{d}p_1 \leq J_1 + J_2, \qquad (10.2.52)$$

in which we have dropped the negative term containing $f(p_2)f(p_3)$.

Now, J_1 can be estimated by using the definition of the Dirac functions $\delta(p_1 - p_2 - p_3)$, $\delta(\mathcal{Z}_{p_1} - \mathcal{Z}_{p_2} - \mathcal{Z}_{p_3})$ and the boundedness of $K^{12}(p_1, p_2, p_1 - p_2)$

$$J_1 = \lambda_1 n_c \iint_{\mathbb{R}^3 \times \mathbb{R}^3} K^{12}(p_1, p_2, p_1 - p_2)\delta(\mathcal{Z}_{p_1} - \mathcal{Z}_{p_2} - \mathcal{Z}_{p_1 - p_2})f(p_1)\mathrm{d}p_1\mathrm{d}p_2$$
$$\lesssim \int_{\mathbb{R}^3} f(p_1)\left(\int_{S^0_{p_1}} \frac{1}{|\nabla H_0^{p_1}|}\mathrm{d}\sigma(p_2)\right)\mathrm{d}p_1,$$

which, by Lemma 10.2.2, can be bounded as

$$J_1 \lesssim \int_{\mathbb{R}^3} f(p_1)|p_1|^2\mathrm{d}p_1.$$

Using the fact that $|p_1|^2$ is dominated by $\mathcal{Z}_{p_1}$ up to a constant, we find

$$J_1 \lesssim \int_{\mathbb{R}^3} f(p_1)\mathcal{Z}_{p_1}\mathrm{d}p_1 \lesssim 1, \qquad (10.2.53)$$

where the last inequality follows from the conservation of energy (10.2.17).

It remains to estimate J_2. By a straightforward use of the definition of the Dirac functions $\delta(p_1 - p_2 - p_3)$ and $\delta(\mathcal{Z}_{p_1} - \mathcal{Z}_{p_2} - \mathcal{Z}_{p_3})$

$$J_2 = 2N_c \iint_{\mathbb{R}^3 \times \mathbb{R}^3} K^{12}(p_1, p_2, p_1 - p_2)$$
$$\times \delta(\mathcal{Z}_{p_1} - \mathcal{Z}_{p_2} - \mathcal{Z}_{p_1 - p_2})f(p_1)f(p_2)\mathrm{d}p_1\mathrm{d}p_2$$
$$= 2N_c \int_{\mathbb{R}^3} f(p_2)\left(\int_{S^2_{p_2}} K^{12}_2(p_1, p_2, p_1 - p_2)f(p_1)\mathrm{d}\sigma(p_1)\right)\mathrm{d}p_2,$$

which, by Lemma 10.2.4, implies

$$J_2 \lesssim \int_{\mathbb{R}^3} f(p_2) \left(\int_{S_{p_2}^2} K_2^{12}(p_1, p_2, p_1 - p_2) f(p_1) d\sigma(p_1) \right) dp_2$$

$$\lesssim \int_{\mathbb{R}^3} f(p_2) \left(\int_{\mathbb{R}^3} \frac{K^{12}(p_1, p_2, p_1 - p_2)}{|p_1||p_2|} f(p_1) dp_1 \right) dp_2.$$

Since $\frac{K^{12}(p_1, p_2, p_1 - p_2)}{|p_1||p_2|}$ is bounded by $|p_1 - p_2|$, up to a constant, J_2 is dominated by

$$J_2 \lesssim \int_{\mathbb{R}^3} f(p_2) \left(\int_{\mathbb{R}^3} f(p_1)|p_1 - p_2| dp_1 \right) dp_2$$

$$\lesssim \int_{\mathbb{R}^3} f(p_2) \left(\int_{\mathbb{R}^3} f(p_1)(|p_1| + |p_2|) dp_1 \right) dp_2$$

$$\lesssim \left(\int_{\mathbb{R}^3} f(p_2) \mathcal{Z}(p_2) dp_2 \right) \left(\int_{\mathbb{R}^3} f(p_1) dp_1 \right),$$

in which we have just used the fact that $|p|$ is bounded by $\mathcal{Z}(p)$ up to a constant, which by the conservation of energy (10.2.17), implies

$$J_2 \lesssim \left(\int_{\mathbb{R}^3} f(p_1) dp_1 \right). \tag{10.2.54}$$

Combining (10.2.52), (10.2.53) and (10.2.54) leads to

$$\frac{\partial}{\partial t} \int_{\mathbb{R}^3} f(p_1) dp_1 = \int_{\mathbb{R}^3} Q[f](p_1) dp_1 \leq C^* \left(1 + \int_{\mathbb{R}^3} f(p_1) dp_1 \right), \tag{10.2.55}$$

for some positive constant C^*, which implies the conclusion of the proposition.

$\square$

10.2.6 Estimating $\mathcal{Q}_{12}$

In this section, we show that the operator $\mathcal{Q}_{12}$ can be bounded by the sum of moments of the solution.

Proposition 10.2.3. *For any positive, radial function* $f(p) = f(|p|)$, *for any* $n \in \mathbb{N}$, *the following bound on the collision operator* $\mathcal{Q}_{12}$ *holds true*

$$\int_{\mathbb{R}^3} \mathcal{Q}_{12}[f](p_1) \mathcal{Z}^n(p_1) dp_1$$

$$\lesssim \sum_{k=1}^{n-1} (m_k[f] + m_{k-1}[f])(m_{n-k-1}[f] + m_{n-k}[f]) - m_{n+1}[f] + m_1[f]).$$

$$\tag{10.2.56}$$

Proof. For the sake of simplicity, we denote $m_k[f]$ by m_k. In view of Lemma 10.2.1,

$$\int_{\mathbb{R}^3} \mathcal{Q}_{12}[f](p_1)\mathcal{Z}^n(p_1)\mathrm{d}p_1 \tag{10.2.57}$$

$$= N_c \iiint_{\mathbb{R}^{3\times3}} K^{12}(p_1,p_2,p_3)\delta(p_1-p_2-p_3)\delta(\mathcal{Z}_{p_1}-\mathcal{Z}_{p_2}-\mathcal{Z}_{p_3})$$

$$\times [f(p_2)f(p_3)-f(p_1)-2f(p_1)f(p_2)][\mathcal{Z}_{p_1}^n-\mathcal{Z}_{p_2}^n-\mathcal{Z}_{p_3}^n]\mathrm{d}p_1\mathrm{d}p_2\mathrm{d}p_3.$$

By the definition of $\delta(\mathcal{Z}_{p_1}-\mathcal{Z}_{p_2}-\mathcal{Z}_{p_3})$, the term $\mathcal{Z}_{p_1}^n-\mathcal{Z}_{p_2}^n-\mathcal{Z}_{p_3}^n$ could be rewritten as $(\mathcal{Z}_{p_2}+\mathcal{Z}_{p_3})^n-\mathcal{Z}_{p_2}^n-\mathcal{Z}_{p_3}^n = \sum_{k=1}^{n-1}\binom{n}{k}\mathcal{Z}_{p_2}^k\mathcal{Z}_{p_3}^{n-k}$, which yields

$$\int_{\mathbb{R}^3} \mathcal{Q}_{12}[f](p_1)\mathcal{Z}^n(p_1)\mathrm{d}p_1$$

$$= N_c\lambda_1 \iiint_{\mathbb{R}^{3\times3}} K^{12}(p_1,p_2,p_3)\delta(p_1-p_2-p_3)\delta(\mathcal{Z}_{p_1}-\mathcal{Z}_{p_2}-\mathcal{Z}_{p_3})$$

$$\times [f(p_2)f(p_3)-f(p_1)-2f(p_1)f(p_2)]\left[\sum_{k=1}^{n-1}\binom{n}{k}\mathcal{Z}_{p_2}^k\mathcal{Z}_{p_3}^{n-k}\right]\mathrm{d}p_1\mathrm{d}p_2\mathrm{d}p_3.$$

Dropping the term containing $-2f(p_1)f(p_2)$, the above quantity could be estimated as

$$\int_{\mathbb{R}^3} \mathcal{Q}_{12}[f](p_1)\mathcal{Z}^n(p_1)\mathrm{d}p_1 \le L_1 + L_2, \tag{10.2.58}$$

where

$$L_1 := N_c \iiint_{\mathbb{R}^{3\times3}} K^{12}(p_1,p_2,p_3)\delta(p_1-p_2-p_3)\delta(\mathcal{Z}_{p_1}-\mathcal{Z}_{p_2}-\mathcal{Z}_{p_3})$$

$$\times f(p_2)f(p_3)\left[\sum_{k=1}^{n-1}\binom{n}{k}\mathcal{Z}_{p_2}^k\mathcal{Z}_{p_3}^{n-k}\right]\mathrm{d}p_1\mathrm{d}p_2\mathrm{d}p_3,$$

$$L_2 := -N_c \iiint_{\mathbb{R}^{3\times3}} K^{12}(p_1,p_2,p_3)\delta(p_1-p_2-p_3)\delta(\mathcal{Z}_{p_1}-\mathcal{Z}_{p_2}-\mathcal{Z}_{p_3})$$

$$\times f(p_1)\left[\sum_{k=1}^{n-1}\binom{n}{k}\mathcal{Z}_{p_2}^k\mathcal{Z}_{p_3}^{n-k}\right]\mathrm{d}p_1\mathrm{d}p_2\mathrm{d}p_3.$$

Let us first estimate L_1. By the definition of $\delta(p_1-p_2-p_3)$,

$$L_1 = N_c\lambda_1 \iint_{\mathbb{R}^{3\times2}} K^{12}(p_2+p_3,p_2,p_3)\delta(\mathcal{Z}_{p_2+p_3}-\mathcal{Z}_{p_2}-\mathcal{Z}_{p_3})$$

$$\times f(p_2)f(p_3)\left[\sum_{k=1}^{n-1}\binom{n}{k}\mathcal{Z}_{p_2}^k\mathcal{Z}_{p_3}^{n-k}\right]\mathrm{d}p_2\mathrm{d}p_3,$$

which, by the boundedness of K^{12}, could be estimated as

$$L_1 \lesssim \iint_{\mathbb{R}^3 \times \mathbb{R}^3} \delta(\mathcal{Z}_{p_2+p_3} - \mathcal{Z}_{p_2} - \mathcal{Z}_{p_3}) f(p_2) f(p_3) \left[\sum_{k=1}^{n-1} \binom{n}{k} \mathcal{Z}_{p_2}^k \mathcal{Z}_{p_3}^{n-k} \right] \mathrm{d}p_2 \mathrm{d}p_3$$

$$\lesssim \sum_{k=1}^{n-1} \int_{\mathbb{R}^3} f(p_2) \mathcal{Z}_{p_2}^k \left[\int_{S_{p_2}^1} f(p_3) \frac{\mathcal{Z}_{p_3}^{n-k}}{|\nabla H_2^{p_2}|} \mathrm{d}\sigma(p_3) \right] \mathrm{d}p_2.$$

Applying Lemma 10.2.3 to the above inequality, we find

$$L_1 \lesssim \sum_{k=1}^{n-1} \int_{\mathbb{R}^3} f(p_2) \frac{\mathcal{Z}_{p_2}^k}{|p_2|} \left[\int_{\mathbb{R}^3} f(p_3) \frac{\mathcal{Z}_{p_3}^{n-k}}{|p_3|} \mathrm{d}p_3 \right] \mathrm{d}p_2.$$

Observe that $\frac{\mathcal{Z}_{p_3}^{n-k}}{|p_3|} \lesssim \left(\mathcal{Z}_{p_3}^{n-k-1} + \mathcal{Z}_{p_3}^{n-k} \right)$ and $\frac{\mathcal{Z}_{p_2}^k}{|p_2|} \lesssim \left(\mathcal{Z}_{p_2}^{k-1} + \mathcal{Z}_{p_2}^k \right)$, which yield

$$L_1 \lesssim \sum_{k=1}^{n-1} \left[\int_{\mathbb{R}^3} f(p_1) \mathcal{Z}_{p_1}^k \mathrm{d}p_1 \right] \left[\int_{\mathbb{R}^3} f(p_1) \mathcal{Z}_{p_1}^{n-k-1} \mathrm{d}p_1 + \int_{\mathbb{R}^3} f(p_1) \mathcal{Z}_{p_1}^{n-k} \mathrm{d}p_1 \right]$$

$$\lesssim \sum_{k=1}^{n-1} [m_k + m_{k-1}][m_{n-k-1} + m_{n-k}].$$

$$(10.2.59)$$

Now, by the definition of $\delta(p_1 - p_2 - p_3)$ and $\delta(\mathcal{Z}_{p_1} - \mathcal{Z}_{p_2} - \mathcal{Z}_{p_1-p_2})$, the second term L_2 can be written as

$$L_2 = -N_c \iint_{\mathbb{R}^{3\times 3}} K^{12}(p_1, p_2, p_1 - p_2) \delta(\mathcal{Z}_{p_1} - \mathcal{Z}_{p_2} - \mathcal{Z}_{p_1-p_2})$$

$$\times f(p_1) \left[\sum_{k=1}^{n-1} \binom{n}{k} \mathcal{Z}_{p_2}^k \mathcal{Z}_{p_1-p_2}^{n-k} \right] \mathrm{d}p_1 \mathrm{d}p_2$$

$$\lesssim -\sum_{k=1}^{n-1} \int_{\mathbb{R}^3} f(p_1) \left[\int_{S_{p_1}^0} K_0^{12}(p_1, p_2, p_1 - p_2) \mathcal{Z}_{p_2}^k \mathcal{Z}_{p_1-p_2}^{n-k} \mathrm{d}\sigma(p_2) \right] \mathrm{d}p_1.$$

Since $\mathcal{Z}_{p_2}^k \mathcal{Z}_{p_1-p_2}^{n-k} \gtrsim \left[|p_2|^k |p_1 - p_2|^{n-k} + |p_2|^{2k} |p_1 - p_2|^{2(n-k)} \right]$, L_2 can be estimated as follows

$$L_2 \lesssim -\sum_{k=1}^{n-1} \int_{\mathbb{R}^3} f(p_1) \times$$

$$\left[\int_{S_{p_1}^0} K_0^{12}(p_1, p_2, p_1 - p_2) \left(|p_2|^k |p_1 - p_2|^{n-k} + |p_2|^{2k} |p_1 - p_2|^{2(n-k)} \right) \mathrm{d}\sigma(p_2) \right] \mathrm{d}p_1,$$

which, due to Lemma 10.2.2, can be bounded by

$$L_2 \lesssim -\int_{\mathbb{R}^3} f(p_1) \left((|p_1| \wedge 1)^{n+7} |p_1|^{n+1} + (|p_1| \wedge 1)^{2n+7} |p_1|^{2n+1} \right) \mathrm{d}p_1.$$

Splitting the integral on $\mathbb{R}^3$ into two integrals on $|p_1| > 1$ and $|p_1| \leq 1$ we find

$$L_2 \lesssim - \int_{|p_1| > 1} f(p_1) \left(|p_1|^{n+1} + |p_1|^{2n+2} \right) \mathrm{d}p_1,$$

where we have used the inequality $-|p_1|^{n+1} > -|p_1|^{n+2}$ for $|p_1| > 1$. Adding and subtracting on the right-hand side of the above inequality an integral with domain $|p_1| \leq 1$, we find

$$L_2 \lesssim - \left[\int_{\mathbb{R}^3} f(p_1) \left(|p_1|^{n+1} + |p_1|^{2n+2} \right) \mathrm{d}_1 - \int_{|p_1| \leq 1} f(p_1) \left(|p_1|^{n+1} + |p_1|^{2n+2} \right) \mathrm{d}p_1 \right]$$

$$\lesssim - \left[\int_{\mathbb{R}^3} f(p_1) \left(|p_1|^{n+1} + |p_1|^{2n+2} \right) \mathrm{d}p_1 - \int_{|p_1| \leq 1} |p_1| f(p_1) \mathrm{d}p_1 \right],$$

where the last inequality is due to the fact that we are integrating on $|p_1| \leq 1$. Bounding the integral on $|p_1| \leq 1$ by the integral on the full space $\mathbb{R}^3$, we obtain

$$L_2 \lesssim - \int_{\mathbb{R}^3} f(p_1) \left(|p_1|^{n+1} + |p_1|^{2n+2} \right) \mathrm{d}p_1 + \int_{\mathbb{R}^3} |p_1| f(p_1) \mathrm{d}p_1.$$

By $|p_1|^{n+1} + |p_1|^{2n+2} \gtrsim \mathcal{Z}_{p_1}^{n+1}$, we find the following estimate on L_2

$$L_2 \lesssim -m_{n+1} + m_1. \tag{10.2.60}$$

Combining (10.2.58), (10.2.59) and (10.2.60), we obtain the conclusion of the proposition. $\square$

10.2.7 Estimating $\mathcal{Q}_{22}$

In this section, we will estimate the operator $\mathcal{Q}_{22}$ by the sum of moments of the solution.

Proposition 10.2.4. *For any positive, radial function $f(p) = f(|p|)$, for any odd $n \in \mathbb{N}$, $n > 2$, the following bound on the collision operator $\mathcal{Q}_{22}$ holds true*

$$\int_{\mathbb{R}^3} \mathcal{Q}_{22}[f](p_1) \mathcal{Z}_{p_1}^n \mathrm{d}p_1 \lesssim$$

$$\sum_{0 \leq i,j,k < n; \ i+j+k=n} \sum_{s=0}^{k} m_{i+s} \left(m_{j+k-s} + m_{j+k-s+1/2} \right)$$

$$+ \sum_{0 \leq i,j,k < n; \ i+j+k=n: \ j,k>0} m_i \left(m_{j-1} + m_{j-1/2} \right) \left(m_{k-1} + m_{k-1/2} \right).$$

$$\tag{10.2.61}$$

Proof. For the sake of simplicity, we denote $m_k[f]$ by m_k. Observe that

$$\int_{\mathbb{R}^3} \mathcal{Q}_{22}[f](p_1)\mathcal{Z}_{p_1}^n \, \mathrm{d}p_1 =$$

$$C\int_{\mathbb{R}^{12}} K^{22}(p_1,p_2,p_3,p_1+p_2-p_3)$$

$$\times \delta(p_1+p_2-p_2-p_4)\delta(\mathcal{Z}_{p_1}+\mathcal{Z}_{p_2}-\mathcal{Z}_{p_3}-\mathcal{Z}_{p_4})$$

$$\times f(p_1)f(p_2)(1+f(p_3)+f(p_4))\Big[\mathcal{Z}_{p_4}^n+\mathcal{Z}_{p_3}^n-\mathcal{Z}_{p_2}^n-\mathcal{Z}_{p_1}^n\Big]\mathrm{d}p_1\mathrm{d}p_2\mathrm{d}p_3\mathrm{d}p_4.$$

Taking into account the fact that p_3 and p_4 are symmetric, and using the definition of the Dirac function to get $\mathcal{Z}_{p_4} = \mathcal{Z}_{p_1} + \mathcal{Z}_{p_2} - \mathcal{Z}_{p_3}$, we obtain

$$\int_{\mathbb{R}^3} \mathcal{Q}_{22}[f](p_1)\mathcal{Z}_{p_1}^n \, \mathrm{d}p_1 =$$

$$C\int_{\mathbb{R}^9} K^{22}(p_1,p_2,p_3,p_1+p_2-p_3) \qquad (10.2.62)$$

$$\times \delta(\mathcal{Z}_{p_1}+\mathcal{Z}_{p_2}-\mathcal{Z}_{p_3}-\mathcal{Z}_{p_1+p_2-p_3})f(p_1)f(p_2)(1+2f(p_3))$$

$$\times \Big[(\mathcal{Z}_{p_1}+\mathcal{Z}_{p_2}-\mathcal{Z}_{p_3})^n+\mathcal{Z}_{p_3}^n-\mathcal{Z}_{p_2}^n-\mathcal{Z}_{p_1}^n\Big]\mathrm{d}p_1\mathrm{d}p_2\mathrm{d}p_3.$$

Since n is an odd number, applying Newton's formula to the term $(\mathcal{Z}_{p_1}+\mathcal{Z}_{p_2}-\mathcal{Z}_{p_3})^n+\mathcal{Z}_{p_3}^n-\mathcal{Z}_{p_2}^n-\mathcal{Z}_{p_1}^n$ yields

$$(\mathcal{Z}_{p_1}+\mathcal{Z}_{p_2}-\mathcal{Z}_{p_3})^n+\mathcal{Z}_{p_3}^n-\mathcal{Z}_{p_2}^n-\mathcal{Z}_{p_1}^n = \qquad (10.2.63)$$

$$\sum_{0\le i,j,k<n;\ i+j+k=n} C_{i,j,k,n}\mathcal{Z}_{p_1}^i\mathcal{Z}_{p_2}^j\mathcal{Z}_{p_3}^k.$$

Plugging (10.2.63) into (10.2.62), we find

$$\int_{\mathbb{R}^3} \mathcal{Q}_{22}[f](p_1)\mathcal{Z}_{p_1}^n \, \mathrm{d}p_1$$

$$\lesssim \int_{\mathbb{R}^9} K^{22}(p_1,p_2,p_3,p_1+p_2-p_3)\delta(\mathcal{Z}_{p_1}+\mathcal{Z}_{p_2}-\mathcal{Z}_{p_3}-\mathcal{Z}_{p_1+p_2-p_3})$$

$$\times \left[\sum_{0\le i,j,k<n;\ i+j+k=n} |C_{i,j,k,n}|\mathcal{Z}_{p_1}^i\mathcal{Z}_{p_2}^j\mathcal{Z}_{p_3}^k\right]\mathrm{d}p_1\mathrm{d}p_2\mathrm{d}p_3$$

$$\lesssim \sum_{0\le i,j,k<n;\ i+j+k=n}\int_{\mathbb{R}^9} K^{22}(p_1,p_2,p_3,p_1+p_2-p_3)$$

$$\times \delta(\mathcal{Z}_{p_1}+\mathcal{Z}_{p_2}-\mathcal{Z}_{p_3}-\mathcal{Z}_{p_1+p_2-p_3})$$

$$\times f(p_1)f(p_2)(1+2f(p_3))\mathcal{Z}_{p_1}^i\mathcal{Z}_{p_2}^j\mathcal{Z}_{p_3}^k\mathrm{d}p_1\mathrm{d}p_2\mathrm{d}p_3.$$

$$(10.2.64)$$

In order to estimate the right-hand side of (10.2.64), we estimate each term containing $f(p_1)f(p_2)$ and $2f(p_1)f(p_2)f(p_3)$ separately. Let us first estimate the term containing $f(p_1)f(p_2)$

$$L_1 := \tag{10.2.65}$$

$$\sum_{0 \le i,j,k < n;\ i+j+k=n} \int_{\mathbb{R}^6} f(p_1)f(p_2) \mathscr{Z}_{p_1}^i \mathscr{Z}_{p_2}^j \int_{\mathcal{S}_{p_1,p_2}} \frac{K^{22} \mathscr{Z}_{p_3}^k}{|\nabla \mathcal{G}^{p_1,p_2}(p_3)|} d\mu(p_3) dp_1 dp_2,$$

which, by the inequality $\mathscr{Z}_{p_1} + \mathscr{Z}_{p_2} \ge \mathscr{Z}_{p_3}$, can be bounded as

$$L_1 \lesssim \sum_{0 \le i,j,k < n;\ i+j+k=n} \int_{\mathbb{R}^6} f(p_1)f(p_2) \chi_{\{|p_1| \ge 1\}} \chi_{\{|p_2| \ge 1\}} \mathscr{Z}_{p_1}^i \mathscr{Z}_{p_2}^j (\mathscr{Z}_{p_1} + \mathscr{Z}_{p_2})^k$$

$$\times \int_{\mathcal{S}_{p_1,p_2}} \frac{K^{22}}{|\nabla \mathcal{G}^{p_1,p_2}(p_3)|} d\mu(p_3) dp_1 dp_2,$$

$$\tag{10.2.66}$$

where $\chi_{\{|p| \ge 1\}}$ is the characteristic function of the set $\{|p| \ge 1\}$. Using Proposition 10.2.1, we can bound the above quantity as

$$L_1 \lesssim \sum_{0 \le i,j,k < n;\ i+j+k=n} \int_{\mathbb{R}^6} f(p_1)f(p_2) \chi_{\{|p_1| \ge 1\}} \chi_{\{|p_2| \ge 1\}} \mathscr{Z}_{p_1}^i \mathscr{Z}_{p_2}^j (\mathscr{Z}_{p_1} + \mathscr{Z}_{p_2})^k$$

$$\times \int_0^{C(|p_1|+|p_2|)} \frac{u}{|p_1 + p_2|} du\, dp_1 dp_2$$

$$\lesssim \sum_{0 \le i,j,k < n;\ i+j+k=n} \int_{\mathbb{R}^6} f(p_1)f(p_2) \chi_{\{|p_1| \ge 1\}} \chi_{\{|p_2| \ge 1\}} \mathscr{Z}_{p_1}^i \mathscr{Z}_{p_2}^j (\mathscr{Z}_{p_1} + \mathscr{Z}_{p_2})^k$$

$$\times \frac{(|p_1| + |p_2|)^2}{|p_1 + p_2|} dp_1 dp_2.$$

$$\tag{10.2.67}$$

Let ς be the angle between p_1, p_2 and rotate the coordinates so that p_1 is the pole. We obtain in spherical coordinates

$$L_1 \lesssim \sum_{0 \le i,j,k < n;\ i+j+k=n} \int_{\mathbb{R}^3} \int_0^\infty \int_0^\pi f(p_1)f(p_2) |p_2|^2 \mathscr{Z}_{p_1}^i \mathscr{Z}_{p_2}^j (\mathscr{Z}_{p_1} + \mathscr{Z}_{p_2})^k$$

$$\times \chi_{\{|p_1| \ge 1\}} \chi_{\{|p_2| \ge 1\}} \frac{(|p_1| + |p_2|)^2 \sin \varsigma}{\sqrt{|p_1|^2 + |p_2|^2 + 2|p_1||p_2| \cos \varsigma}} d\varsigma\, d|p_2| dp_1.$$

$$\tag{10.2.68}$$

Now, since

$$\frac{(|p_1| + |p_2|)^2 \sin \varsigma}{\sqrt{|p_1|^2 + |p_2|^2 + 2|p_1||p_2| \cos \varsigma}} \lesssim \frac{(|p_1|^2 + |p_2|^2) \sin(\varsigma)}{\sqrt{|p_1||p_2|(1 + \cos \varsigma)}} \lesssim \frac{|p_1|^2 + |p_2|^2}{\sqrt{|p_1||p_2|}},$$

we learn from (10.2.68) that

$$L_1 \lesssim \sum_{0 \le i,j,k < n; i+j+k=n} \int_{\mathbb{R}_+^2} |p_1|^{\frac{3}{2}} |p_2|^{\frac{3}{2}} (|p_1|^2 + |p_2|^2) f(p_1)f(p_2)$$

$$\times \chi_{\{|p_1| \ge 1\}} \chi_{\{|p_2| \ge 1\}} \mathscr{Z}_{p_1}^i \mathscr{Z}_{p_2}^j (\mathscr{Z}_{p_1} + \mathscr{Z}_{p_2})^k d|p_1| d|p_2|.$$

$$\tag{10.2.69}$$

Again, by Newton's formula, we find $(\mathcal{Z}_{p_1} + \mathcal{Z}_{p_2})^k = \sum_0^k \binom{k}{s} \mathcal{Z}_{p_1}^s \mathcal{Z}_{p_2}^{k-s}$, which, together with (10.2.69), leads to

$$
\begin{aligned}
L_1 \lesssim \sum_{0 \leq i,j,k < n;\ i+j+k=n} \sum_{s=0}^k \int_{\mathbb{R}_+^2} &|p_1|^{\frac{3}{2}} |p_2|^{\frac{3}{2}} (|p_1|^2 + |p_2|^2) f(p_1) f(p_2) \\
&\times \chi_{\{|p_1| \geq 1\}} \chi_{\{|p_2| \geq 1\}} \mathcal{Z}_{p_1}^{i+s} \mathcal{Z}_{p_2}^{j+k-s} \mathrm{d}|p_1| \mathrm{d}|p_2|.
\end{aligned}
\tag{10.2.70}
$$

By the symmetry of p_1 and p_2, we deduce from (10.2.70) that

$$
\begin{aligned}
L_1 \lesssim \sum_{0 \leq i,j,k < n;\ i+j+k=n} \sum_{s=0}^k \int_{|p_1| \leq |p_2|} &|p_1|^{\frac{3}{2}} |p_2|^{\frac{7}{2}} f(p_1) f(p_2) \\
&\times \chi_{\{|p_1| \geq 1\}} \chi_{\{|p_2| \geq 1\}} \mathcal{Z}_{p_1}^{i+s} \mathcal{Z}_{p_2}^{j+k-s} \mathrm{d}|p_1| \mathrm{d}|p_2|.
\end{aligned}
\tag{10.2.71}
$$

Note that the integrals of $\mathrm{d}|p_1|$ and $\mathrm{d}|p_2|$ in (10.2.70) are separated. It is straightforward that the integral of $\mathrm{d}|p_1|$ can be computed, by a spherical coordinate change of variables, as

$$
\int_{\mathbb{R}_+} |p_1|^{\frac{3}{2}} \chi_{\{|p_1| \geq 1\}} f(p_1) \mathcal{Z}_{p_1}^{i+s} \mathrm{d}|p_1| \lesssim \int_{\mathbb{R}_+} |p_1|^2 f(p_1) \mathcal{Z}_{p_1}^{i+s} \mathrm{d}|p_1| = m_{i+s}.
\tag{10.2.72}
$$

Now, for the second integral concerning $\mathrm{d}|p_2|$, by the inequality $\mathcal{Z}_{p_2} \lesssim |p_2| + |p_2|^2$, one gets

$$
\int_{\mathbb{R}_+} |p_2|^{\frac{7}{2}} f(p_2) \mathcal{Z}_{p_2}^{j+k-s} \mathrm{d}|p_2| \lesssim m_{j+k-s} + m_{j+k-s+1/2}.
\tag{10.2.73}
$$

Combining (10.2.70), (10.2.72) and (10.2.73) yields

$$
L_1 \lesssim \sum_{0 \leq i,j,k < n;\ i+j+k=n} \sum_{s=0}^k m_{i+s} \left(m_{j+k-s} + m_{j+k-s+1/2} \right).
\tag{10.2.74}
$$

Now, for the term containing $2f(p_1) f(p_2) f(p_3)$,

$$
L_2 :=
\tag{10.2.75}
$$

$$
\sum_{0 \leq i,j,k < n;\ i+j+k=n} \int_{\mathbb{R}^6} f(p_1) f(p_2) \mathcal{Z}_{p_1}^k \mathcal{Z}_{p_2}^j \int_{S_{p_1,p_2}} \frac{K^{22} \mathcal{Z}_{p_3}^i f(p_3)}{|\nabla \mathcal{G}^{p_1,p_2}(p_3)|} \mathrm{d}\mu(p_3) \mathrm{d}p_1 \mathrm{d}p_2.
$$

There are only two cases: $i, j, k > 0$ and one of i, j, k is 0. Due to the constraints $i+j+k = n$ and $0 \leq i, j, k < n$, the case in which two of the indices i, j, k are 0 does happen. Hence, we can suppose without loss of generality that $i \geq 0$ and $j, k > 0$. The terms on the right-hand side of (10.2.75) can be estimated, by Proposition 10.2.1, as

$$L_2 \lesssim \tag{10.2.76}$$

$$\sum_{0 \le i,j,k<n;\ i+j+k=n} \int_{\mathbb{R}^6} f(p_1)f(p_2)\mathcal{Z}_{p_1}^k \mathcal{Z}_{p_2}^j \int_0^\infty \frac{K^{22}\mathcal{Z}_{p_3}^i f(p_3)|p_3|}{|p_1+p_2|} \mathrm{d}|p_3|\mathrm{d}p_1\mathrm{d}p_2.$$

By the same argument used to obtain (10.2.69), as well as the inequality $K^{22}|p_3| \lesssim K^{22}|p_3|^2$, we get

$$L_2 \lesssim \tag{10.2.77}$$

$$\sum_{0 \le i,j,k<n;\ i+j+k=n} \int_{\mathbb{R}^6} f(p_1)f(p_2)\frac{\mathcal{Z}_{p_1}^k}{\sqrt{|p_1|}}\frac{\mathcal{Z}_{p_2}^j}{\sqrt{|p_2|}} \int_0^\infty \mathcal{Z}_{p_3}^i f(p_3)|p_3|^2 \mathrm{d}|p_3|\mathrm{d}p_1\mathrm{d}p_2,$$

which yields

$$L_2 \lesssim \sum_{0 \le i,j,k<n;\ i+j+k=n:\ j,k>0} m_i\left(m_{j-1}+m_{j-1/2}\right)\left(m_{k-1}+m_{k-1/2}\right). \tag{10.2.78}$$

From (10.2.64), (10.2.74) and (10.2.78), we get

$$\int_{\mathbb{R}^3} Q_{22}[f](p_1)\mathcal{Z}_{p_1}^n \mathrm{d}p_1 \lesssim \sum_{0 \le i,j,k<n;\ i+j+k=n} \sum_{s=0}^k m_{i+s}\left(m_{j+k-s}+m_{j+k-s+1/2}\right)$$

$$+ \sum_{0 \le i,j,k<n;\ i+j+k=n;\ j,k>0} m_i\left(m_{j-1}+m_{j-1/2}\right)\left(m_{k-1}+m_{k-1/2}\right).$$

$$\square$$

10.2.8 Finite Time Moment Estimates

The proposition below shows that moments of all orders greater than 1 are bounded on any time interval $[0, T]$.

Proposition 10.2.5. *Suppose that $f_0(p) = f_0(|p|)$ is positive and radial. Moreover,*

$$\int_{\mathbb{R}^3} f_0(p)\mathcal{Z}_p \mathrm{d}p < \infty, \qquad \int_{\mathbb{R}^3} f_0(p)\mathrm{d}p < \infty,$$

then for any finite time interval $[0, T]$, and for any $n \ge 1$, the positive radial solution $f(t, p) = f(t, |p|)$ of (10.2.1) satisfies

$$\sup_{t \in [\tau, T]} \int_{\mathbb{R}^3} f(t, p)\mathcal{Z}_p^n \mathrm{d}p < C_\tau, \quad \forall\, 0 < \tau \le T,$$

where C_τ is a constant depending on τ.
    If

$$\int_{\mathbb{R}^3} f_0(p)\mathcal{Z}_p^n \mathrm{d}p < \infty,$$

then

$$\sup_{t\in[0,T]}\int_{\mathbb{R}^3} f(t,p)\mathcal{Z}_p^n \mathrm{d}p < \infty.$$

In order to prove Proposition 10.2.5, we need the following Hölder inequality.

Lemma 10.2.5. *Let f be a function in $L^1(\mathbb{R}^3) \cap L_n^1(\mathbb{R}^3)$, then*

$$\|f\|_{L_k^1} \lesssim \|f\|_{L_n^1}^{\frac{k}{n}},$$

where the constant on the right-hand side depends on $\|f\|_{L^1}$, k and n.

Proof. This can be shown easily as follows

$$\int_{\mathbb{R}^3} |p|^k f(p)\mathrm{d}p \le \left(\int_{\mathbb{R}^3} |f(p)|\mathrm{d}p\right)^{\frac{n-k}{n}} \left(\int_{\mathbb{R}^3} |p|^n |f(p)|\mathrm{d}p\right)^{\frac{k}{n}}$$

$$\lesssim \left(\int_{\mathbb{R}^3} |p|^n f(p)\mathrm{d}p\right)^{\frac{k}{n}}.$$

$\square$

Proof of Proposition 10.2.5. Fix a time interval $[0,T]$. It is sufficient to prove Proposition 10.2.5 for $n \in \mathbb{N}$, n odd. Using $\mathcal{Z}_{p_1}^n$ as a test function in (10.2.1), we adapt the classical moment estimates [4, 27, 156, 161] for the classical Boltzmann equation. In view of Lemma 10.2.1, we obtain

$$\frac{\partial}{\partial t}\int_{\mathbb{R}^3} f(p_1)\mathcal{Z}_{p_1}^n \mathrm{d}p_1 = \int_{\mathbb{R}^3} \mathcal{Q}_{12}[f](p_1)\mathcal{Z}_{p_1}^n \mathrm{d}p_1 + \int_{\mathbb{R}^3} \mathcal{Q}_{22}[f](p_1)\mathcal{Z}_{p_1}^n \mathrm{d}p_1.$$
$$(10.2.79)$$

For the sake of simplicity, we denote $m_k[f(t)]$ as $m_k(t)$. First, let us estimate the $\mathcal{Q}_{12}$ collision operator. By Proposition 10.2.3

$$\int_{\mathbb{R}^3} \mathcal{Q}_{12}[f](p_1)\mathcal{Z}^n(p_1)\mathrm{d}p_1 \times$$

$$\sum_{k=1}^{n-1}(m_k(t) + m_{k-1}(t))(m_{n-k-1}(t) + m_{n-k}(t)) - m_{n+1}(t) + m_1(t).$$

Since, by Proposition 10.2.2, $m_0(t)$ is bounded by a constant C on $[0,T]$, we deduce from Lemma 10.2.5 that

$$m_k(t) \lesssim m_n(t)^{\frac{k}{n}}, \quad m_{k-1}(t) \lesssim m_n(t)^{\frac{k-1}{n}}, \quad m_{n-k-1}(t) \lesssim m_n(t)^{\frac{n-k-1}{n}},$$
$$m_{n-k}(t) \lesssim m_n(t)^{\frac{n-k}{n}}, \quad m_{n+1}(t) \gtrsim m_n(t)^{\frac{n+1}{n}}, \quad m_1(t) \lesssim m_n(t)^{\frac{1}{n}},$$

where the constants on the right-hand side depend on n, k, and the bound of the mass on $[0,T]$ in Proposition 10.2.2. As a consequence, we obtain the following estimate

$$\int_{\mathbb{R}^3} \mathcal{C}_{12}[f](p_1)\mathcal{Z}^n(p_1)\mathrm{d}p_1 \lesssim m_n(t) + m_n(t)^{\frac{n-1}{n}} + m_n(t)^{\frac{n-2}{n}} + m_n(t)^{\frac{1}{n}} - m_n(t)^{\frac{n+1}{n}}.$$

$$(10.2.80)$$

Now, for the $\mathcal{C}_{22}$ collision operator, according to Proposition 10.2.4,

$$\int_{\mathbb{R}^3} \mathcal{C}_{22}[f](p_1)\mathcal{Z}^n_{p_1}\,\mathrm{d}p_1 \lesssim$$

$$\sum_{0 \le i,j,k<n;\ i+j+k=n} \sum_{s=0}^{k} \left(m_{i+s}(t) + m_{j+k-s}(t) + m_{j+k-s+1/2}(t)\right)$$

$$+ \sum_{0 \le i,j,k<n;\ i+j+k=n:\ j,k>0} m_i(t)\left(m_{j-1}(t) + m_{j-1/2}(t)\right)\left(m_{k-1}(t) + m_{k-1/2}(t)\right).$$

Again, by Proposition 10.2.2, and Lemma 10.2.5,

$$m_{i+s}(t) \lesssim m_n(t)^{\frac{i+s}{n}}, \qquad m_{j+k-s}(t) \lesssim m_n(t)^{\frac{j+k-s}{n}},$$

$$m_{j+k-s+1/2}(t) \lesssim m_n(t)^{\frac{j+k-s+1/2}{n}}, \qquad m_i(t) \lesssim m_n(t)^{\frac{i}{n}},$$

$$m_{j-1}(t) \lesssim m_n(t)^{\frac{j-1}{n}}, \qquad m_{j-1/2}(t) \lesssim m_n(t)^{\frac{j-1/2}{n}},$$

$$m_{k-1}(t) \lesssim m_n(t)^{\frac{k-1}{n}}, \qquad m_{k-1/2}(t) \lesssim m_n(t)^{\frac{k-1/2}{n}},$$

we find

$$\int_{\mathbb{R}^3} \mathcal{C}_{22}[f](p_1)\mathcal{Z}^n_{p_1}\,\mathrm{d}p_1 \lesssim$$

$$\sum_{0 \le i,j,k<n,\ i+j+k=n} \sum_{s=0}^{k+1} m_n(t)^{\frac{i+s}{n}} \left(m_n(t)^{\frac{j+k-s}{n}} + m_n(t)^{\frac{j+k-s+1/2}{n}}\right)$$

$$+ \sum_{0 \le i,j,k<n;\ i+j+k=n:\ j,k>0} m_n(t)^{\frac{i}{n}} \left(m_n(t)^{\frac{j-1}{n}} + m_n(t)^{\frac{j-1/2}{n}}\right)\left(m_n(t)^{\frac{k-1}{n}} + m_n(t)^{\frac{k-1/2}{n}}\right).$$

$$(10.2.81)$$

Combining (10.2.79), (10.2.80) and (10.2.81) leads to

$$\frac{\mathrm{d}}{\mathrm{d}t} m_n(t) \lesssim m_n(t) + m_n(t)^{\frac{n-1}{n}} + m_n(t)^{\frac{n-2}{n}} + m_n(t)^{\frac{1}{n}} - m_n^{\frac{n+1}{n}}$$

$$+ \sum_{0 \le i,j,k<n;\ i+j+k=n} \sum_{s=0}^{k+1} m_n(t)^{\frac{i+s}{n}} \left(m_n(t)^{\frac{j+k-s}{n}} + m_n(t)^{\frac{j+k-s+1/2}{n}}\right)$$

$$+ \sum_{0 \le i,j,k<n;\ i+j+k=n;\ j,k>0} m_n(t)^{\frac{i}{n}} \left(m_n(t)^{\frac{j-1}{n}} + m_n(t)^{\frac{j-1/2}{n}}\right)\left(m_n(t)^{\frac{k-1}{n}} + m_n(t)^{\frac{k-1/2}{n}}\right),$$

$$(10.2.82)$$

where the constant on the right-hand side depends on n, k, and the bound of the mass on $[0,T]$ in Proposition 10.2.2. Notice that $-m_n(t)^{\frac{n+1}{n}}$ has the highest order on the right-hand side of (10.2.82). By the same argument as in [123], the conclusion of the theorem then follows. □

10.2.9 Hölder Estimates for the Collision Operators

In this section, we will establish Hölder estimates for the two collision operators Q_{12} and Q_{22}. We split Q_{22} into two operators

$$Q_{22}^1[f](p_1) = \kappa_3 \iiint_{\mathbb{R}^9} K^{22}(p_1, p_2, p_3, p_4)\delta(p_1 + p_2 - p_3 - p_4) \quad (10.2.83)$$
$$\times \delta(\mathcal{Z}_{p_1} + \mathcal{Z}_{p_2} - \mathcal{Z}_{p_3} - \mathcal{Z}_{p_4})[f(p_3)f(p_4) - f(p_1)f(p_2)]\mathrm{d}p_2\mathrm{d}p_3\mathrm{d}p_4$$

and

$$Q_{22}^2[f](p_2) = \kappa_3 \iiint_{\mathbb{R}^9} K^{22}(p_1, p_2, p_3, p_4)\delta(p_1 + p_2 - p_3 - p_4) \quad (10.2.84)$$
$$\times \delta(\mathcal{Z}_{p_1} + \mathcal{Z}_{p_2} - \mathcal{Z}_{p_3} - \mathcal{Z}_{p_4})[f(p_3)f(p_4)(f(p_1) + f(p_2))$$
$$- f(p_1)f(p_2)(f(p_3) + f(p_4))]\mathrm{d}p_2\mathrm{d}p_3\mathrm{d}p_4.$$

We will show in Proposition 10.2.6, Proposition 10.2.7 and Proposition 10.2.8 that Q_{12}, Q_{22}^1 and Q_{22}^2 are Hölder continuous.

Proposition 10.2.6. *Let f and g be two functions in $L_{n+3}^1(\mathbb{R}^3) \cap L^1(\mathbb{R}^3)$, $n \in \mathbb{R}_+$, $n > 0$. Then the following estimate holds true*

$$\|Q_{12}[f] - Q_{12}[g]\|_{L_n^1} \lesssim \|f - g\|_{L_{n+3}^1} + \|f - g\|_{L^1}. \quad (10.2.85)$$

If $\|f\|_{L_{n+4}^1}, \|g\|_{L_{n+4}^1} < C_0$, then

$$\|Q_{12}[f] - Q_{12}[g]\|_{L_n^1} \lesssim \|f - g\|_{L^1}^{\frac{1}{n+4}} + \|f - g\|_{L^1}. \quad (10.2.86)$$

Proof. First, let us estimate the L_n^1 norm of the difference $Q_{12}[f] - Q_{12}[g]$. In view of Lemma 10.2.1,

$$\|Q_{12}[f] - Q_{12}[g]\|_{L_n^1} = \int_{\mathbb{R}^3} |p_1|^n |Q_{12}[f] - Q_{12}[g]|\mathrm{d}p_1$$
$$\lesssim N_c \iiint_{\mathbb{R}^{3\times3}} K^{12}(p_1, p_2, p_3)\delta(p_1 - p_2 - p_3)\delta(\mathcal{Z}_{p_1} - \mathcal{Z}_{p_2} - \mathcal{Z}_{p_3})$$
$$\times |f(p_2)f(p_3) - 2f(p_3)f(p_1) - f(p_1) - g(p_2)g(p_3)$$
$$+ 2g(p_3)g(p_1) + g(p_1)|[|p_1|^n + |p_2|^n + |p_3|^n]\,\mathrm{d}p_1\mathrm{d}p_2\mathrm{d}p_3.$$
$$(10.2.87)$$

Therefore, $\|Q_{12}[f] - Q_{12}[g]\|_{L_n^1}$ can be bounded by the sum of the following three terms

$$N_1 = N_c \iiint_{\mathbb{R}^{3\times3}} K^{12}(p_1, p_2, p_3)\delta(p_1 - p_2 - p_3)\delta(\mathcal{Z}_{p_1} - \mathcal{Z}_{p_2} - \mathcal{Z}_{p_3})$$
$$\times |f(p_2)f(p_3) - g(p_2)g(p_3)|[|p_1|^n + |p_2|^n + |p_3|^n]\,\mathrm{d}p_1\mathrm{d}p_2\mathrm{d}p_3,$$

$$N_2 = 2N_c \iiint_{\mathbb{R}^{3\times3}} K^{12}(p_1, p_2, p_3)\delta(p_1 - p_2 - p_3)\delta(\mathcal{Z}_{p_1} - \mathcal{Z}_{p_2} - \mathcal{Z}_{p_3})$$
$$\times |f(p_3)f(p_1) - g(p_3)g(p_1)| \left[|p_1|^n + |p_2|^n + |p_3|^n\right] \mathrm{d}p_1 \mathrm{d}p_2 \mathrm{d}p_3,$$

and

$$N_3 = N_c \iiint_{\mathbb{R}^{3\times3}} K^{12}(p_1, p_2, p_3)\delta(p_1 - p_2 - p_3)\delta(\mathcal{Z}_{p_1} - \mathcal{Z}_{p_2} - \mathcal{Z}_{p_3})$$
$$\times |f(p_1) - g(p_1)| \left[|p_1|^n + |p_2|^n + |p_3|^n\right] \mathrm{d}p_1 \mathrm{d}p_2 \mathrm{d}p_3.$$

We will estimate N_1, N_2, N_3 in three steps.

Step 1: Estimating N_1.

By the definition of $\delta(p_1 - p_2 - p_3)$, N_1 can be rewritten as

$$N_1 = N_c\lambda_1 \iint_{\mathbb{R}^{3\times2}} K^{12}(p_2 + p_3, p_2, p_3)\delta(\mathcal{Z}_{p_2+p_3} - \mathcal{Z}_{p_2} - \mathcal{Z}_{p_3})$$
$$\times |f(p_2)f(p_3) - g(p_2)g(p_3)| \left[|p_2 + p_3|^n + |p_2|^n + |p_3|^n\right] \mathrm{d}p_2 \mathrm{d}p_3.$$

By the triangle inequality,

$$|f(p_2)f(p_3) - g(p_2)g(p_3)| \leq |f(p_2) - g(p_2)||f(p_3)| + |f(p_3) - g(p_3)||g(p_2)|,$$

the first quantity N_1 can be bounded

$$N_1 \lesssim \iint_{\mathbb{R}^{3\times2}} K^{12}(p_2 + p_3, p_2, p_3)\delta(\mathcal{Z}_{p_2+p_3} - \mathcal{Z}_{p_2} - \mathcal{Z}_{p_3})$$
$$\times |f(p_2) - g(p_2)||f(p_3)| \left[|p_2 + p_3|^n + |p_2|^n + |p_3|^n\right] \mathrm{d}p_2 \mathrm{d}p_3$$
$$+ \iint_{\mathbb{R}^{3\times2}} K^{12}(p_2 + p_3, p_2, p_3)\delta(\mathcal{Z}_{p_2+p_3} - \mathcal{Z}_{p_2} - \mathcal{Z}_{p_3})$$
$$\times |f(p_3) - g(p_3)||g(p_2)| \left[|p_2 + p_3|^n + |p_2|^n + |p_3|^n\right] \mathrm{d}p_2 \mathrm{d}p_3.$$

Again, by the triangle inequality

$$|p_2 + p_3|^n \leq (|p_2| + |p_3|)^n \leq 2^{n-1}(|p_2|^n + |p_3|^n),$$

one can estimate N_1

$$N_1 \lesssim \iint_{\mathbb{R}^{3\times2}} \delta(\mathcal{Z}_{p_2+p_3} - \mathcal{Z}_{p_2} - \mathcal{Z}_{p_3})K^{12}(p_2 + p_3, p_2, p_3)$$
$$\times |f(p_2) - g(p_2)||f(p_3)| \left[|p_2|^n + |p_3|^n\right] \mathrm{d}p_2 \mathrm{d}p_3$$
$$+ \iint_{\mathbb{R}^{3\times2}} \delta(\mathcal{Z}_{p_2+p_3} - \mathcal{Z}_{p_2} - \mathcal{Z}_{p_3})K^{12}(p_2 + p_3, p_2, p_3)$$
$$\times |f(p_3) - g(p_3)||g(p_2)| \left[|p_2|^n + |p_3|^n\right] \mathrm{d}p_2 \mathrm{d}p_3.$$

The above estimate can be rewritten, taking into account the definition of $\delta(\mathcal{Z}_{p_2+p_3} - \mathcal{Z}_{p_2} - \mathcal{Z}_{p_3})$. We then find

$$N_1 \lesssim$$

$$\int_{\mathbb{R}^3} \int_{S^1_{p_3}} K_1^{12}(p_2 + p_3, p_2, p_3) |f(p_2) - g(p_2)| |f(p_3)| \left[|p_2|^n + |p_3|^n \right] d\sigma(p_3) dp_2$$

$$+ \int_{\mathbb{R}^3} \int_{S^1_{p_2}} K_1^{12}(p_2 + p_3, p_2, p_3) |f(p_3) - g(p_3)| |g(p_2)| \left[|p_2|^n + |p_3|^n \right] d\sigma(p_2) dp_3.$$

By Lemma 10.2.3, one can estimate N_1

$$N_1 \lesssim$$

$$\iint_{\mathbb{R}^{3\times2}} |f(p_2) - g(p_2)| |f(p_3)| \frac{K^{12}(p_2 + p_3, p_2, p_3)}{|p_2||p_3|} \left[|p_2|^n + |p_3|^n \right] dp_3 dp_2$$

$$+ p \iint_{\mathbb{R}^{3\times2}} |f(p_3) - g(p_3)| |g(p_2)| \frac{K^{12}(p_2 + p_3, p_2, p_3)}{|p_2||p_3|} \left[|p_2|^n + |p_3|^n \right] dp_2 dp_3.$$

Since $\frac{K^{12}(p_2+p_3,p_2,p_3)}{|p_2||p_3|}$ and $\frac{K^{12}(p_2+p_3,p_2,p_3)}{|p_2||p_3|}$ are bounded, N_1 is bounded as

$$N_1 \lesssim \iint_{\mathbb{R}^{3\times2}} |f(p_2) - g(p_2)| |f(p_3)| \left[|p_2|^n + |p_3|^n \right] dp_3 dp_2$$

$$+ \iint_{\mathbb{R}^{3\times2}} |f(p_3) - g(p_3)| |g(p_2)| \left[|p_2|^n + |p_3|^n \right] dp_2 dp_3,$$

which leads to the following estimates on N_1

$$\begin{aligned}
N_1 \lesssim\ & \int_{\mathbb{R}^3} |f(p_2) - g(p_2)| |p_2|^n d_2 \int_{\mathbb{R}^3} |f(p_3)| dp_3 \\
& + \int_{\mathbb{R}^3} |f(p_2) - g(p_2)| dp_2 \int_{\mathbb{R}^3} |f(p_3)| |p_3|^n dp_3 \\
& + \int_{\mathbb{R}^3} |f(p_3) - g(p_3)| |p_3|^n d_3 \int_{\mathbb{R}^3} |f(p_2)| dp_2 \qquad (10.2.88) \\
& + \int_{\mathbb{R}^3} |f(p_3) - g(p_3)| dp_3 \int_{\mathbb{R}^3} |f(p_2)| |p_2|^n dp_2 \\
\lesssim\ & \int_{\mathbb{R}^3} |f(p_1) - g(p_1)| |p_1|^n dp_1 + \int_{\mathbb{R}^3} |f(p_1) - g(p_1)| dp_1.
\end{aligned}$$

Step 2: Estimating N_2.
By the definition of $\delta(p_1 - p_2 - p_3)$, N_2 can be rewritten as

$$\begin{aligned}
N_2 =\ & 2N_c \iint_{\mathbb{R}^{3\times2}} K^{12}(p_1, p_1 - p_3, p_3) \delta(\mathscr{Z}_{p_1} - \mathscr{Z}_{p_1-p_3} - \mathscr{Z}_{p_3}) \\
& \times |f(p_3)f(p_1) - g(p_3)g(p_1)| \left[|p_1|^n + |p_1 - p_3|^n + |p_3|^n \right] dp_1 dp_3,
\end{aligned}$$

which, since $|p_1 - p_3|^n \leq (|p_1| + |p_3|)^n \leq 2^{n-1}(|p_1|^n + |p_3|^n)$, can be bounded from above

$$N_2 \;\lesssim\; \iint_{\mathbb{R}^{3\times 2}} K^{12}(p_1, p_1 - p_3, p_3)\delta(\mathcal{Z}_{p_1} - \mathcal{Z}_{p_1 - p_3} - \mathcal{Z}_{p_3})$$
$$\times |f(p_3) - g(p_3)||f(p_1)| \left[|p_1|^n + |p_3|^n\right] \mathrm{d}p_1 \mathrm{d}p_3$$
$$+ \iint_{\mathbb{R}^{3\times 2}} K^{12}(p_1, p_1 - p_3, p_3)\delta(\mathcal{Z}_{p_1} - \mathcal{Z}_{p_1 - p_3} - \mathcal{Z}_{p_3})$$
$$\times |f(p_1) - g(p_1)||g(p_3)| \left[|p_1|^n + |p_3|^n\right] \mathrm{d}p_1 \mathrm{d}p_3.$$

Applying the definition of $\delta(\mathcal{Z}_{p_1} - \mathcal{Z}_{p_1 - p_3} - \mathcal{Z}_{p_3})$, one can estimate N_2

$$N_2 \;\lesssim\;$$
$$\int_{\mathbb{R}^3} \int_{S^0_{p_1}} K^{12}_0(p_1, p_1 - p_3, p_3)|f(p_3) - g(p_3)||f(p_1)| \left[|p_1|^n + |p_3|^n\right] \mathrm{d}\sigma(p_3)\mathrm{d}p_1$$
$$+ \int_{\mathbb{R}^3} \int_{S^0_{p_1}} K^{12}_0(p_1, p_1 - p_3, p_3)|f(p_1) - g(p_1)||g(p_3)| \left[|p_1|^n + |p_3|^n\right] \mathrm{d}\sigma(p_3)\mathrm{d}p_1,$$

which, by Lemma 10.2.2, yields

$$N_2 \;\lesssim\;$$
$$\int_{\mathbb{R}^3} \int_0^{|p_1|} K^{12}(p_1, p_1 - p_3, p_3)|f(p_1) - g(p_1)||g(p_3)| \left[|p_1|^n + |p_3|^n\right] |p_3|\mathrm{d}|p_3|\mathrm{d}p_1$$
$$+ \int_{\mathbb{R}^3} \int_0^{|p_1|} K^{12}(p_1, p_1 - p_3, p_3)|f(p_3) - g(p_3)||f(p_1)| \left[|p_1|^n + |p_3|^n\right] |p_3|\mathrm{d}|p_3|\mathrm{d}p_1.$$

Bounding the integral from 0 to $|p_1|$ by an integral from 0 to ∞ leads to

$$N_2 \;\lesssim\;$$
$$\int_{\mathbb{R}^3} \int_0^{\infty} K^{12}(p_1, p_1 - p_3, p_3)|f(p_1) - g(p_1)||g(p_3)| \left[|p_1|^n + |p_3|^n\right] |p_3|\mathrm{d}|p_3|\mathrm{d}p_1$$
$$+ \int_{\mathbb{R}^3} \int_0^{\infty} K^{12}(p_1, p_1 - p_3, p_3)|f(p_3) - g(p_3)||f(p_1)| \left[|p_1|^n + |p_3|^n\right] |p_3|\mathrm{d}|p_3|\mathrm{d}p_1.$$

We now switch the integral from $\mathrm{d}|p_3|$ to $\mathrm{d}p_3$ in the above inequality to obtain

$$N_2 \;\lesssim\;$$
$$\iint_{\mathbb{R}^{3\times 2}} \frac{K^{12}(p_1, p_1 - p_3, p_3)}{|p_3|}|f(p_1) - g(p_1)||g(p_3)| \left[|p_1|^n + |p_3|^n\right] \mathrm{d}p_3\mathrm{d}p_1$$
$$+ \iint_{\mathbb{R}^{3\times 2}} (1 + |p_1|)\frac{K^{12}(p_1, p_1 - p_3, p_3)}{|p_3|}|f(p_3) - g(p_3)||f(p_1)| \left[|p_1|^n + |p_3|^n\right] \mathrm{d}p_3\mathrm{d}p_1.$$

Applying the inequality $|p_1|^n + |p_3|^n \lesssim 1 + |p_1|^n + |p_3|^n$ to the above estimate, we find

$$N_2 \;\lesssim\;$$
$$\iint_{\mathbb{R}^{3\times 2}} \frac{K^{12}(p_1, p_1 - p_3, p_3)}{|p_3|}|f(p_3) - g(p_3)||f(p_1)| \left[1 + |p_1|^n + |p_3|^n\right] \mathrm{d}p_3\mathrm{d}p_1$$
$$+ \iint_{\mathbb{R}^{3\times 2}} \frac{K^{12}(p_1, p_1 - p_3, p_3)}{|p_3|}|f(p_1) - g(p_1)||g(p_3)| \left[1 + |p_1|^n + |p_3|^n\right] \mathrm{d}p_3\mathrm{d}p_1.$$

The same argument used to obtain (10.2.88) yields

$$N_2 \lesssim \int_{\mathbb{R}^3} |f(p_1) - g(p_1)||p_1|^n dp_1 + \int_{\mathbb{R}^3} |f(p_1) - g(p_1)| dp_1. \qquad (10.2.89)$$

Step 3: Estimating N_3.
By the definition of $\delta(p_1 - p_2 - p_3)$, N_3 can be rewritten as

$$N_3 = N_c \iint_{\mathbb{R}^{3 \times 2}} K^{12}(p_1, p_2, p_1 - p_2) \delta(\mathcal{Z}_{p_1} - \mathcal{Z}_{p_2} - \mathcal{Z}_{p_1 - p_2})$$
$$\times |f(p_1) - g(p_1)| \left[|p_1|^n + |p_2|^n + |p_1 - p_2|^n \right] dp_1 dp_2,$$

which, thanks to the inequality $|p_1 - p_2|^n \leq (|p_1| + |p_2|)^n \leq 2^{n-1}(|p_1|^n + |p_2|^n)$, can be bounded as

$$N_3 \lesssim \int_{\mathbb{R}^3} \int_{S_{p_1}} K^{12}(p_1, p_2, p_1 - p_2) |f(p_1) - g(p_1)| \left[|p_1|^n + |p_2|^n \right] d\sigma(p_2) dp_1.$$

Now, as an application of Lemma 10.2.2,

$$\int_{S_{p_1}} (|p_1|^n + |p_2|^n) d\sigma(p_2) \lesssim |p_1|^{n+2} + \int_{S_{p_1}} |p_2|^n d\sigma(p_2)$$
$$\lesssim |p_1|^{n+2} + \int_0^{|p_1|} |p_2|^{n+1} d|p_2|$$
$$\lesssim 1 + |p_1|^{n+3},$$

which together with the fact that $K^{12}(p_1, p_2, p_1 - p_2)$ is bounded, yields

$$N_3 \lesssim \int_{\mathbb{R}^3} |f(p_1) - g(p_1)| \left[|p_1|^{n+3} + 1 \right] dp_1. \qquad (10.2.90)$$

Combining (10.2.88), (10.2.89), and (10.2.90) leads to

$$\|\mathcal{Q}_{12}[f] - \mathcal{Q}_{12}[g]\|_{L_n^1} \lesssim$$
$$\int_{\mathbb{R}^3} |f(p_1) - g(p_1)| \left[|p_1|^{n+3} + |p_1|^{n+1} + |p_1|^n + 1 \right] dp_1. \qquad (10.2.91)$$

Since $|p|^n \lesssim (|p|^{n+3} + 1)$ and $|p|^{n+1} \lesssim (|p|^{n+3} + 1)$, inequality (10.2.85) follows from (10.2.91). Inequality (10.2.86) is a consequence of inequality (10.2.85), Lemma 10.2.5 and

$$\|f - g\|_{L_{n+3}^1} \lesssim \|f - g\|_{L^1}^{\frac{1}{n+4}} \left(\|f\|_{L_{n+4}^1} + \|g\|_{L_{n+4}^1} \right)^{\frac{n+3}{n+4}}.$$

$\square$

Proposition 10.2.7. *Let f and g be two functions in $L_n^1(\mathbb{R}^3) \cap L^1(\mathbb{R}^3)$, $n \in \mathbb{N}$, $n/2$ an odd number. Then the following inequality*

$$\|Q_{22}^1[f] - Q_{22}^1[g]\|_{L_n^1} \lesssim \|f - g\|_{L_{n+1}^1} + \|f - g\|_{L^1} \qquad (10.2.92)$$

holds true. If $\|f\|_{L_{n+2}^1}, \|g\|_{L_{n+2}^1} < \mathcal{C}_0$, then

$$\|Q_{22}^1[f] - Q_{22}^1[g]\|_{L_n^1} \lesssim \|f - g\|_{L^1}^{\frac{1}{n+2}} + \|f - g\|_{L^1}. \qquad (10.2.93)$$

Proof. Let us estimate the L_n^1 norm of the difference $Q_{22}^1[f] - Q_{22}^1[g]$. In view of Lemma 10.2.1

$$\int_{\mathbb{R}^3} \left| Q_{22}^1[f](p_1) - Q_{22}^1[g](p_1) \right| |p_1|^n \mathrm{d}p_1$$

$$\lesssim \iiiint_{\mathbb{R}^{12}} K^{22}(p_1, p_2, p_3, p_4) \delta(p_1 + p_2 - p_3 - p_4) \delta(\mathcal{Z}_{p_1} + \mathcal{Z}_{p_2} - \mathcal{Z}_{p_3} - \mathcal{Z}_{p_4})$$

$$\times |f(p_1)f(p_2) - g(p_1)g(p_2)| \Big[|p_4|^n + |p_3|^n + |p_2|^n + |p_1|^n \Big] \mathrm{d}p_1 \mathrm{d}p_2 \mathrm{d}p_3 \mathrm{d}p_4.$$

Thanks to the inequality $|p|^n \lesssim \mathcal{Z}_p^{n/2}$, one gets $|p_4|^n + |p_3|^n + |p_2|^n + |p_1|^n \lesssim \mathcal{Z}_{p_4}^{n/2} + \mathcal{Z}_{p_3}^{n/2} + \mathcal{Z}_{p_2}^{n/2} + \mathcal{Z}_{p_1}^{n/2}$, which yields

$$\int_{\mathbb{R}^3} \left| Q_{22}^1[f](p_1) - Q_{22}^1[g](p_1) \right| |p_1|^n \mathrm{d}p_1$$

$$\lesssim \iiiint_{\mathbb{R}^{12}} K^{22}(p_1, p_2, p_3, p_4) \delta(p_1 + p_2 - p_3 - p_4) \delta(\mathcal{Z}_{p_1} + \mathcal{Z}_{p_2} - \mathcal{Z}_{p_3} - \mathcal{Z}_{p_4})$$

$$\times |f(p_1)f(p_2) - g(p_1)g(p_2)| \Big[\mathcal{Z}_{p_4}^{n/2} + \mathcal{Z}_{p_3}^{n/2} + \mathcal{Z}_{p_2}^{n/2} + \mathcal{Z}_{p_1}^{n/2} \Big] \mathrm{d}p_1 \mathrm{d}p_2 \mathrm{d}p_3 \mathrm{d}p_4.$$

Now, thanks to the Dirac function $\delta(\mathcal{Z}_{p_1} + \mathcal{Z}_{p_2} - \mathcal{Z}_{p_3} - \mathcal{Z}_{p_4})$, one can write $\mathcal{Z}_{p_4}$ as $\mathcal{Z}_{p_1} + \mathcal{Z}_{p_2} - \mathcal{Z}_{p_3}$, which implies

$$\int_{\mathbb{R}^3} \left| Q_{22}^1[f](p_1) - Q_{22}^1[g](p_1) \right| |p_1|^n \mathrm{d}p_1$$

$$\lesssim \iiint_{\mathbb{R}^9} K^{22}(p_1, p_2, p_3, p_1 + p_2 - p_3) \delta(\mathcal{Z}_{p_1} + \mathcal{Z}_{p_2} - \mathcal{Z}_{p_3} - \mathcal{Z}_{p_1 + p_2 - p_3})$$

$$\times |f(p_1)f(p_2) - g(p_1)g(p_2)|$$

$$\times \Big[(\mathcal{Z}_{p_1} + \mathcal{Z}_{p_2} - \mathcal{Z}_{p_3})^{n/2} + \mathcal{Z}_{p_3}^{n/2} + \mathcal{Z}_{p_2}^{n/2} + \mathcal{Z}_{p_1}^{n/2} \Big] \mathrm{d}p_1 \mathrm{d}p_2 \mathrm{d}p_3.$$

Since $n/2$ is an odd number, by Newton's formula

$$(\mathcal{Z}_{p_1} + \mathcal{Z}_{p_2} - \mathcal{Z}_{p_3})^{n/2} + \mathcal{Z}_{p_3}^{n/2} + \mathcal{Z}_{p_2}^{n/2} + \mathcal{Z}_{p_1}^{n/2}$$

$$= \sum_{0 \le i,j,k \; ; \; i+j+k=n/2} B_{i,j,k,n} \mathcal{Z}_{p_1}^i \mathcal{Z}_{p_2}^j \mathcal{Z}_{p_3}^k$$

and the inequality $\mathcal{Z}_{p_3} \le \mathcal{Z}_{p_1} + \mathcal{Z}_{p_2}$ for $\mathcal{Z}_{p_1} + \mathcal{Z}_{p_2} = \mathcal{Z}_{p_1 + p_2 - p_3} + \mathcal{Z}_{p_3}$, we find

$$\int_{\mathbb{R}^3} \left| Q_{22}^1[f](p_1) - Q_{22}^1[g](p_1) \right| |p_1|^n \mathrm{d}p_1 \lesssim X, \qquad (10.2.94)$$

where

$$X := \iiint_{\mathbb{R}^{12}} K^{22}(p_1, p_2, p_3, p_1 + p_2 - p_3)\delta(\mathcal{Z}_{p_1} + \mathcal{Z}_{p_2} - \mathcal{Z}_{p_3} - \mathcal{Z}_{p_1 + p_2 - p_3})$$

$$\times \left| f(p_1)f(p_2) - g(p_1)g(p_2) \right|$$

$$\times \left[\sum_{0 \leq i,j,k \;;\; i+j+k=n/2} |B_{i,j,k,n}| \mathcal{Z}_{p_1}^i \mathcal{Z}_{p_2}^j (\mathcal{Z}_{p_1} + \mathcal{Z}_{p_2})^k \right] \mathrm{d}p_1 \mathrm{d}p_2 \mathrm{d}p_3.$$

The same argument used to obtain (10.2.69) can be applied to X, to get

$$X \lesssim \iint_{\mathbb{R}^6} \frac{|p_1|^2 + |p_2|^2}{|p_1||p_2|} \left| f(p_1)f(p_2) - g(p_1)g(p_2) \right|$$

$$\times \left[\sum_{0 \leq i,j,k \;;\; i+j+k=n/2} \mathcal{Z}_{p_1}^i \mathcal{Z}_{p_2}^j (\mathcal{Z}_{p_1} + \mathcal{Z}_{p_2})^k \right] \mathrm{d}p_1 \mathrm{d}p_2.$$

Using Newton's formula again, we find

$$X \lesssim \iint_{\mathbb{R}^6} \frac{|p_1|^2 + |p_2|^2}{|p_1||p_2|} \left| f(p_1)f(p_2) - g(p_1)g(p_2) \right|$$

$$\times \left[\sum_{0 \leq i,j,k \;;\; i+j+k=n/2} \sum_{s=0}^{k} \binom{k}{s} \mathcal{Z}_{p_1}^{i+s} \mathcal{Z}_{p_2}^{k+1+j-s} \right] \mathrm{d}p_1 \mathrm{d}p_2.$$

Let us bound X by estimating the terms inside the sum

$$\iint_{\mathbb{R}^6} \frac{|p_1|^2 + |p_2|^2}{|p_1||p_2|} \left| f(p_1)f(p_2) - g(p_1)g(p_2) \right| \mathcal{Z}_{p_1}^{i+s} \mathcal{Z}_{p_2}^{k+j-s} \mathrm{d}p_1 \mathrm{d}p_2$$

$$\lesssim \iint_{\mathbb{R}^6} \frac{|p_1|^2 + |p_2|^2}{|p_1||p_2|} |f(p_1) - g(p_1)||g(p_2)| \mathcal{Z}_{p_1}^{i+s} \mathcal{Z}_{p_2}^{k+j-s} \mathrm{d}p_1 \mathrm{d}|p_2$$

$$+ \iint_{\mathbb{R}^6} \frac{|p_1|^2 + |p_2|^2}{|p_1||p_2|} |f(p_2) - g(p_2)||f(p_1)| \mathcal{Z}_{p_1}^{i+s} \mathcal{Z}_{p_2}^{k+j-s} \mathrm{d}p_1 \mathrm{d}p_2,$$

in which the following triangle inequality has been used

$$|f(p_1)f(p_2) - g(p_1)g(p_2)| \leq |f(p_1) - g(p_1)||g(p_2)| + |f(p_2) - g(p_2)||f(p_1)|.$$

Since $0 < i + s \leq n/2$ and $0 \leq k + j - s \leq n/2$, we have $\mathcal{Z}_{p_1}^{i+s} \lesssim |p_1| + |p_1|^n$, and $\mathcal{Z}_{p_2}^{k+1+j-s} \lesssim |p_2| + |p_2|^n$, which yield

$$\iint_{\mathbb{R}^6} \frac{|p_1|^2 + |p_2|^2}{|p_1||p_2|} |f(p_1)f(p_2) - g(p_1)g(p_2)| \mathcal{Z}_{p_1}^{i+s} \mathcal{Z}_{p_2}^{k+j-s} \mathrm{d}p_1 \mathrm{d}p_2$$

$$\lesssim \iint_{\mathbb{R}^6} (|p_1|^2 + |p_2|^2)|f(p_1) - g(p_1)||g(p_2)|(1 + |p_1|^{n-1})(1 + |p_2|^{n-1})\mathrm{d}p_1 \mathrm{d}p_2$$

$$+ \iint_{\mathbb{R}^6} (|p_1|^2 + |p_2|^2)|f(p_2) - g(p_2)||f(p_1)|(1 + |p_1|^{n-1})(1 + |p_2|^{n-1})\mathrm{d}p_1 \mathrm{d}p_2.$$

We then find

$$\int_{\mathbb{R}^3} |C_{22}^1[f](p_1) - C_{22}^1[g](p_1)| \, |p_1|^n \mathrm{d}_1 \lesssim \|f - g\|_{L^1} + \|f - g\|_{L_{n+1}^1}. \quad (10.2.95)$$

Inequality (10.2.93) is a consequence of inequality (10.2.92), Lemma 10.2.5 and

$$\|f - g\|_{L_{n+1}^1} \lesssim \|f - g\|_{L^1}^{\frac{1}{n+2}} \left(\|f\|_{L_{n+2}^1} + \|g\|_{L_{n+2}^1} \right)^{\frac{n+1}{n+2}}.$$

$\square$

Proposition 10.2.8. *Let f and g be two functions in $L_n^1(\mathbb{R}^3) \cap L^1(\mathbb{R}^3)$, $n/2 \in \mathbb{N}$, $n > 0$, then the following inequality*

$$\|Q_{22}^2[f] - Q_{22}^2[g]\|_{L_n^1} \lesssim (\|f - g\|_{L_n^1} + \|f - g\|_{L^1}) \quad (10.2.96)$$

holds true. If $\|f\|_{L_{n+1}^1}, \|g\|_{L_{n+1}^1} < \mathcal{C}_0$, then

$$\|Q_{22}^2[f] - Q_{22}^2[g]\|_{L_n^1} \lesssim \left(\|f - g\|_{L^1}^{\frac{1}{n+1}} + \|f - g\|_{L^1} \right). \quad (10.2.97)$$

Proof. Proceeding similarly as the proof of Proposition 10.2.7, we have

$$\int_{\mathbb{R}^3} |Q_{22}^2[f](p_1) - Q_{22}^2[g](p_1)| \, |p_1|^n \mathrm{d}p_1 \lesssim Y, \quad (10.2.98)$$

where

$$Y := \iiint_{\mathbb{R}^9} K^{22}(p_1, p_2, p_3, p_1 + p_2 - p_3)\delta(\mathcal{Z}_{p_1} + \mathcal{Z}_{p_2} - \mathcal{Z}_{p_3} - \mathcal{Z}_{p_1+p_2-p_3})$$

$$\times \left| f(p_1)f(p_2)f(p_3) - g(p_1)g(p_2)g(p_3) \right|$$

$$\times \left[\sum_{0 \leq i,j,k \; ; \; i+j+k=n/2} \mathcal{Z}_{p_1}^i \mathcal{Z}_{p_2}^j \mathcal{Z}_{p_3}^k \right] \mathrm{d}p_1 \mathrm{d}p_2 \mathrm{d}p_3.$$

Applying the triangle inequality

$$\begin{aligned}
&|f(p_1)f(p_2)f(p_3) - g(p_1)g(p_2)g(p_3)| \\
&\leq |f(p_1) - g(p_1)||f(p_2)||f(p_3)| + |f(p_2) - g(p_2)||g(p_1)||f(p_3)| \\
&\quad + |f(p_3) - g(p_3)||g(p_1)||g(p_3)|,
\end{aligned}$$

to the previous estimate gives

$$Y \lesssim \iiint_{\mathbb{R}^9} K^{22}(p_1, p_2, p_3, p_1 + p_2 - p_3)\delta(\mathcal{Z}_{p_1} + \mathcal{Z}_{p_2} - \mathcal{Z}_{p_3} - \mathcal{Z}_{p_1+p_2-p_3})$$

$$\times |f(p_1) - g(p_1)||f(p_2)||f(p_3)| \left[\sum_{0 \leq i,j,k \ ; \ i+j+k=n/2} \mathcal{Z}_{p_1}^i \mathcal{Z}_{p_2}^j \mathcal{Z}_{p_3}^k \right] \mathrm{d}p_1 \mathrm{d}p_2 \mathrm{d}p_3$$

$$+ \iiint_{\mathbb{R}^9} K^{22}(p_1, p_2, p_3, p_1 + p_2 - p_3)\delta(\mathcal{Z}_{p_1} + \mathcal{Z}_{p_2} - \mathcal{Z}_{p_3} - \mathcal{Z}_{p_1+p_2-p_3})$$

$$\times |f(p_2) - g(p_2)||g(p_1)||f(p_3)| \left[\sum_{0 \leq i,j,k \ ; \ i+j+k=n/2} \mathcal{Z}_{p_1}^i \mathcal{Z}_{p_2}^j \mathcal{Z}_{p_3}^k \right] \mathrm{d}p_1 \mathrm{d}p_2 \mathrm{d}p_3$$

$$+ \iiint_{\mathbb{R}^9} K^{22}(p_1, p_2, p_3, p_1 + p_2 - p_3)\delta(\mathcal{Z}_{p_1} + \mathcal{Z}_{p_2} - \mathcal{Z}_{p_3} - \mathcal{Z}_{p_1+p_2-p_3})$$

$$\times |f(p_3) - g(p_3)||g(p_1)||g(p_3)| \left[\sum_{0 \leq i,j,k \ ; \ i+j+k=n/2} \mathcal{Z}_{p_1}^i \mathcal{Z}_{p_2}^j \mathcal{Z}_{p_3}^k \right] \mathrm{d}p_1 \mathrm{d}p_2 \mathrm{d}p_3.$$

Denote the three terms on the right-hand side by Y_1, Y_2, Y_3. Let us estimate Y_1. Observe that $\mathcal{Z}_p^i \lesssim |p| + |p|^n$, $\mathcal{Z}_p^j \lesssim |p| + |p|^n$ and $\mathcal{Z}_p^k \lesssim |p| + |p|^n$. By the argument used to obtain (10.2.69), we find that

$$Y_1 \lesssim \sum_{0 \leq i,j,k \ ; \ i+j+k=n/2} \int_{\mathbb{R}^{3 \times 3}} |f(p_1) - g(p_1)||f(p_2)||f(p_3)|$$

$$\times \frac{|p_1| + |p_1|^n}{\sqrt{|p_1|}} \frac{|p_2| + |p_2|^n}{\sqrt{|p_2|}} \frac{|p_3| + |p_3|^n}{|p_3|} \mathrm{d}p_1 \mathrm{d}p_2 \mathrm{d}p_3$$

$$\lesssim \|f - g\|_{L^1} + \|f - g\|_{L_n^1}.$$

Since the same estimates can also be proved for Y_2, Y_3, we finally get

$$\int_{\mathbb{R}^3} |Q_{22}^2[f](p_1) - Q_{22}^2[g](p_1)| \, |p_1|^n \mathrm{d}p_1 \lesssim \|f - g\|_{L^1} + \|f - g\|_{L_n^1}.$$

Inequality (10.2.97) is a consequence of inequality (10.2.96), Lemma 10.2.5 and

$$\|f - g\|_{L_n^1} \leq \|f - g\|_{L^1}^{\frac{1}{n+1}} \left(\|f\|_{L_{n+1}^1} + \|g\|_{L_{n+1}^1} \right)^{\frac{n}{n+1}}.$$

$\square$

10.2.10 Proof of Theorem 10.2.1

In order to prove Theorem 10.2.1, we will need the following lemma, introduced first in [5], inspired by a previous unpublished work by Bressan [32]. The authors would like to thank Ricardo Alonso and Irene M. Gamba for the fruitful discussions on the lemma.

Lemma 10.2.6. *Let $[0,T]$ be a time interval, $E := (E, \|\cdot\|)$ be a Banach space, $\mathcal{V}_T$ be a bounded, convex and closed subset of E, and $Q : \mathcal{V}_T \to E$ be an operator satisfying the following properties:*

(𝔍)   *Let $\|\cdot\|_*$ be a different norm of E, satisfying $\|\cdot\|_* \leq C_E\|\cdot\|$ for some universal constant C_E, and let*

$$|\cdot|_* : E \longrightarrow \mathbb{R}$$
$$u \longrightarrow |u|_*$$

be a function satisfying $|u+v|_* \leq |u|_* + |v|_*$, and $|\alpha u|_* = \alpha|u|_*$ for all u, v in E and $\alpha \in \mathbb{R}_+$.
Moreover, suppose

$$|u|_* = \|u\|_*, \forall u \in \mathcal{V}_T, \quad |u|_* \leq \|u\|_* \leq C_E\|u\|, \forall u \in E,$$

$$|Q(u)|_* \leq C_*(1+|u|_*), \forall u \in \mathcal{V}_T,$$

and $\mathcal{V}_T$ is a subset of

$$\overline{B_*\left(O, (2R_*+1)e^{(C_*+1)T}\right)} := \overline{\left\{u \in E \,\middle|\, \|u\|_* \leq (2R_*+1)e^{(C_*+1)T}\right\}},$$

for some positive constant $R_* \geq 1$.

(𝔍𝔍)   *Sub-tangent condition*

$$\liminf_{h\to 0^+} h^{-1}\mathrm{dist}\left(u + hQ[u], \mathcal{V}_T\right) = 0,$$

$$\forall u \in \mathcal{V}_T \cap B_*\left(O, (2R_*+1)e^{(C_*+1)T}\right).$$

(𝔍𝔍𝔍)   *Hölder continuity condition*

$$\|Q[u] - Q[v]\| \leq C\|u-v\|^\beta, \quad \beta \in (0,1), \quad \forall u,v \in \mathcal{V}_T.$$

(𝔍𝔙)   *One-sided Lipschitz condition*

$$[Q[u] - Q[v], u-v] \leq C\|u-v\|, \quad \forall u,v \in \mathcal{V}_T,$$

where

$$[\varphi, \phi] := \lim_{h\to 0^-} h^{-1}\left(\|\phi + h\varphi\| - \|\phi\|\right).$$

Then the equation

$$\partial_t u = Q[u] \ on \ [0,T] \times E, \qquad u(0) = u_0 \in \mathcal{V}_T \cap B_*(O, R_*) \qquad (10.2.99)$$

has a unique solution in $C^1((0,T), E) \cap C([0,T], \mathcal{V}_T)$.

Let us set $E = \mathbb{L}^1_{2n}(\mathbb{R}^3)$, and define $|f|_* = \int_{\mathbb{R}^3} f(p) dp$ and $\|f\|_* = \int_{\mathbb{R}^3} |f(p)| dp$. It is proved in (10.2.55) that for all $f \geq 0$, $f \in E$, the inequality $|Q[f]|_* \leq C^* (1 + \|f\|_*)$ holds true, where C^* depends on $\|f\|_{\mathcal{L}^1_2(\mathbb{R}^3)}$. We then choose C_* in Theorem 10.2.6 to be C^*.

We now define the set $\mathcal{S}_T$ as follows

$$\mathcal{V}_T := \Big\{ f \in \mathbb{L}^1_{2n}(\mathbb{R}^3) \mid (V_1) \ f \geq 0, \quad f(p) = f(|p|), \quad (V_2) \int_{\mathbb{R}^3} f(|p|) dp \leq \mathfrak{c}_0,$$

$$(V_3) \int_{\mathbb{R}_+} f(|p|) \mathcal{Z}_p dp = \mathfrak{c}_1, \quad (V_4) \int_{\mathbb{R}_+} f(|p|) \mathcal{Z}_p^{n^*} dp \leq \mathfrak{c}_{n^*} \Big\},$$

$$(10.2.100)$$

in which the constants have the explicit formulations $\mathfrak{c}_0 := (2\mathcal{R} + 1) e^{(C^*+1)T}$ and $\mathfrak{c}_{n^*} = \frac{3\rho_{n_*}}{2}$, with ρ_{n_*} defined in (10.2.102). It is clear that $\mathcal{V}_T$ is a bounded, convex and closed subset of $\mathbb{L}^1_{2n}(\mathbb{R}^3)$. Moreover for all f in $\mathcal{V}_T$, it is straightforward that $|f|_* = \|f\|_*$.

Theorem 10.2.1 will follow as an application of Theorem 10.2.6, if the four conditions $(\mathfrak{I})$, $(\mathfrak{II})$, $(\mathfrak{III})$ and $(\mathfrak{IV})$ are checked.

Condition $(\mathfrak{I})$ We choose the constant R_* to be $\mathcal{R} + 1$, then for all u in $\mathcal{V}_T$, $\|u\|_* \leq (2R_* + 1) e^{(C^*+1)T}$. Condition $(\mathfrak{I})$ is automatically satisfied.

Condition $(\mathfrak{II})$ First, the same lines of argument used to obtained (10.2.82) gives

$$\int_{\mathbb{R}^3} Q[f] \mathcal{Z}_p^{n^*} dp \leq \mathbb{P}[m_{n^*}(f)] :=$$

$$C m_{n^*}(f) + C m_{n^*}(f)^{\frac{n^*-1}{n^*}} + C m_{n^*}(f)^{\frac{n^*-2}{n^*}} + C m_{n^*}(f)^{\frac{1}{n^*}} - C m_{n^*}(f)^{\frac{n^*+1}{n^*}}$$

$$+ C \sum_{0 \leq i,j,k < n^*; \ i+j+k=n^*} \sum_{s=0}^{k} m_{n^*}(f)^{\frac{i+s}{n^*}} \left(m_{n^*}(f)^{\frac{j+k-s}{n^*}} + m_{n^*}(f)^{\frac{j+k-s+1/2}{n^*}} \right)$$

$$+ C \sum_{0 \leq i,j,k < n^*; \ i+j+k=n^*; \ j,k>0} m_{n^*}(f)^{\frac{i}{n^*}} \left(m_{n^*}(f)^{\frac{j-1}{n^*}} + m_{n^*}(f)^{\frac{j-1/2}{n^*}} \right)$$

$$\times \left(m_{n^*}(f)^{\frac{k-1}{n^*}} + m_{n^*}(f)^{\frac{k-1/2}{n^*}} \right), \qquad \forall f \in \mathcal{V}_T,$$

$$(10.2.101)$$

in which C is a positive constant depending on $\mathfrak{c}_0$.

Let ρ_{n^*} be the solution of $\mathbb{P}(\rho) = 0$ so that:
If $0 < \rho < \rho_{n^*}$, $\mathbb{P}(\rho) < 0$; if $\rho > \rho_{n^*}$, $\mathbb{P}(\rho) > 0$.

$$(10.2.102)$$

Notice that ρ_{n^*} depends on $\mathfrak{c}_0$. Let f be an arbitrary element of the set $\mathcal{V}_T \cap B_* \Big(O, (2R_* + 1) e^{(C_*+1)T} \Big)$ and consider the function $f + hQ[f]$. We will prove that for all $\epsilon > 0$, there exists a strictly positive constant h_* depending on f and ϵ such that the set $B(f + hQ[f], h\epsilon) \cap \mathcal{V}_T$ is not empty for all $0 < h < h_*$. To this extent, we define $\chi_r(p)$ to be the characteristic function of the ball $B(O, r)$, centered at the origin with radius r. We set $f_r(p) = \chi_r(p)f(p)$ and $w_r = f + hQ[f_r]$. Notice that $Q[f_r] \in \mathbb{L}^1_{2n}(\mathbb{R}^3)$. It follows that $w_r \in \mathbb{L}^1_{2n}(\mathbb{R}^3)$. We will prove that for r large enough and h_* small enough, w_r belongs to the set $\mathcal{V}_T$. We now check the four conditions (V_1), (V_2), (V_3) and (V_4).

Condition (V_1): Observe that f_r is compactly supported, it follows that $Q^-[f_r]$, defined in (10.2.15), is bounded by $C(f, r, \mathfrak{c}_0, \mathfrak{c}_{n^*})$, a strictly positive constant depending on f, r, $\mathfrak{c}_0$, $\mathfrak{c}_{n^*}$. Therefore, $w_r \geq f - hf_rQ^-[f_r] \geq f(1 - hQ^-[f_r]) \geq 0$, for $h < C(f, r, \mathfrak{c}_0, \mathfrak{c}_{n^*})^{-1}$.

Condition (V_2): Since $\|f\|_* < (2R_*+1)e^{(C_*+1)T}$, and $\lim_{h\to 0}\|f-w_r\|_* = 0$, h_* can be chosen small enough such that $\|w_r\|_* < (2R_* + 1)e^{(C_*+1)T}$.

Condition (V_3): This follows from the conservation of energy

$$\int_{\mathbb{R}^3} w_r \mathcal{Z}_p \mathrm{d}p = \int_{\mathbb{R}^3} (f + hQ[f_r])\mathcal{Z}_p \mathrm{d}p = \int_{\mathbb{R}^3} f\mathcal{Z}_p \mathrm{d}p = \mathfrak{c}_1.$$

Condition (V_4): Now, we show that r and h_* can be chosen such that $\int_{\mathbb{R}^3} w_r \mathcal{Z}_p^{n^*} \mathrm{d}p < \frac{3\rho_{n^*}}{2}$. To this extent, we consider two cases. If $\int_{\mathbb{R}^3} f\mathcal{Z}_p^{n^*} \mathrm{d}p < \frac{3\rho_{n^*}}{2}$, we deduce from $\lim_{h\to 0}\int_{\mathbb{R}^3} |w_r - f|\mathcal{Z}_p^{n^*} \mathrm{d}p = 0$ that h_* can be chosen small enough such that $\int_{\mathbb{R}^3} w_r \mathcal{Z}_p^{n^*} \mathrm{d}p < \frac{3\rho_{n^*}}{2}$. If, on the other hand, $\int_{\mathbb{R}^3} f\mathcal{Z}_p^{n^*} \mathrm{d}p = \frac{3\rho_{n^*}}{2}$, r can be chosen large enough such that $\int_{\mathbb{R}^3} f_r\mathcal{Z}_p^{n^*} \mathrm{d}p > \rho_{n^*}$, which implies, by (10.2.102), that $\int_{\mathbb{R}^3} Q[f_r]\mathrm{d}p < 0$. As a consequence, $\int_{\mathbb{R}^3} w_r\mathcal{Z}_p^{n^*} \mathrm{d}p < \int_{\mathbb{R}^3} f\mathcal{Z}_p^{n^*} \mathrm{d}p = \frac{3\rho_{n^*}}{2}$.

Therefore, $w_r \in \mathcal{V}_T$ for all $0 < h < h_*$. Now since

$$\lim_{r\to\infty} \frac{1}{h}\|w_r - f - hQ[f_r]\|_{\mathbb{L}^1_{2n}(\mathbb{R}^3)} = \lim_{r\to\infty} \|Q[f] - Q[f_r]\|_{\mathbb{L}^1_{2n}(\mathbb{R}^3)} = 0,$$

then for r large enough, $w_r \in B(f + hQ[f], h\epsilon)$, which implies $B(f + hQ[f], h\epsilon) \cap \mathcal{V}_T\backslash\{f + hQ[f]\}$. Condition $(\mathfrak{II})$ is verified.

Condition $(\mathfrak{III})$ Condition $(\mathfrak{III})$ follows from Propositions 10.2.6, 10.2.7, and 10.2.8.

Condition $(\mathfrak{IV})$ By the Lebesgue dominated convergence theorem,

$$\left[\varphi, \phi\right] \leq \int_{\mathbb{R}^3} \varphi(p)\mathrm{sign}(\phi(p))(1 + \mathcal{Z}_p^n)\mathrm{d}p, \tag{10.2.103}$$

which means that Condition $(\mathfrak{IV})$ is satisfied if the following inequality holds true

$$\mathcal{P}_0 := \int_{\mathbb{R}^3} [Q[f](p) - Q[g](p)]\mathrm{sign}((f - g)(p))(1 + \mathcal{Z}_p^n)\mathrm{d}p \lesssim \|f - g\|_{\mathrm{L}_{2n}^1}.$$

(10.2.104)

Since $Q = \mathcal{Q}_{12} + \mathcal{Q}_{22}$, let us split $\mathcal{P}_0 = \mathcal{P}_1 + \mathcal{P}_2$, where

$$\mathcal{P}_1 := \int_{\mathbb{R}^3} [\mathcal{Q}_{12}[f](p) - \mathcal{Q}_{12}[g](p)]\mathrm{sign}((f - g)(p))(1 + \mathcal{Z}_p^n)\mathrm{d}p,$$

and

$$\mathcal{P}_2 := \int_{\mathbb{R}^3} [\mathcal{Q}_{22}[f](p) - \mathcal{Q}_{22}[g](p)]\mathrm{sign}((f - g)(p))(1 + \mathcal{Z}_p^n)\mathrm{d}p.$$

We will now estimate $\mathcal{P}_1$ and $\mathcal{P}_2$ separately.

Estimating $\mathcal{P}_1$.
Define $\varphi_k(p) = \mathrm{sign}((f - g)(p))\mathcal{Z}_p^k$, $k \in \mathbb{Z}, k \geq 0$, $k \neq 1$. Let us consider the following generalized term of $\mathcal{P}_1$

$$\mathcal{H}_0 := \int_{\mathbb{R}^3} [\mathcal{Q}_{12}[f](p) - \mathcal{Q}_{12}[g](p)]\varphi_k(p)\mathrm{d}p, \qquad (10.2.105)$$

which, thanks to Lemma 10.2.1, can be rewritten as

$$
\begin{aligned}
\mathcal{H}_0 := \ & \int_{\mathbb{R}^{3 \times 3}} [R_{12}[f](p_1) - R_{12}[g](p_1)][\varphi_k(p_1) - \varphi_k(p_2) - \varphi_k(p_3)]\mathrm{d}p_1\mathrm{d}p_2\mathrm{d}p_3 \\
= \ & \int_{\mathbb{R}^{3 \times 2}} \bar{K}^{12}(p_2 + p_3, p_2, p_3)\delta(\mathcal{Z}_{p_2+p_3} - \mathcal{Z}_{p_2} - \mathcal{Z}_{p_3}) \\
& \times [(f(p_2)f(p_3) - g(p_2)g(p_3)) \\
& - 2(f(p_2)f(p_2 + p_3) - g(p_2)g(p_2 + p_3)) - (f(p_2 + p_3) - g(p_2 + p_3))] \\
& \times [\varphi_k(p_2 + p_3) - \varphi_k(p_2) - \varphi_k(p_3)]\mathrm{d}p_2\mathrm{d}p_3.
\end{aligned}
$$

(10.2.106)

Split $\mathcal{H}_0$ into the sum of three terms

$$
\begin{aligned}
\mathcal{H}_1 := \ & \int_{\mathbb{R}^{3 \times 2}} \bar{K}^{12}(p_2 + p_3, p_2, p_3)\delta(\mathcal{Z}_{p_2+p_3} - \mathcal{Z}_{p_2} - \mathcal{Z}_{p_3}) \qquad (10.2.107) \\
& \times [f(p_2)f(p_3) - g(p_2)g(p_3)][\varphi_k(p_2 + p_3) - \varphi_k(p_2) - \varphi_k(p_3)]\mathrm{d}p_2\mathrm{d}p_3,
\end{aligned}
$$

$$
\begin{aligned}
\mathcal{H}_2 := \ & -2\int_{\mathbb{R}^{3 \times 2}} \bar{K}^{12}(p_2 + p_3, p_2, p_3)\delta(\mathcal{Z}_{p_2+p_3} - \mathcal{Z}_{p_2} - \mathcal{Z}_{p_3}) \qquad (10.2.108) \\
& \times [f(p_2)f(p_2 + p_3) - g(p_2)g(p_2 + p_3)] \\
& \times [\varphi_k(p_2 + p_3) - \varphi_k(p_2) - \varphi_k(p_3)]\mathrm{d}p_2\mathrm{d}p_3,
\end{aligned}
$$

and

$$\mathcal{H}_3 := -\int_{\mathbb{R}^{3 \times 2}} \bar{K}^{12}(p_2 + p_3, p_2, p_3)\delta(\mathcal{Z}_{p_2+p_3} - \mathcal{Z}_{p_2} - \mathcal{Z}_{p_3}) \qquad (10.2.109)$$

$$\times [f(p_2 + p_3) - g(p_2 + p_3)]$$
$$\times [\varphi_k(p_2 + p_3) - \varphi_k(p_2) - \varphi_k(p_3)] \mathrm{d}p_2 \mathrm{d}p_3.$$

The same arguments used to obtained (10.2.88) and (10.2.89) give

$$\mathcal{H}_1 \lesssim \|f - g\|_{L^1_{2k}(\mathbb{R}^3)} \tag{10.2.110}$$

and

$$\mathcal{H}_2 \lesssim \|f - g\|_{L^1_{2k}(\mathbb{R}^3)}. \tag{10.2.111}$$

The third term $\mathcal{H}_3$ can be estimated as

$$\mathcal{H}_3 = -\int_{\mathbb{R}^{3\times 2}} \bar{K}^{12}(p_2 + p_3, p_2, p_3)\delta(\mathcal{Z}_{p_2+p_3} - \mathcal{Z}_{p_2} - \mathcal{Z}_{p_3}) \tag{10.2.112}$$
$$\times [f(p_2 + p_3) - g(p_2 + p_3)]$$
$$\times [\mathcal{Z}^k_{p_2+p_3}\mathrm{sign}((f(p_2 + p_3) - g(p_2 + p_3)) - \mathcal{Z}^k_{p_2}\mathrm{sign}((f(p_2) - g(p_2))$$
$$- \mathcal{Z}^k_{p_3}\mathrm{sign}((f(p_3) - g(p_3))]\mathrm{d}p_2\mathrm{d}p_3$$
$$\leq \int_{\mathbb{R}^{3\times 2}} \bar{K}^{12}(p_2 + p_3, p_2, p_3)\delta(\mathcal{Z}_{p_2+p_3} - \mathcal{Z}_{p_2} - \mathcal{Z}_{p_3})$$
$$\times |f(p_2 + p_3) - g(p_2 + p_3)|$$
$$\times [\mathcal{Z}^k_{p_2} + \mathcal{Z}^k_{p_3} - \mathcal{Z}^k_{p_2+p_3}]\mathrm{d}p_2\mathrm{d}p_3.$$

Now, let us consider the two cases $k = 0$ and $k > 1$.

- If $k = 0$,

$$\mathcal{H}_3 \leq \int_{\mathbb{R}^{3\times 2}} \bar{K}^{12}(p_2 + p_3, p_2, p_3)\delta(\mathcal{Z}_{p_2+p_3} - \mathcal{Z}_{p_2} - \mathcal{Z}_{p_3}) \tag{10.2.113}$$
$$\times |f(p_2 + p_3) - g(p_2 + p_3)|\mathrm{d}p_2\mathrm{d}p_3,$$

 which, by the same arguments that led to (10.2.90), can be bounded as

$$\mathcal{H}_3 \lesssim \|f - g\|_{L^1_3(\mathbb{R}^3)}. \tag{10.2.114}$$

- If $k > 1$, since $\mathcal{Z}_{p_2+p_3} = \mathcal{Z}_{p_2} + \mathcal{Z}_{p_3}$, it is straightforward that $\mathcal{Z}^k_{p_2} + \mathcal{Z}^k_{p_3} - \mathcal{Z}^k_{p_2+p_3} = \mathcal{Z}^k_{p_2} + \mathcal{Z}^k_{p_3} - (\mathcal{Z}_{p_2} + \mathcal{Z}_{p_3})^k \leq -k\mathcal{Z}_{p_2}\mathcal{Z}^{k-1}_{p_3} \leq 0$. As a consequence, we can estimate $\mathcal{H}_3$ as

$$\mathcal{H}_3 \leq -\int_{\mathbb{R}^{3\times 2}} \bar{K}^{12}(p_2 + p_3, p_2, p_3)\delta(\mathcal{Z}_{p_2+p_3} - \mathcal{Z}_{p_2} - \mathcal{Z}_{p_3})$$
$$\times |f(p_2 + p_3) - g(p_2 + p_3)|k\mathcal{Z}_{p_2}\mathcal{Z}^{k-1}_{p_3}\mathrm{d}p_2\mathrm{d}p_3$$
$$\leq -\int_{\mathbb{R}^3}\int_{S^0_{p_1}} \bar{K}^{12}_0(p_1, p_2, p_1 - p_2)|f(p_1) - g(p_1)|k\mathcal{Z}_{p_2}\mathcal{Z}^{k-1}_{p_1-p_2}\mathrm{d}\sigma(p_2)\mathrm{d}p_1. \tag{10.2.115}$$

 In view of Lemma 10.2.2, we find the following bound on $\mathcal{H}_3$

$$\mathcal{H}_3 \lesssim -\int_{\mathbb{R}^3} |f(p) - g(p)| \left(|p|^{2k+1} \min\{1, |p|\}^{2k+7}\right) dp. \qquad (10.2.116)$$

Combining (10.2.110), (10.2.111), (10.2.114) and (10.2.116) yields

$$\mathcal{P}_1 \lesssim \qquad\qquad\qquad\qquad\qquad\qquad\qquad\qquad\qquad\qquad (10.2.117)$$

$$\int_{\mathbb{R}^3} |f(p) - g(p)| \left(1 + |p| + |p|^3 + |p|^{2n} + |p|^{2n+1} - |p|^{2n+1} \min\{1, |p|\}^{2n+7}\right) dp.$$

Estimating $\mathcal{P}_2$.
We can estimate $\mathcal{P}_2$ using Propositions 10.2.7 and 10.2.8, as follows

$$\mathcal{P}_2 \lesssim \int_{\mathbb{R}^3} |f(p) - g(p)| \left(1 + |p| + |p|^{2n} + |p|^{2n+1}\right) dp. \qquad (10.2.118)$$

Estimating $\mathcal{P}_0$.
Combining (10.2.117) and (10.2.118) yields

$$\mathcal{P}_0 \lesssim \int_{\mathbb{R}^3} |f(p) - g(p)| \left(1 + |p| + |p|^3 + |p|^{2n} - |p|^{2n+1} \min\{1, |p|\}^{2n+7}\right) dp.$$
$$\qquad\qquad\qquad\qquad\qquad\qquad\qquad\qquad\qquad\qquad\qquad (10.2.119)$$

Since the weight $1 + |p| + |p|^3 + |p|^{2n} - |p|^{2n+1} \min\{1, |p|\}^{2n+7}$ of (10.2.119) is bounded uniformly in p by a universal positive constant C, inequality (10.2.119) implies $\mathcal{P}_0 \lesssim \int_{\mathbb{R}^3} |f(p) - g(p)| dp$, which concludes the proof of (10.2.104).

References

1. A.I. Akhiezer and S.V. Peletminskii. *Methods of statistical physics.* Oxford: Pergamon, 1981.
2. T. Allemand. Derivation of a two-fluids model for a Bose gas from a quantum kinetic system. *Kinetic and Related Models*, (2):379–402, 2009.
3. A.J. Allen, C.F. Barenghi, N.P. Proukakis, and E. Zaremba. A dynamical self-consistent finite temperature kinetic theory: The ZNG scheme. *arXiv preprint arXiv:1206.0145*, 2012.
4. R. Alonso, J.A. Canizo, I.M. Gamba, and C. Mouhot. A new approach to the creation and propagation of exponential moments in the Boltzmann equation. *Communications in Partial Differential Equations*, 38(1):155–169, 2013.
5. R. Alonso, I.M. Gamba, and M.-B. Tran. The Cauchy problem and BEC stability for the quantum Boltzmann–Gross–Pitaevskii system for bosons at very low temperature. *arXiv preprint arXiv:1609.07467*, 2016.
6. D.F. Anderson. Global asymptotic stability for a class of nonlinear chemical equations. *SIAM J. Appl. Math.*, 68(5):1464–1476, 2008.
7. D.F. Anderson. A proof of the global attractor conjecture in the single linkage class case. *SIAM J. Appl. Math.*, 71(4):1487–1508, 2011.
8. M.H. Anderson, J.R. Ensher, M.R. Matthews, C.E. Wieman, and E.A. Cornell. Observation of Bose–Einstein Condensation in a dilute atomic vapor. *Science*, 269(5221):198–201, 1995.
9. D.F. Anderson and A. Shiu. The dynamics of weakly reversible population processes near facets. *SIAM J. Appl. Math.*, 70(6):1840–1858, 2010.
10. M.R. Andrews, C.G. Townsend, H.-J. Miesner, D.S. Durfee, D.M. Kurn, and W. Ketterle. Observation of interference between two Bose condensates. *Science*, 275 (5300):637–641, 1997.
11. N. Angelescu, A. Verbeure, and V.A. Zagrebnov. On Bogoliubov's model of superfluidity. *Journal of Physics A: Mathematical and General*, 25(12):3473, 1992.
12. D. Angeli, P. De Leenheer, and E.D. Sontag. Persistence results for chemical reaction networks with time-dependent kinetics and no global conservation laws. *SIAM J. Appl. Math.*, 71(1):128–146, 2011.
13. J.R. Anglin and W. Ketterle. Bose–Einstein condensation of atomic gases. *Nature*, 416(6877):211–218, 2002.
14. L. Arkeryd and A. Nouri. Bose condensates in interaction with excitations: a kinetic model. *Comm. Math. Phys.*, 310(3):765–788, 2012.
15. L. Arkeryd and A. Nouri. A Milne problem from a Bose condensate with excitations. *Kinetic and Related Models*, 6(4):671–686, 2013.
16. L. Arkeryd and A. Nouri. Bose condensates in interaction with excitations: a two-component space-dependent model close to equilibrium. *J. Stat. Phys.*, 160(1):209–238, 2015.

© Springer Nature Switzerland AG 2019
Y. Pomeau, M.-B. Tran, *Statistical Physics of Non Equilibrium Quantum Phenomena*,
Lecture Notes in Physics 967, https://doi.org/10.1007/978-3-030-34394-1

17. L. Arlotti and M. Lachowicz. Euler and Navier–Stokes limits of the Uehling–Uhlenbeck quantum kinetic equations. *J. Math. Phys.*, 38(7):3571–3588, 1997.

18. R. Balescu. *Equilibrium and nonequilibrium statistical mechanics*. New York: Wiley, 1975.

19. J. Bandyopadhyay and J.J.L. Velázquez. Blow-up rate estimates for the solutions of the bosonic Boltzmann–Nordheim equation. *Journal of Mathematical Physics*, 56(6):063302, 2015.

20. D. Benedetto, C. Castella, R. Esposito and M. Pulvirenti. Some considerations on the derivation of the nonlinear quantum Boltzmann equation. *Journal of statistical physics*, 116(1-4):381–410, 2004.

21. D. Benedetto, C. Castella, R. Esposito and M. Pulvirenti. On the weak-coupling limit for bosons and fermions. *Mathematical Models and Methods in Applied Sciences*, 15(12):1811–1843, 2005.

22. D. Benedetto, C. Castella, R. Esposito and M. Pulvirenti. From the N-body Schrödinger equation to the quantum Boltzmann equation: a term-by-term convergence result in the weak coupling regime. *Comm. Math. Phys.*, 277(1):1–44, 2008.

23. N. Bernhoff. Half-space problems for a linearized discrete quantum kinetic equation. *J. Stat. Phys.*, 159(2):358–379, 2015.

24. N. Bernhoff. Boundary layers for discrete kinetic models: multicomponent mixtures, polyatomic molecules, bimolecular reactions, and quantum kinetic equations. *Kinetic and Related Models*, 10(4):925–955, 2017.

25. N. Bernhoff. Discrete quantum Boltzmann equation. *AIP Conference Proceedings*, Vol. 2132. No. 1. AIP Publishing, 2019.

26. M.J. Bijlsma, E. Zaremba, and H.T.C. Stoof. Condensate growth in trapped Bose gases. *Physical Review A*, 62(6):063609, 2000.

27. A.V. Bobylev, I.M. Gamba, and V.A. Panferov. Moment inequalities and high-energy tails for Boltzmann equations with inelastic interactions. *Journal of Statistical Physics*, 116(5–6):1651–1682, 2004.

28. N. Bogoliubov. On the theory of superfluidity. *J. Phys*, 11(1):23, 1947.

29. N.N. Bogoliubov. *Lectures on quantum statistics*, volume 2. CRC Press, 1970.

30. D. Bohm and D. Pines. A collective description of electron interactions. I. Magnetic interactions. *Physical Review*, 82(5):625, 1951.

31. F. Bouchut, F. Golse, and M. Pulvirenti. *Kinetic equations and asymptotic theory*, volume 4 of *Series in Applied Mathematics (Paris)*. Edited by and with a foreword by Benoît Perthame and Laurent Desvillettes. Gauthier-Villars, Éditions Scientifiques et Médicales Elsevier, Paris, 2000.

32. D. Bressan. Notes on the Boltzmann equation. *Lecture notes for a summer course, S.I.S.S.A. Trieste*, 2005.

33. J.B. Bru and V.A. Zagrebnov. Exact solution of the Bogoliubov Hamiltonian for weakly imperfect Bose gas. *Journal of Physics A: Mathematical and General*, 31(47):9377, 1998.

34. T. Buckmaster, P. Germain, Z. Hani, and J. Shatah. Effective dynamics of the nonlinear Schrödinger equation on large domains. *Communications on Pure and Applied Mathematics*, 71(7):1407–1460, 2018.

35. T. Buckmaster, P. Germain, Z. Hani, and J. Shatah. Onset of the wave turbulence description of the longtime behavior of the nonlinear Schrödinger equation. *arXiv preprint arXiv:1907.03667*, 2019.

36. R.E. Caflisch. The fluid dynamic limit of the nonlinear Boltzmann equation. *Comm. Pure Appl. Math.*, 33(5):651–666, 1980.

37. S. Cai and X. Lu. The spatially homogeneous Boltzmann equation for Bose–Einstein particles: Rate of strong convergence to equilibrium. *arXiv preprint arXiv:1808.04038*, 2018.

38. T. Carleman. Sur la théorie de l'équation intégrodifférentielle de Boltzmann. *Acta Math.*, 60(1):91–146, 1933.

39. T. Castella. From the von Neumann equation to the quantum Boltzmann equation in a deterministic framework. *Journal of Statistical Physics*, 104(1-2):387–447, 2001.

40. T. Castella. From the von-Neumann equation to the quantum Boltzmann equation II: Identifying the Born series. *Journal of Statistical Physics*, 106(5-6):1197–1220, 2002.

41. C. Cercignani. *The Boltzmann equation and its applications*, volume 67 of *Applied Mathematical Sciences*. Springer-Verlag, New York, 1988.

42. C. Cercignani, R. Illner, and M. Pulvirenti. *The mathematical theory of dilute gases*, volume 106 of *Applied Mathematical Sciences*. Springer-Verlag, New York, 1994.

43. X. Chen and Y. Guo. On the weak coupling limit of quantum many-body dynamics and the quantum Boltzmann equation. *Kinetic and Related Models*, 8(3):443–465, 2015.

44. C. Connaughton, A.C. Newell, and Y. Pomeau. Non-stationary spectra of local wave turbulence. *Physica D: Nonlinear Phenomena*, 184(1):64–85, 2003.

45. C. Connaughton and Y. Pomeau. Kinetic theory and Bose–Einstein condensation. *Comptes Rendus Physique*, 5(1):91–106, 2004.

46. E. Cortés and M. Escobedo. On a system of equations for the normal fluid-condensate interaction in a Bose gas. *arXiv:1803.09964*, 2018.

47. G. Craciun. Toric differential inclusions and a proof of the global attractor conjecture. *Submitted.*

48. G. Craciun, A. Dickenstein, A. Shiu, and B. Sturmfels. Toric dynamical systems. *J. Symbolic Comput.*, 44(11):1551–1565, 2009.

49. G. Craciun, F. Nazarov, and C. Pantea. Persistence and permanence of mass-action and power-law dynamical systems. *SIAM J. Appl. Math.*, 73(1):305–329, 2013.

50. G. Craciun and M.-B. Tran. A reaction network approach to the convergence to equilibrium of quantum Boltzmann equations for Bose gases. *arXiv preprint arXiv:1608.05438*, 2016.

51. F. Dalfovo, S. Giorgini, L.P. Pitaevskii, and S. Stringari. Theory of Bose–Einstein condensation in trapped gases. *Reviews of Modern Physics*, 71(3):463, 1999.

52. H.G. Dehmelt. Proposed $10^{14} \Delta v$ v laser fluorescence spectroscopy on $T1+$ mono-ion oscillator II (spontaneous quantum jumps). *Bull. Am. Phys. Soc.* 20:60, 1975.

53. H.G. Dehmelt, Mono-ion oscillator as potential ultimate laser frequency standard, *IEEE Trans. Instrum. Meas.* 31:83, 1982, and *Nature* 325: 581, 1987.

54. P.A.M. Dirac. The quantum theory of the emission and absorption of radiation. *Proc. Roy. Soc.* A114: 243, 1927.

55. U. Eckern. Relaxation processes in a condensed Bose gas. *J. Low Temp. Phys.*, 54:333–359, 1984.

56. A. Einstein. Uber einen Erzeugung und Verwandlung des Lichtes betreffenden heuristischen Gesichtspunkt. *Ann. d. Phys.* 17:132, 1905, and Zur Theorie der Lichterzeugung und Lichtabsorption. *Ann. d. Phys.* 20:199, 1906.

57. A. Einstein. On the quantum theory of radiation. *Physikalishe Zeitschrift* 18:121–128, 1917.

58. L. Erdos. Lecture Notes on Quantum Brownian Motion. *arXiv preprint arXiv:1009.0843*, 2010.

59. L. Erdos, M. Salmhofer and H.-T. Yau. On the quantum Boltzmann equation. *Journal of statistical physics*, 116(1-4):367–380, 2004.

60. M. Escobedo, S. Mischler, and J.J.L. Velazquez. On the fundamental solution of a linearized Uehling–Uhlenbeck equation. *Archive for Rational Mechanics and Analysis*, 186(2):309–349, 2007.

61. M. Escobedo, S. Mischler, and J.J.L. Velazquez. Singular solutions for the Uehling–Uhlenbeck equation. *Proceedings of the Royal Society of Edinburgh Section A: Mathematics*, 138(1):67–107, 2008.

62. M. Escobedo, F. Pezzotti, and M. Valle. Analytical approach to relaxation dynamics of condensed Bose gases. *Ann. Physics*, 326(4):808–827, 2011.

63. M. Escobedo and M.-B. Tran. Convergence to equilibrium of a linearized quantum Boltzmann equation for bosons at very low temperature. *Kinetic and Related Models*, 8(3):493–531, 2015.

64. M. Escobedo and J.J.L. Velázquez. On the blow up and condensation of supercritical solutions of the Nordheim equation for bosons. *Communications in Mathematical Physics*, 330(1):331–365, 2014.

65. M. Escobedo and J.J.L. Velázquez. Finite time blow-up and condensation for the bosonic Nordheim equation. *Invent. Math.*, 200(3):761–847, 2015.

66. M. Escobedo and J.J.L. Velázquez. On the theory of weak turbulence for the nonlinear Schrödinger equation. *Mem. Amer. Math. Soc.*, 238(1124):v+107, 2015.

67. H. Everett. Relative State Formulation of Quantum Mechanics. *Reviews of Modern Physics*, 29:454–462, 1957.

68. E. Faou, P. Germain, and Z. Hani. The weakly nonlinear large-box limit of the 2d cubic nonlinear Schrödinger equation. *Journal of the American Mathematical Society*, 29(4):915–982, 2016.

69. M. Feinberg. Complex balancing in general kinetic systems. *Arch. Rational Mech. Anal.*, 49:187–194, 1972/73.

70. M. Feinberg. The existence and uniqueness of steady states for a class of chemical reaction networks. *Arch. Rational Mech. Anal.*, 132(4):311–370, 1995.

71. I.M. Gamba, L.M. Smith, and M.-B. Tran. On the wave turbulence theory for stratified flows in the ocean. *arXiv preprint arXiv:1709.08266*, 2017.

72. C. Gardiner and P. Zoller. *Quantum kinetic theory. A quantum kinetic master equation for condensation of a weakly interacting Bose gas without a trapping potential.* Volume 55 of *Phys. Rev. A*. 1997.

73. C. Gardiner and P. Zoller. *Quantum kinetic theory. III. Quantum kinetic master equation for strongly condensed trapped systems.* Volume 58 of *Phys. Rev. A*. 1998.

74. C.W. Gardiner, P. Zoller, R.J. Ballagh, and M.J. Davis. Kinetics of Bose–Einstein condensation in a trap. *Physical review letters*, 79(10):1793, 1997.

75. P. Germain, A.D. Ionescu, and M.-B. Tran. Optimal local well-posedness theory for the kinetic wave equation. *arXiv preprint arXiv:1711.05587*, 2017.

76. A. Griffin, T. Nikuni, and E. Zaremba. *Bose-condensed gases at finite temperatures.* Cambridge University Press, Cambridge, 2009.

77. E.D. Gust. *Characteristic Relaxation Rates of a Bose Gas in the Classical, Quantum and Condensed Regimes.* PhD Thesis, The University of Texas at Austin, August 2011.

78. E.D. Gust and L.E. Reichl. Collision integrals in the kinetic equations of dilute Bose–Einstein condensates. *arXiv:1202.3418*, 2012.

79. E.D. Gust and L.E. Reichl. Relaxation rates and collision integrals for Bose–Einstein condensates. *Phys. Rev. A*, 170:43–59, 2013.

80. E.D. Gust and L.E. Reichl. Transport coefficients from the boson Uehling–Uhlenbeck equation. *Physical Review E*, 87(4):042109, 2013.

81. W. Heisenberg. *The physical principles of the quantum theory.* Dover, New York, 1949.

82. F. Horn. Necessary and sufficient conditions for complex balancing in chemical kinetics. *Arch. Rational Mech. Anal.*, 49:172–186, 1972/73.

83. F. Horn. The dynamics of open reaction systems. In *Mathematical aspects of chemical and biochemical problems and quantum chemistry (Proc. SIAM-AMS Sympos. Appl. Math., New York, 1974)*, pages 125–137. SIAM–AMS Proceedings, Vol. VIII. Amer. Math. Soc., Providence, R.I., 1974.

84. F. Horn and R. Jackson. General mass action kinetics. *Arch. Rational Mech. Anal.*, 47:81–116, 1972.

85. K. Huang. *Introduction to statistical physics.* Chapman and Hall/CRC, 2009.

86. K. Huang and C.N. Yang. Quantum-mechanical many-body problem with hard-sphere interaction. *Physical review*, 105(3):767, 1957.

87. K. Huang, C.N. Yang, and J.M. Luttinger. Imperfect Bose gas with hard-sphere interaction. *Physical Review*, 105(3):776, 1957.

88. N.M. Hugenholtz and D. Pines. Ground-state energy and excitation spectrum of a system of interacting bosons. *Physical Review*, 116(3):489, 1959.

89. D.A.W. Hutchinson, E. Zaremba, and A. Griffin. Finite temperature excitations of a trapped bose gas. *Physical review letters*, 78(10):1842, 1997.

90. M. Imamovic-Tomasovic and A. Griffin. Quasiparticle kinetic equation in a trapped Bose gas at low temperatures. *J. Low Temp. Phys.*, 122:617–655, 2001.

91. D. Jaksch, C. Gardiner, and P. Zoller. *Quantum kinetic theory. II. Simulation of the quantum Boltzmann master equation.* Volume 56 of *Phys. Rev. A.* 1997.

92. S. Jin and M.-B. Tran. Quantum hydrodynamic approximations to the finite temperature trapped bose gases. *Physica D: Nonlinear Phenomena*, 380-381:45–57, 1 October 2018.

93. C. Josserand and Y. Pomeau. Nonlinear aspects of the theory of Bose–Einstein condensates. *Nonlinearity*, 14(5):R25, 2001.

94. C. Josserand, Y. Pomeau, and S. Rica. Self-similar singularities in the kinetics of condensation. *Journal of Low Temperature Physics*, 145(1):231–265, 2006.

95. Y.M. Kagan, B.V. Svistunov, and G.V. Shlyapnikov. Erratum: Kinetics of Bose condensation in an interacting Bose gas [sov. physics-jetp 74, 279–285 (February 1992)]. *JETP*, 75(2):387, 1992.

96. W. Ketterle, D.S. Durfee, and D.M. Stamper-Kurn. Making, probing and understanding Bose–Einstein condensates. *arXiv preprint cond-mat/9904034*, 1999.

97. T.R. Kirkpatrick and J.R. Dorfman. Transport theory for a weakly interacting condensed Bose gas. *Phys. Rev. A (3)*, 28(4):2576–2579, 1983.

98. T.R. Kirkpatrick and J.R. Dorfman. Transport coefficients in a dilute but condensed Bose gas. *J. Low Temp. Phys.*, 58:399–415, 1985.

99. T.R. Kirkpatrick and J.R. Dorfman. Transport in a dilute but condensed nonideal Bose gas: Kinetic equations. *J. Low Temp. Phys.*, 58:301–331, 1985.

100. A.N. Kolmogorov. Uber die analytischen Methoden in der Wahrscheinlichkeitsrechnung (On Analytical Methods in the Theory of Probability). *Math. Ann.* 104:415, 1931.

101. R. Lacaze, P. Lallemand, Y. Pomeau, and S. Rica. Dynamical formation of a Bose–Einstein condensate. Advances in nonlinear mathematics and science. *Phys. D*, 152/153:779–786, 2001.

102. L.D. Landau. Two-fluid model of liquid helium II. *J. Phys.(USSR)*, 5:71, 1941.

103. T.D. Lee and C.N. Yang. Question of parity conservation in weak interactions. *Physical Review*, 104(1):254, 1956.

104. T.D. Lee and C.N. Yang. Low-temperature behavior of a dilute Bose system of hard spheres. I. Equilibrium properties. *Physical Review*, 112(5):1419, 1958.

105. W. Li and X. Lu. Global existence of solutions of the Boltzmann equation for Bose–Einstein particles with anisotropic initial data. *Journal of Functional Analysis*, 276(1):231–283, 2019.

106. E.M. Lifshitz and L.D. Landau. *Fluid mechanics: Volume 6 (course of theoretical physics).* Elsevier, 1987.

107. X. Lu. On isotropic distributional solutions to the Boltzmann equation for Bose–Einstein particles. *J. Statist. Phys.*, 116(5-6):1597–1649, 2004.

108. X. Lu. The Boltzmann equation for Bose–Einstein particles: velocity concentration and convergence to equilibrium. *J. Stat. Phys.*, 119(5-6):1027–1067, 2005.

109. X. Lu. The Boltzmann equation for Bose–Einstein particles: condensation in finite time. *J. Stat. Phys.*, 150(6):1138–1176, 2013.

110. X. Lu. The Boltzmann equation for Bose–Einstein particles: regularity and condensation. *J. Stat. Phys.*, 156(3):493–545, 2014.

111. X. Lu. Long time convergence of the Bose–Einstein condensation. *J. Stat. Phys.*, 162(3):652–670, 2016.

112. X. Lu. Long time strong convergence to Bose–Einstein distribution for low temperature. *Kinetic and Related Models*, 11(4):715–734, 2018.

113. J. Lukkarinen and M. Marcozzi. Wick polynomials and time-evolution of cumulants. *Journal of Mathematical Physics* 57(8), 083301.

114. J. Lukkarinen, M. Peng and H. Spohn. Global well-posedness of the spatially homogeneous Hubbard–Boltzmann equation. *Communications on Pure and Applied Mathematics*, 168(5):758–807, 2015.

115. J. Lukkarinen and H. Spohn. Kinetic limit for wave propagation in a random medium. *Arch. Ration. Mech. Anal.* 183(1):93–162 , 2007.

116. J. Lukkarinen and H. Spohn. Not to normal order – notes on the kinetic limit for weakly interacting quantum fluids. *Journal of Statistical Physics*, 134(5-6):1133–1172, 2009.

117. J. Lukkarinen and H. Spohn. Weakly nonlinear Schrödinger equation with random initial data. *Invent. Math.*, 183(1):79–188, 2011.

118. S. Métens, Y. Pomeau, M.A. Brachet and S. Rica. Théorie cinétique d'un gaz de bose dilué avec condensat. *C. R. Acad. Sci. Paris S'er. IIb M'ec. Phys. Astr.*, 327:791–798, 1999.

119. H.J. Miesner, D.M. Stamper-Kurn, M.R. Andrews, D.S. Durfee, S. Inouye, and W. Ketterle. Bosonic stimulation in the formation of a Bose–Einstein condensate. *Science*, 279(5353):1005–1007, 1998.

120. W. Nagourney, J. Sanderg and H.G. Dehmelt. Shelved optical electron amplifier: observation of quantum jumps. *Phys. Rev. Lett.*, 56:2797, 1986.

121. S. Nazarenko. *Wave turbulence.* Springer Science & Business Media, 83, 2013.

122. S. Nazarenko, A. Soffer and M.-B. Tran. On the wave turbulence theory for the nonlinear Schrödinger equation with random potentials. *arXiv preprint arXiv:1905.06323*, 2019.

123. T.T. Nguyen and M.-B. Tran. On the kinetic equation in Zakharov's wave turbulence theory for capillary waves. *SIAM J. Math. Anal.*, 50(2):2020–2047, 2018.

124. T.T. Nguyen and M.-B. Tran. Uniform in time lower bound for solutions to a quantum Boltzmann equation of bosons. *Archive for Rational Mechanics and Analysis*, January 2019, Volume 231, Issue 1, pp. 63–89.

125. L.W. Nordheim. Transport phenomena in Einstein–Bose and Fermi–Dirac gases. *Proc. Roy. Soc. London A*, 119:689, 1928.

126. S. Peletminskii and A. Yatsenko. Contribution to the quantum theory of kinetic and relaxation processes. *Soviet Physics JETP*, 26(773), 1968.

127. C.J. Pethick and H. Smith. *Bose–Einstein condensation in dilute gases.* Cambridge University Press, 2002.

128. L. Pitaevskii and S. Stringari. *Bose–Einstein condensation and superfluidity.* International Series of Monographs on Physics, volume 164, Oxford University Press, 2016.

129. M. Planck, 1900. Über des Gestzes der Energieverteilung in Normal spectrum. *Verh. Dtsch. Phys. Ges.* **2**: 237–245. Reprinted 1958 in *Max Planck Physikalisch Abhandlungen und Vortrage, Friedr. Vieweg und Sohn*, Braunschweig Vol. 1: 698–706. *Verh. Dtsch. Phys. Ges.* **2**: 237–245. Reprinted 1958 in *Max Planck Physikalisch Abhandlungen und Vortrage, Friedr. Vieweg und Sohn*, Braunschweig Vol. 1: 698–706.

130. M.B. Plenio and P.L. Knight. The quantum-jump approach to dissipative dynamics in quantum optics. *Reviews of Modern Physics*, 70(1):101, 1998.

131. Y. Pomeau. Asymptotic time behaviour of nonlinear classical field equations. *Nonlinearity*, 5(3):707–720, 1992.

132. Y. Pomeau. Kinetic theory of quantum gases with condensate: irreversibility resulting from the coupling between coherent and incoherent parts. *Chaos, Solitons & Fractals*, 12(14-15):2675–2682, 2001.

133. Y. Pomeau, M.-E. Brachet, S. Métens, and S. Rica. Théorie cinétique d'un gaz de Bose dilué avec condensat. *Comptes Rendus de l'Académie des Sciences-Series IIB-Mechanics-Physics-Astronomy*, 327(8):791–798, 1999.

134. Y. Pomeau, M. Le Berre and J. Ginibre. Ultimate statistical physics: Fluorescence of a single atom. *Journal of Stat. Phys.*, 26:104002, 2016.

135. Y. Pomeau and S. Rica. Thermodynamics of a dilute Bose–Einstein gas with repulsive interactions. *Journal of Physics A: Mathematical and General*, 33(4):691, 2000.

136. N.P. Proukakis and B. Jackson. Finite-temperature models of Bose–Einstein condensation. *Journal of Physics B: Atomic, Molecular and Optical Physics*, 41(20):203002, 2008.

137. M. Pulvirenti. The weak-coupling limit of large classical and quantum systems. *Proceedings of the International Congress of Mathematicians: Madrid*, August 22–30, 2006: Invited Lectures (pp. 229–256).

138. I.I. Rabi. Space quantization in a gyrating magnetic field. *Physical Review*, 51:652, 1937.
139. L.E. Reichl. Microscopic modes in a Fermi superfluid. I. Linearized kinetic equations. *Journal of Statistical Physics*. 23(1):83–110, 1980.
140. L.E. Reichl. Microscopic modes in a Fermi superfluid. II. Dispersion relations. *Journal of Statistical Physics*. 23(1):111–125, 1980.
141. L.E. Reichl. *A modern course in statistical physics*. A Wiley-Interscience Publication. John Wiley & Sons, Inc., New York, fourth edition, 2016.
142. L.E. Reichl and E.D. Gust. Transport theory for a dilute Bose–Einstein condensate. *J. Low Temp. Phys.*, 88:053603, 2013.
143. L.E. Reichl and M.-B. Tran. A kinetic model for very low temperature dilute Bose gases. *Journal of Physics A: Mathematical and Theoretical*, Volume 52, Number 6, 063001, January 2019.
144. P. Résibois, and M. de Leener. Chapter VII-4 *Classical Kinetic theory of fluids*, Wiley, New York, 1976.
145. D.V. Semikoz and I.I. Tkachev. Kinetics of Bose condensation. *Phys. Rev. Lett.*, 74:3093–3097, 1995.
146. D.V. Semikoz and I.I. Tkachev. Condensation of bosons in the kinetic regime. *Phys. Rev. D*, 55:489–502, 1997.
147. I. Shammass, S. Rinott, A. Berkovitz, R. Schley, and J. Steinhauer. Phonon dispersion relation of an atomic Bose–Einstein condensate. *Physical review letters*, 109(19):195301, 2012.
148. A. Sinatra, Y. Castin and E. Witkowska. Nondiffusive phase spreading of a Bose–Einstein condensate at finite temperature. *Physical Review A*, 75(3):033616, 2007.
149. A. Soffer and M.-B. Tran. On coupling kinetic and Schrodinger equations. *Journal of Differential Equations*, 265(5):2243–2279, 2018.
150. A. Soffer and M.-B. Tran. On the dynamics of finite temperature trapped bose gases. *Advances in Mathematics*, 325:533–607, 2018.
151. Y. Sone. *Kinetic theory and fluid dynamics*. Modeling and Simulation in Science, Engineering and Technology. Birkhäuser Boston, Inc., Boston, MA, 2002.
152. H. Spohn. Kinetic equations for quantum many-particle systems. *arXiv preprint arXiv:0706.0807*, 2007.
153. H. Spohn. Kinetics of the Bose–Einstein condensation. *Physica D: Nonlinear Phenomena*, 239(10):627–634, 2010.
154. H. Spohn. Weakly nonlinear wave equations with random initial data. *Proceedings of the International Congress of Mathematicians 2010 (ICM 2010)*, (In 4 Volumes) Vol. I: Plenary Lectures and Ceremonies Vols. II–IV: Invited Lectures (pp. 2128–2143).
155. B.V. Svistunov. Highly nonequilibrium Bose condensation in a weakly interacting gas. *J. Moscow Phys. Soc*, 1:373, 1991.
156. M. Taskovic, R.J. Alonso, I.M. Gamba, N. Pavlovic. On Mittag–Leffler moments for the Boltzmann equation for hard potentials without cutoff. *SIAM Journal on Mathematical Analysis* 50(1):834–869, 2019.
157. M.-B. Tran, G. Craciun, L.M. Smith, and S. Boldyrev. A reaction network approach to the theory of acoustic wave turbulence. *arXiv preprint arXiv:1901.03005*, 2019.
158. C. Truesdell and R.G. Muncaster. *Fundamentals of Maxwell's kinetic theory of a simple monatomic gas – Treated as a branch of rational mechanics.*. Volume 83 of *Pure and Applied Mathematics*. Academic Press, Inc. [Harcourt Brace Jovanovich, Publishers], New York-London, 1980.
159. E.A. Uehling and G.E. Uhlenbeck. On the kinetic methods in the new statistics and its applications in the electron theory of conductivity. *I Phys. Rev.*, 43:552–561, 1933.
160. C. Villani. A review of mathematical topics in collisional kinetic theory. In *Handbook of mathematical fluid dynamics, Vol. I*, pages 71–305. North-Holland, Amsterdam, 2002.
161. B. Wennberg. Entropy dissipation and moment production for the Boltzmann equation. *Journal of Statistical Physics*, 86(5-6):1053–1066, 1997.
162. E. Wigner. On the quantum correction for thermodynamic equilibrium. *Physical review*, 40(5):749, 1932.

163. V.E. Zakharov, V.S. L'vov, and G. Falkovich. *Kolmogorov spectra of turbulence I: Wave turbulence.* Springer Science & Business Media, 2012.

164. E. Zaremba, T. Nikuni, and A. Griffin. Two-fluid hydrodynamics for a trapped weakly interacting Bose gas. *Physical Review A*, 57(6):46–95, 1998.

165. E. Zaremba, T. Nikuni, and A. Griffin. Dynamics of trapped Bose gases at finite temperatures. *J. Low Temp. Phys.*, 116:277–345, 1999.

166. E. Zaremba, T. Nikuni, and A. Griffin. Two-fluid hydrodynamics in trapped Bose gases and in superfluid helium. *Journal of low temperature physics*, 121(5-6):247–256, 2000.

167. Y.B. Zel'dovich. Hydrodynamics of the Universe. *Review of Fluid Mechanics*, 9(1):215–228, 1977.

Index

annihilation operator, 69, 92, 93, 98, 102, 154–157
approximated coupling system, 167
approximation
  Euler, 119, 122, 123, 147
  first-order, 95, 135
  hydrodynamic, 119, 147
  Popov, 67, 100

Bogoliubov
  assumption, 97
  dispersion relation, 166
  renormalization, 61, 149, 153, 155, 161
  theory, 153, 154
  transformation, 61, 102, 158–160
bogolon, 64–67, 102, 103, 120, 137, 142
  energy, 66, 140, 142
  kinetic equation, 102, 138, 142
  momentum, 103, 140, 142
  number, 65, 140
  spectrum, 147
Boltzmann constant, 59, 64, 67, 85, 150
Boltzmann–Nordheim
  equation, 59, 62, 68, 69, 88, 90, 94, 96,
    105, 108, 115, 125, 128, 133, 134, 163
  kinetic theory, 115, 162
Bose–Einstein
  condensate, v, 60, 62, 88, 94, 107, 137
  distribution, 67, 103, 129, 134, 166
boson, viii, 59–61, 67, 69, 73, 78, 91–94,
    96, 105, 127
  commutation relations, 96

canonical partition function, 150, 157
chemical potential, 63, 98, 103, 106, 114,
    121, 125, 150, 151, 157–159, 161
chemical reaction networks, 170
classical Hamiltonian, 74, 96
closed to equilibrium, 68

coherent collective behavior, 61
collision operator, 64, 65, 84, 86, 87, 89,
    91–95, 116, 120, 121, 126, 131, 132,
    137–140, 142, 143, 145, 167, 168,
    173–175, 190, 191, 194, 199–201
condensate growth, 61, 64, 91, 94
constant
  Boltzmann, 59, 64, 67, 85, 150
  Planck, 10, 59, 60, 71
contact potential, 96
coupling condensate-thermal cloud system,
    132, 163
creation operator, 69, 92, 93, 98, 102,
    154–157

Dehmelt, 6, 35, 48
density
  function, 61, 75, 105, 167
  operator, 96, 97
Discrete Velocity Models, 170

eigenfunctions, 95, 139, 142, 168
eigenstates, 29–31, 139, 140, 143, 144
energy
  bogolon, 66, 140, 142
  Hartree–Fock, 64
  kinetic, vii, 115, 153, 154, 158
  operator, 30, 149, 154, 156–158, 161
  thermal particle, 89
entropy production, 89, 110, 112, 113
equation
  bogolon kinetic, 102, 138, 142
  Boltzmann–Nordheim, 59, 62, 68, 69, 88,
    90, 96, 105, 108, 115, 125, 128, 133,
    134, 163
  Euler, 147
  Gross–Pitaevskii, viii, 61, 62, 67, 123,
    167
  Hugenholtz–Pines, 67, 96, 103

© Springer Nature Switzerland AG 2019
Y. Pomeau, M.-B. Tran, *Statistical Physics of Non Equilibrium Quantum Phenomena*,
Lecture Notes in Physics 967, https://doi.org/10.1007/978-3-030-34394-1

Printed in the United States
By Bookmasters